The Minerals, Metals & Materials Series

Aaron P. Stebner · Gregory B. Olson
Editors

Peter M. Anderson · Mohsen Asle Zaeem · Othmane Benafan
Amy Brice · Emmanuel De Moor · Ibrahim Karaman
Ricardo Komai · Valery I. Levitas · Michael J. Mills
Peter Müllner · Alan R. Pelton · David J. Rowenhorst
Avadh Saxena
Co-editors

Proceedings of the International Conference on Martensitic Transformations: Chicago

Editors
Aaron P. Stebner
Colorado School of Mines
Golden, CO
USA

Gregory B. Olson
Northwestern University
Evanston, IL
USA

ISSN 2367-1181   ISSN 2367-1696   (electronic)
The Minerals, Metals & Materials Series
ISBN 978-3-319-76967-7   ISBN 978-3-319-76968-4   (eBook)
https://doi.org/10.1007/978-3-319-76968-4

Library of Congress Control Number: 2018933500

Printed on acid-free paper

This Springer imprint is published by the registered company Springer International Publishing AG
part of Springer Nature
The registered company address is: Gewerbestrasse 11, 6330 Cham, Switzerland

Preface

The International Conference on Martensitic Transformations (ICOMAT) was held on July 9–14, 2017, at the Hyatt Regency in Chicago, Illinois, USA. The event was hosted by the Chicago-based CHiMaD Center for Hierarchical Materials Design in downtown Chicago.

Located on the shores of Lake Michigan in the heart of the Great Lakes Region of North America, the city of Chicago is truly an epitome of the "melting pot" reputation of the United States, evidenced by a diverse array of cultural, economic, and social experiences that draw tourism, science and technology, arts, and music to its streets year-round. The "Chicago School" represents a unique tradition of innovation across many fields, notably including materials, as the birthplace of Materials Science in the 1950s and Materials Design in the 1990s.

ICOMAT 2017 was organized around the central theme of "Martensite by Design." Symposia included design for microstructures, properties, advanced manufacturing, and performance. Topics included:

- Theory & Methods for Martensite Design
- Interactions of Phase Transformations and Plasticity
- Quenching and Partitioning of Martensite and Other Advancements in Steels
- Novel Shape Memory Alloys
- Novel Functional Behaviors: Beyond Shape Memory Effect & Superelasticity
- Martensitic Transformations in Non-Metallic Materials
- Size Effects in Martensitic Transformations
- Advanced Characterization of Martensite—3D & High Resolution
- Quasimartensitic Modulations
- Advanced Processing Techniques: Additive, Porous, and Others
- Engineering Applications and Devices
- MSMnet: Magnetomechanics of Magnetic Shape Memory Alloys

The next ICOMAT conference will be held in 2020 in Jeju Island, South Korea.

Aaron P. Stebner
Gregory B. Olson

Contents

Part VIII   Advanced Processing Techniques: Additive, Porous, and Others

Part IX   Advanced Characterization of Martensite—3D and High Resolution

Part X   MSMnet: Magnetomechanics of Magnetic Shape Memory Alloys

Part XI   Novel Shape Memory Alloys

Part XII   Quasimartensitic Modulations

Part I

International Conference on Martensitic Transformations Plenary

Phase Transformations Under High Pressure and Large Plastic Deformations: Multiscale Theory and Interpretation of Experiments

Valery I. Levitas

Abstract

It is known that superposition of large plastic shear at high pressure in a rotational diamond anvil cell (RDAC) or high-pressure torsion leads to numerous new phenomena, including drastic reduction in phase transformation (PT) pressure and appearance of new phases. Here, our four-scale theory and corresponding simulations are reviewed. Molecular dynamic simulations were used to determine lattice instability conditions under six components of the stress tensor, which demonstrate strong reduction of PT pressure under nonhydrostatic loading. At nanoscale, nucleation at various evolving dislocation configurations is studied utilizing a developed phase field approach. The possibility of reduction in PT pressure by an *order of magnitude* due to stress concentration at the shear-generated dislocation pileup is proven. At microscale, a strain-controlled kinetic equation is derived and utilized in large-strain macroscopic theory for coupled PTs and plasticity. At macroscale, the behavior of the sample in DAC and RDAC is studied using a finite-element approach. A comprehensive computational study of the effects of different material and geometric parameters is performed, and various experimental effects are reproduced. Possible misinterpretation of experimental PT pressure is demonstrated. The obtained results offer new methods for controlling PTs and searching for new high-pressure phases (HPPs), as well as methods for characterization of high-pressure PTs in traditional DAC and RDAC.

V. I. Levitas (✉)
Departments of Aerospace Engineering, Mechanical Engineering, and Material Science and Engineering, Iowa State University, Ames, IA 50011, USA
e-mail: vlevitas@iastate.edu

V. I. Levitas
Ames Laboratory, Division of Materials Science & Engineering, Ames, IA 50011, USA

Keywords

Strain-induced phase transformations • High pressure
Four-scale theory • Nucleation at dislocation pile-up
Rotational diamond anvil cell

Introduction

In situ studies of material behavior, including phase transformations (PTs), under high pressure up to several hundred GPa are performed in a diamond anvil cell (DAC). PTs are usually characterized in terms of pressure for initiation and completion of direct and reverse PTs, and in some cases for the concentration of phases versus pressure. It is recognized that the entire process of producing high pressure is accompanied by large nonhydrostatic (deviatoric) stresses and large plastic deformations that drastically affect PTs. A specially-designed device [1, 2], the RDAC (Fig. 4a), is utilized to study the strong, multifaceted, and unique effects of large plastic shear on PTs:

(a) *Plastic shear under high pressure leads to the formation of new phases* that may not be producible without shear [1–7].
(b) *Plastic shear under high pressure reduces the transformation pressure by a factor between 2 and 10* for some PTs [1–4, 6–9], e.g., for transformations from highly disordered hexagonal hBN into superhard wurtzitic wBN from 52.5 to 6.7 GPa [8].
(c) The concentration of the high-pressure phase is an increasing function of plastic shear [1–8]. Plastic strain is a time-like parameter, so one needs to consider strain-controlled kinetics instead of time-controlled kinetics.

Here, we review our results on the development of four-scale theory and corresponding simulations for understanding and description of the reasons for the phenomena mentioned above in (a)–(c).

Pressure-Induced Versus Strain-Induced Phase Transformations Under High Pressure

Note that classical macroscopic thermodynamics fails to explain a strong reduction in transformation pressure due to nonhydrostatic stresses. A thermodynamic treatment leads to a simplified condition in which the transformation work W reaches some critical value k, given by $W = -p\varepsilon_{0t} + \tau\gamma_t = k$, where $\varepsilon_{0t} < 0$ and γ_t are the volumetric and shear transformation strains, and p and τ are the pressure and the shear stress. Under hydrostatic conditions $-p_h\varepsilon_{0t} = k$, which determines k. Then the maximum difference between transformation pressure under hydrostatic and nonhydrostatic conditions is $p_h - p = \tau_y\gamma_t/|\varepsilon_{0t}|$, taking into account that maximum shear stress is limited by the yield strength in shear τ_y. Let us assume as a simple estimate that $\tau_y = 1$ GPa and $\gamma_t/|\varepsilon_{0t}| = 1 - 5$. Then the reduction in transformation pressure is just 1–5 GPa. If PT pressure under hydrostatic conditions is 15 or 50 GPa, then maximum pressure reduction is 33 or 10%, considerably below than observed in experiments, e.g., in [8]

The first step in understanding the PTs in RDAC was recognition in [6, 7] that there is a basic difference between plastic strain-induced PTs under high pressure in RDAC and pressure- or stress-induced PTs in DAC under quasi-hydrostatic conditions. *Pressure- and stress-tensor-induced PTs* occur predominantly at pre-existing defects (e.g., dislocations and various tilt boundaries) at stress levels below the yield strength. These defects represent stress (pressure) concentrators. The number of nucleation sites is limited, so one has to increase pressure to activate less potent defects, i.e., defects with smaller stress concentration.

Strain-induced PTs occur by nucleation at new defects, e.g., dislocation pileups generated during plastic flow. That is why it is possible to increase local stresses and promote PTs near the new defects by increasing plastic shear at constant pressure. Since concentration of all components of a stress tensor is proportional to the number of dislocations in a pileup, which can be as large as 10–100, new defects may be much stronger than the pre-existing defects. Such high local pressure and deviatoric stresses may cause nucleation of the HPP at an external pressure much below that under hydrostatic conditions. Note that shear stresses within small nanoscale regions may be limited by the theoretical shear strength that could range from one to two orders of magnitude larger that τ_y. It was concluded in [6, 7] that strain-induced PTs require a completely different experimental characterization as well as new thermodynamic and kinetic descriptions.

Nanoscale Continuum Treatment: Phase Field Approach

A simple analytical model of nucleation at the tip of the dislocation pileup [6, 7] demonstrated that, with a sufficient number of dislocations, nucleation of HPP can indeed occur at external pressures an order of magnitude smaller than under hydrostatic conditions. This model, however, contained a number of simplifying assumptions. For a more precise and advanced proof of the complex, the *first phase field approach (PFA) for the interaction* of PTs and dislocations was developed [10–13], which synergistically combines the most advanced fully geometrically nonlinear theories for martensitic PTs [14–16] and dislocations [17, 18], with nontrivial interactions and inheritance of dislocations during PTs. This approach combined with the finite-element method (FEM) was applied for the first simulation of PT under pressure and shear (Fig. 1) [11–13] in a nanograined bicrystal. A model material with phase equilibrium pressure $p_{eq} = 10$ GPa and instability pressure of 20 GPa was considered; transformation strains were $\varepsilon_{0t} = -0.1$ and $\gamma_t = 0.2$. Under hydrostatic pressure and single dislocation in the grain, PT occurs at 15.75 GPa. Under a compressive stress $\sigma_n = 3.05$ GPa (corresponding to an averaged pressure of 2.0 GPa) and shear, dislocations are generated in the left grain and pile up against grain boundaries. Strong concentration of all components of the stress tensor near the pileup tip leads to a combination of stresses that satisfy local lattice instability criterion and lead to barrierless nucleation of a HPP. For the configuration shown in Fig. 1a, HPP is observed at $\gamma = 0.15$ and pressure $\bar{p} = 1.2$ GPa averaged over the transforming grain. When shear increases to 0.2 (Fig. 1 b), a major part of a grain transforms and, in the stationary state due to volume decrease, pressure drops to 0.8 GPa (averaged over both grains) and 0.06 GPa averaged over the transformed grain. Such averaged pressures (rather than σ_n) are usually reported by experimentalists. Thus, a *dislocation pileup can indeed cause a reduction in transformation pressure more than an order of magnitude below that under hydrostatic conditions.*

The above results have been obtained for neglected plasticity in the transforming grain. When plasticity is included (Fig. 1c), the volume fraction of the HPP reduces in comparison with cases without plasticity. Thus, dislocations play a dual role: they promote PT by producing strong stress concentrators and also suppress PT by relaxing the stresses. Some combinations of normal and shear stresses have been found in [12] where the volume fraction of HPP either with or without plasticity in the transformed grain

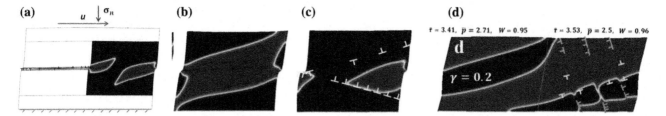

Fig. 1 Stationary nanostructures for coupled evolution of dislocations and high-pressure phase (red) in a two-grain sample under pressure and prescribed shear $\gamma = 0.15$ (**a**), and in the right grain without (**b**) and with (**c**) dislocations for prescribed shear of 0.2. Due to dislocation pileup, PT occurs at applied pressure more than an order of magnitude lower than for hydrostatic conditions with one dislocation. Plasticity plays a dual role: it promotes PT by stress concentrators (**a, b**) but suppresses it by relaxing these stresses (**c**). **d** Complex phase and dislocation nanostructure obtained under compression and shear, which includes grain rotation and switching between slip systems. Reproduced with permission from [12]

does not essentially differ. Figure 1d shows a nanostructure obtained under compression and shear that includes grain rotation and switching between slip systems. Major parts of both grains are transformed despite the fact that dislocations in the left grain disappeared. Some additional studies without [11] and with normal stress are presented in [13].

Lattice Instability Criteria: PFA Approach and MD Simulations

Since shear (deviatoric) stresses near the tip of a dislocation pile are not limited by the macroscopic yield strength but are limited by the theoretical strength, unique stress states not achievable in bulk can be produced. Such stresses may cause PTs into stable or metastable phases that are not or even could not be achieved in bulk under hydrostatic or nonhydrostatic conditions without significant plastic deformation. The next step is to find the maximum possible effect of deviatoric stresses, i.e., their effect on PT (lattice instability) criteria of an ideal (defect-free) single crystal under complex loading.

A general lattice instability (or PT) criterion was derived under complex stress states and large strains utilizing PFA [15]. To test and specify this criterion for silicon (Si), molecular dynamics (MD) simulations [19, 20] were performed for direct and reverse Si I $\leftrightarrow$ Si II PTs (Fig. 2). Lattice instability criteria are thus linear in terms of normal to cubic faces stresses σ_i and they are independent of shear stresses, at least below 3 GPa. The effect of nonhydrostatic stresses is drastic. Thus, under hydrostatic conditions, the PT pressure is 80 GPa; under uniaxial loading, the stress is 12 GPa, i.e., the mean pressure is 4 GPa, so reduction in PT pressure is by a factor of 20. While the uniaxial stress of 12 GPa definitely cannot be applied to the real (defective) bulk sample because it is much higher than the macroscopic yield strength, it can be obtained at the tip of the dislocation pileup and drastically reduce the external pressure required for PT.

Microscale Strain Controlled Kinetic Equations

To produce high pressure in DAC, one must compress a sample plastically. Thus, not only in RDAC, but in most cases in DAC without a hydrostatic medium, PTs should be treated as plastic strain-induced PTs. It was suggested in [6, 7] that strain-induced PTs should be described theoretically and characterized experimentally with the help of strain-controlled kinetic equations for concentrations of high-pressure phases c_k and parameters D_m characterizing the defect structure of the type

$$dc_k/dq = f_k(\boldsymbol{\sigma}, E_p, q, c_k, D_m), \tag{1}$$

where q is the accumulated plastic strain that plays a role of a time-like parameter and $\boldsymbol{\sigma}$ and E_p are the stress and plastic strain tensors. A simplified version of this equation for a single PT was derived in [6, 7], taking into account conceptually all information learned from the nanoscale model in [6, 7]:

$$\frac{dc}{dq} = A\frac{(1-c)\bar{p}_d H(\bar{p}_d)M - c\bar{p}_r H(\bar{p}_r)}{c + (1-c)M};$$
$$\bar{p}_d = \frac{p - p_\varepsilon^d}{p_h^d - p_\varepsilon^d} \text{ and } \bar{p}_r = \frac{p - p_\varepsilon^r}{p_h^r - p_\varepsilon^r}; \ M = \frac{\sigma_{y2}}{\sigma_{1y}}. \tag{2}$$

Here, four characteristic pressures are utilized: p_ε^d is the minimum pressure below which a direct plastic strain-induced PT to HPP does not take place, p_ε^r is the maximum pressure above which a reverse strain-induced PT to low pressure phase cannot occur, and p_h^d and p_h^r are the pressures for direct and reverse PTs under hydrostatic loading, respectively. These pressures are combined into two dimensionless characteristic pressures for direct and reverse PTs, $\bar{p}_d$ and $\bar{p}_r$, respectively. In addition, A is the kinetic parameter, σ_{1y} and σ_{1y} are the yield strengths in compression of low- and high-pressure phases, respectively, and H is

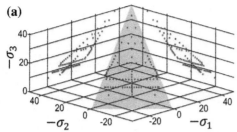

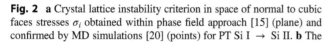

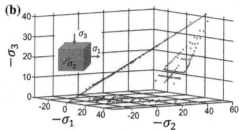

Fig. 2 a Crystal lattice instability criterion in space of normal to cubic faces stresses σ_i obtained within phase field approach [15] (plane) and confirmed by MD simulations [20] (points) for PT Si I → Si II. **b** The

same plot but twisted until plane is visible as a line, to demonstrate closeness of MD and phase field results. Reproduced with permission from [20]

the Heaviside function ($H(y) = 1$ for $y \geq 0$ and $H(y) = 0$ for $y < 0$). The barrierless character of nucleation at defects results in time not being a parameter in Eq. (2). Since the number of pileups and the number of dislocations in each grow with increasing plastic strain, and when plastic straining stops, PT stops as well, the accumulated plastic strain q is a time-like parameter. Defects like dislocation pileups generate both compressive and tensile stresses of the same magnitude but in different regions. Consequently, strain-induced defects simultaneously promote both direct and reverse PTs, as reflected in the two opposite terms in Eq. (2). In addition, due to different yield strengths in different phases, plastic strain localizes in the weaker phase, explaining the ratio of the yield strength of phases in Eq. (2).

Equation (2) was generalized for multiple PTs in [21] and analysed in [6, 7, 21] for interpretation of experimental phenomena. Some of the general results are:

(1) If $p_\varepsilon^d < p_\varepsilon^r$ (i.e., direct PT can start at pressure below the maximum pressure for the reverse PT) and $p_\varepsilon^d < p < p_\varepsilon^r$; for $q \to \infty$ there is a stationary solution $0 < c < 1$ of Eq. (2). This explains incomplete PT and existence of the stationary concentration, see [22].

(2) The solution explains the zero-pressure hysteresis observed in [1] for B_1 to B_2 PT in KCl. However, this PT pressure is not the phase equilibrium pressure that cannot be determined from a macroscopic plastic strain-induced experiment because it is not present in Eq. (2). At the same time, as was found in [11, 12] with the PFA, the phase equilibrium stress tensor governs the stationary morphology of the nanostructure, both locally and in terms of stresses averaged over the transformed region or grain.

(3) The stationary concentration of the HPP increases with the ratio σ_{y2}/σ_{y1}, i.e., a strong HPP is promoted more by plastic strain than a weak one. Similarly, hard (soft) inert particles promote (decelerate) plastic strain-induced chemical reactions [23, 24] that can also be described by Eq. (2).

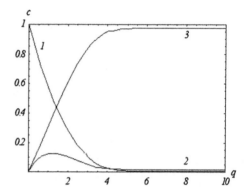

Fig. 3 Change in volume fraction of Si I (1), Si II (2), and Si III (3) with plastic straining at pressure of 7 GPa. Reproduced with permission from [21]

Figure 3 presents a solution for strain-controlled kinetics for PTs between Si I, Si II, and Si III at a pressure of 7 GPa. Note that under hydrostatic pressure of 7 GPa, Si I is stable. For strain-induced PTs, Si III, which is not present on the equilibrium phase diagram, monotonically increases in concentration until completion. Si II appears and disappears with straining. If one assumes that Si II is an unknown phase to be discovered, then traditional wisdom—the larger the plastic strain, the larger the promotion for PT—fails. Plastic strain should be optimal, as also found in [25] for PT in boron nitride.

Macroscale Modeling Coupled Phase Transformation and Plastic Flow of a Sample in DAC and RDAC

Kinetic Eq. (2) was incorporated into *macroscale* theory for stress- and plastic strain-induced PTs for the description of coupled plastic flow and PTs in a sample treated in DAC and RDAC. An initial simplified version of the theory for small elastic and transformational strains (but large plastic strains), linear elasticity, and pressure-independent yield strength was developed and applied in [26–34]. Strain hardening was neglected based on experimental and theoretical results given in [35]. Even for a model material, the results of a parametric study changed the fundamental understanding of the interpretation of experimentally observed effects and measurements and the extraction of information on material behavior from sample behavior. Some results for processes in RDAC both without and with a gasket are presented in Figs. 4 and 5, respectively. In particular, the results in Fig. 4 reproduce the following experimental phenomena.

(a) A pressure self-multiplication effect [1, 3, 4], i.e., pressure growth during PT during torsion at fixed force, despite the volume reduction due to PT. Even though this sounds like violation of the Le Shatelie principle, as mentioned above, strain-induced PTs require non-traditional thermodynamic treatment. This effect was automatically reproduced when the yield strength grew during PTs, similar to that described in a simplified analytical treatment in [6, 7]. It represents positive mechanochemical feedback in promotion of PTs by torsion of an anvil.

(b) Note that even when the pressure exceeds p_ε^d by a factor of two (line 3 in Fig. 4b and c), PT is not complete. Experimentalists would say that PT is spread over the range from p_ε^d to more than $2p_\varepsilon^d$. However, according to Eq. (2), this is not kinetic property of the PT, which could occur until completion at pressure slightly above p_ε^d under large plastic strains. While not required for PT, such high pressure appears because of a coupled PT and plastic flow under such loadings, i.e., as system (sample) behavior rather than material properties. Pressure growth can be suppressed by using a sample-gasket system, see [25, 33] and Fig. 5. Thus, for synthesis of HPP at minimal pressure, one must design the sample-gasket system and loading program. The results shown in Fig. 5 show smaller heterogeneity of all parameters and a smaller change in pressure during anvil rotation.

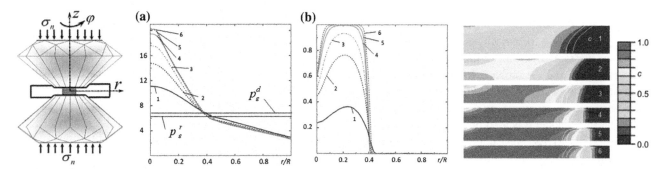

Fig. 4 a Schematics of RDAC; **b** and **c** distributions of pressure and concentration of high-pressure phase, respectively, along the radius of the contact surface of a sample; **d** distribution of concentration of high-pressure phase within a quarter of a sample. Simulations are for the case without gasket, with the yield strength $\sigma_{2y} = 5\ \sigma_{1y}$, for dimensionless applied axial force $F = 4.44$, and different values of angle of rotation φ. 1: $\varphi = 0.09$, 2: $\varphi = 0.38$, 3: $\varphi = 0.61$, 4: $\varphi = 0.94$, 5: $\varphi = 1.10$, and 6: $\varphi = 1.30$ rad [27]. White lines in the concentration distribution in (**d**) correspond to pressure $p = p_\varepsilon^d$. Reproduced with permission from [27]

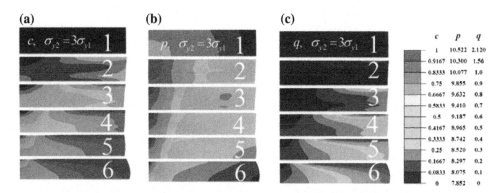

Fig. 5 Evolution of distributions of **a** concentration of the high-pressure phase c, **b** pressure p, and **c** accumulated plastic strain q in the quarter of a sample within a gasket at a fixed compressive force $F = 6.19$ and different values of angle of rotation φ. 1: $\varphi = 0$, 2: $\varphi = 0.1$, 3: $\varphi = 0.3$, 4: $\varphi = 0.5$, 5: $\varphi = 0.8$, and 6: $\varphi = 1.0$ rad. Reproduced with permission from [33]

(c) Steps (horizontal plateaus) at the experimental pressure distribution [1, 3, 4]. Since the pressure gradient is proportional to the frictional stress at the contact surface between the anvil and a sample, for intense plastic flow equal to the yield strength in shear [35], these steps led to the conclusion that the yield strength reduces to zero in the PT region [1]. This, however, could not be true, because, e.g., TRIP steels show high strength during PT. In Fig. 4b, these steps appear automatically as a result of the solution when PT kinetics is fast enough. Also, the pressure at the steps lies between two characteristic pressures p_ε^d and p_ε^r, allowing experimental determination of their range.

(d) For relatively large rotations, HPP is detectable in the region where $p < p_\varepsilon^d$ (Fig. 4d), where transformation is by definition impossible. HPP is brought into this region by radial plastic flow that occurs due to reduction of sample thickness during rotation of an anvil. Since pressure is usually measured after each rotation-PT increment, an experimentalist may report pressure where HPP is found as the PT pressure, which would be a misinterpretation.

In [36], a fully geometrically and physically nonlinear model for elastoplasticity under high pressure without PTs was developed, and the problem for compression of a rhenium in DAC was solved, producing results in good correspondence with experiments in [37] for pressures up to 300 GPa. The same model was used in [38, 39] for compression and torsion in RDAC, revealing a pressure-self-focusing effect, and suggesting how to utilize it to increase maximum achievable pressure. In [40], this model was generalized by including strain-induced PTs and subsequently used to study PT in BN.

Concluding Remarks

The main mysterious experimental results obtained under compression and torsion of materials in RDAC are the reduction in PT pressure by up to an order of magnitude in comparison with that under quasi-hydrostatic loading, and appearance of new phases that were not or could not be obtained under quasi-hydrostatic conditions. Classical macroscopic thermodynamics fails in explaining these phenomena. In this paper, we review our four-scale theory and simulations, demonstrating how events at each scale contribute to the explanation of these phenomena. MD simulations demonstrated reduction in PT (lattice instability) pressure for uniaxial loading of perfect Si by a factor of 20 in

comparison with that from hydrostatic loading. However, such uniaxial loading cannot be applied to an actual defective sample because it significantly exceeds the macroscopic yield strength. At the same time, such stresses can be obtained at the tip of strong defects as dislocation pileups. Concentration of all components of the stress tensor is proportional to the number of dislocations in a pileup that can be as many as 10–100. Plastic deformation generates dislocation pileups with high concentration of both pressure and deviatoric stresses near their tips, satisfying the lattice instability criterion, and cause barrierless nucleation of an HPP. Because of the large increase in local stresses, applied pressure required for nucleation can be significantly reduced. The possibility of reduction of PT pressure by more than an order of magnitude was first confirmed by a simple analytical model and then by advanced PFA. Plasticity plays a dual role: it promotes PT by producing stress concentrators but also suppresses PT by relaxing these stresses. Thus, some combinations of compression and shear were found for which PT wins this competition. Since such unique stress states with extremely high deviatoric components, limited by the theoretical strength (rather than the macroscopic strength, which is one to two orders of magnitude smaller), cannot be obtained in a bulk real (defective) material, they can lead to the appearance of new phases that were not or even could not be obtained without plastic deformation.

Such an understanding may lead to new directions in design and synthesis of high-pressure novel phases and compounds under much lower pressure than under hydrostatic conditions. Instead of increasing external pressure, successively filling the material with strain-induced defects causing unique stress tensor states at the limit of lattice stability and barrierless PT near their tips is suggested. Classical thermodynamics and kinetics should be conceptually advanced to multiscale thermodynamics and kinetics in 12 + D parameter space (six components each of the stress and plastic strain tensors, and quantitative characteristics of the strain-induced defect structures). However, nanoscale understanding alone is not sufficient for achieving this goal. Micro- and macroscale theories and measurements give important information on how to characterize PTs in terms of strain-controlled kinetic equations of the type given in Eqs. (1) and (2) and how to design a sample-gasket system and loading programs to produce desired phases at lowest pressure in RDAC or, for larger scale, during high-pressure torsion. The key points are related to the heterogeneities of stress and strain fields at both microscale and macroscale. At microscale, they are caused by specifics of the stress field of the dislocation pileup that generate both compressive and tensile stresses of the same magnitude in different regions and thus simultaneously promote both direct and reverse PTs. Also, at both scales they are related to essentially different two- or multiphase system behaviors depending on the ratio of the yield strength of the high- and low-pressure phases, as well as that of a gasket.

Acknowledgements The support of ARO (W911NF-12-1-0340), NSF (DMR-1434613 and CMMI-1536925), and Iowa State University (Schafer 2050 Challenge Professorship and Vance Coffman Faculty Chair Professorship) is gratefully acknowledged.

References

1. Blank VD, Estrin EI (2014) Phase transitions in solids under high pressure. CRC Press, New York
2. Edalati K, Horita Z (2016) A review on high-pressure torsion (HPT) from 1935 to 1988. Mat Sci Eng A 652:325–352
3. Novikov NV, Polotnyak SB, Shvedov LK, Levitas VI (1999) Phase transitions under compression and shear in diamond anvils: experiment and theory. Superhard Mat 3:39–51
4. Blank VD et al (1994) Is C_{60} fullerite harder than diamond? Phys Lett A 188:281–286
5. Levitas VI, Ma Y, Selvi E, Wu J, Patten J (2012) High-density amorphous phase of silicon carbide obtained under large plastic shear and high pressure. Phys Rev B 85:054114
6. Levitas VI (2004) Continuum mechanical fundamentals of Mechanochemistry. In: Gogotsi Y, Domnich V (eds) High pressure surface science and engineering. Inst. of Physics, Bristol, Section 3, p 159–292
7. Levitas VI (2004) High-pressure mechanochemistry: conceptual multiscale theory and interpretation of experiments. Phys Rev B 70:184118
8. Ji C, Levitas VI, Zhu H, Chaudhuri J, Marathe A, Ma Y (2012) Shear-induced phase transition of nanocrystalline hexagonal boron nitride to wurtzitic structure at room temperature and low pressure. Proc Natl Acad Sci USA 109:19108–19112
9. Levitas VI, Shvedov LK (2002) Low pressure phase transformation from rhombohedral to cubic BN: experiment and theory. Phys Rev B 65:104109
10. Levitas VI, Javanbakht M (2015) Interaction between phase transformations and dislocations at the nanoscale. Part 1. General phase field approach. J Mech Phys Solids 82:287–319
11. Javanbakht M, Levitas VI (2015) Interaction between phase transformations and dislocations at the nanoscale. Part 2. Phase field simulation examples. J Mech Phys Solids 82:164–185
12. Levitas VI, Javanbakht M (2014) Phase transformations in nanograin materials under high pressure and plastic shear: nanoscale mechanisms. Nanoscale 6:162–166
13. Javanbakht M, Levitas VI (2016) Phase field simulations of plastic strain-induced phase transformations under high pressure and large shear. Phys Rev B 94:214104
14. Levitas VI, Levin VA, Zingerman KM, Freiman EI (2009) Displacive phase transitions at large strains: phase-field theory and simulations. Phys Rev Lett 103:025702
15. Levitas VI (2013) Phase-field theory for martensitic phase transformations at large strains. Int J Plast 49:85–118
16. Levitas VI (2014) Phase field approach to martensitic phase transformations with large strains and interface stresses. J Mech Phys Solids 70:154–189
17. Levitas VI, Javanbakht M (2015) Thermodynamically consistent phase field approach to dislocation evolution at small and large strains. J Mech Phys Solids 82:345–366
18. Javanbakht M, Levitas VI (2016) Phase field approach to dislocation evolution at large strains: computational aspects. Int J Solids Struct 82:95–110
19. Levitas VI, Chen H, Xiong L (2017) Triaxial-stress-induced homogeneous hysteresis-free first-order phase transformations with stable intermediate phases. Phys Rev Lett 118:025701
20. Levitas VI, Chen H, Xiong L (2017) Lattice instability during phase transformations under multiaxial stress: modifed transformation work criterion. Phys Rev B 96:054118
21. Levitas VI, Zarechnyy OM (2006) Kinetics of strain-induced structural changes under high pressure. J Phys Chem B 110:16035–16046
22. Straumal BB, Kilmametov AR, Ivanisenko Y et al (2015) Phase transitions induced by severe plastic deformation: steady-state and equifinality. Int J Mat Res 106:657–664
23. Zharov A (1984) The polymerisation of solid monomers under conditions of deformation at a high pressure. Usp Khim 53:236–250
24. Zharov A (1994) High pressure chemistry and physics of polymers, Kovarskii AL (ed). CRC Press, Boca Raton, Chapter 7, pp 267–301
25. Levitas VI, Ma Y, Hashemi J, Holtz M, Guven N (2006) Strain-induced disorder, phase transformations and transformation induced plasticity in hexagonal boron nitride under compression and shear in a rotational diamond anvil cell: in-situ X-ray diffraction study and modeling. J Chem Phys 25:044507
26. Levitas VI, Zarechnyy OM (2010) Modeling and simulation of strain-induced phase transformations under compression in a diamond anvil cell. Phy Rev B 82:174123
27. Levitas VI, Zarechnyy OM (2010) Modeling and simulation of strain-induced phase transformations under compression and torsion in a rotational diamond anvil cell. Phys Rev B 82:174124
28. Feng B, Levitas VI, Zarechnyy OM (2013) Plastic flows and phase transformations in materials under compression in diamond anvil cell: effect of contact sliding. J Appl Phys 114:043506
29. Feng B, Zarechnyy OM, Levitas VI (2013) Strain-induced phase transformations under compression, unloading, and reloading in a diamond anvil cell. J Appl Phys 113:173514
30. Feng B, Levitas VI (2013) Coupled phase transformations and plastic flows under torsion at high pressure in rotational diamond anvil cell: effect of contact sliding. J Appl Phys 114:213514
31. Feng B, Levitas VI, Zarechnyy OM (2014) Strain-induced phase transformations under high pressure and large shear in a rotational diamond anvil cell: simulation of loading, unloading, and reloading. Comput Mater Sci 84:404–416
32. Feng B, Levitas VI, Ma Y (2014) Strain-induced phase transformation under compression in a diamond anvil cell: simulations of a sample and gasket. J Appl Phys 115:163509
33. Feng B, Levitas VI (2016) Effects of the gasket on coupled plastic flow and strain-induced phase transformations under high pressure and large torsion in a rotational diamond anvil cell. J Appl Phys 119:015902
34. Feng B, Levitas VI (2017) Plastic flows and strain-induced alpha to omega phase transformation in zirconium during compression in a diamond anvil cell: finite element simulations. Mater Sci Eng A 680:130–140
35. Levitas VI (1996) Large deformation of materials with complex rheological properties at normal and high pressure. Nova Science Publishers, New York
36. Feng B, Levitas VI, Hemley RJ (2016) Large elastoplasticity under static megabar pressures: formulation and application to compression of samples in diamond anvil cells. Int J Plast 84:33–57

37. Hemley RJ, Mao HK, Shen GY, Badro J, Gillet P, Hanfland M, Hausermann D (1997) X-ray imaging of stress and strain of diamond, iron, and tungsten at megabar pressures. Science 276:1242–1245

38. Feng B, Levitas VI (2017) Pressure self-focusing effect and novel methods for increasing the maximum pressure in traditional and rotational diamond anvil cells. Sci Reports 7:45461

39. Feng B, Levitas VI (2017) Large elastoplastic deformation of a sample under compression and torsion in a rotational diamond anvil cell under megabar pressures. Int J Plast 92:79–95

40. Feng B, Levitas VI (2017) Coupled elastoplasticity and strain-induced phase transformation under high pressure and large strains: formulation and application to BN sample compressed in a diamond anvil cell. Int J Plast 96:156–181

Activation Energy of Time-Dependent Martensite Formation in Steel

Matteo Villa and Marcel A. J. Somers

Abstract

The kinetics of $\{5\,5\,7\}_\gamma$ lath martensite formation in (wt %) 17Cr-7Ni-1Al-0.09C and 15Cr-7Ni-2Mo-1Al-0.08C steels was assessed with magnetometry at sub-zero Celsius temperatures. Samples were cooled to 77 K by immersion in boiling nitrogen to suppress martensite formation. Thereafter, thermally activated martensite formation was monitored during: (i) isochronal (re) heating at different heating rates; (ii) isothermal holding at temperatures between 120 and 310 K. The activation energy, E_A, of thermally activated martensite formation was quantified from the results of both isochronal and isothermal tests by applying a Kissinger-like method. In addition, the isothermal data was interpreted applying the approach presented by Borgenstam and Hillert. The results of the independent quantification methods were consistent and indicated an E_A in the range 9–13 kJ mol^{-1}. Thereafter, the two methods were applied to evaluate the data available in the literature. The overall analysis showed that E_A varies in the range 2–27 kJ mol^{-1} and increases logarithmically with the total fraction of interstitials in the steel.

Keywords

Isothermal martensite • Transformation kinetics Martensitic steel

Introduction

The design of martensitic steels requires models to accurately describe the kinetics of the austenite-to-martensite transformation in this class of materials. In early work, martensitic transformations were considered athermal, meaning that the degree of transformation is determined exclusively by the lowest temperature reached, independent of time [1]. This approach has remained [2] and, in the large majority of cases, allows a consistent description of the transformation kinetics. Nevertheless, martensite formation can also proceed isothermally or, rather, time-dependent, particularly at sub-zero Celsius temperatures.

The first evidence of time-dependent martensite formation was reported in 1948 [3]. Extensive evidence followed (see Refs. [4, 5]) and in the 1990s the isothermal behaviour was brought to a rationalization as a common characteristic of martensite formation in ferrous alloys [6]. This implies that martensite formation in steel can be suppressed by sufficiently fast cooling to a temperature where the transformation proceeds (virtually) infinitely slowly. Consistently, it was shown in several cases (see Refs. [4, 5, 7]) that martensite formation can be *partly* suppressed by fast cooling to temperatures T $\leq$ 77 K, and the transformation can continue on subsequent (re)heating. Also, the transformation can be *fully* suppressed, as firstly demonstrated in 1953 for Fe-Ni-Mn alloys [8], in 1960 for stainless steel [9] and in 1990 for Fe-Ni alloys [10].

Conversely, the possibility to form martensite at temperatures as low as 4 K was demonstrated already in 1950 [11] and is evidence that the transformation in the investigated alloys is un-suppressible (i.e., intrinsically athermal). Martensite formation at 4 K also indicates that the growth of martensite units does not determine the overall rate of the isothermal process. It was established as early as the 1930–1950s [12–14] that the duration of an austenite-to-martensite transformation event can be of the order of a small fraction of a second and that the growth rate of the martensite units

M. Villa (✉) · M. A. J. Somers
Department of Mechanical Engineering, Technical University of Denmark, 2800 Kongens Lyngby, Denmark
e-mail: matv@mek.dtu.dk

M. A. J. Somers
e-mail: somers@mek.dtu.dk

can be independent of temperature within a significantly large temperature interval (i.e., growth is athermal) [14, 15]. Nevertheless, time-dependent growth of martensite has been observed several times (see Ref. [16]).

To reconcile the above experimental observations, it has been suggested that the martensite sub-structure controls the kinetics of martensite formation [13, 17–19]: athermal martensite is internally twinned and time-dependent martensite is internally slipped. This straightforward description is, unfortunately, not consistent with all experiments. Slow growth of martensite always involves slipped sub-structures; however, slipped martensite can also grow instantaneously [20, 21]. Furthermore, the transformation of austenite into slipped martensite can be suppressed by fast cooling [22–24]; on the other hand, slipped martensite can form at 4 K (see, for example, Ref. [11]). Evidently, the kinetics of martensitic transformations, the roles of nucleation and growth and the significance of the martensite substructure are incompletely understood.

Following Huizing and Klostermann [25], we recently suggested that the products of martensitic transformations in steel should be classified into two groups [26]:

i. *Schiebung, S,* martensite corresponds to internally slipped $\{557\}_\gamma$ lath martensite and to the internally slipped product growing on $\{225\}_\gamma$ plate and $\{259\}_\gamma$ lenticular martensites. *S* martensite is *s*uppressible and cannot form at an observable rate at temperatures approaching absolute zero. The growth of *S* martensite can be time dependent.

ii. *U*mklapp, *U,* martensite corresponds to twinned $\{3\,10\,15\}_\gamma$ thin plate martensite, to the twinned parts of the $\{225\}_\gamma$ plate and $\{259\}_\gamma$ lenticular martensites, as well as to internally slipped strain induced and $\{112\}_\gamma$ martensites. *U* martensite is *u*n-suppressible and can form at 4 K. The growth of *U* martensite is instantaneous.

Additionally, we suggested that the existing kinetics models, which typically describe the kinetics of martensite formation in steel as nucleation-controlled (i.e., implicitly assuming instantaneous growth) and define nucleation as *u*n-suppressible upon reaching a certain critical driving force for transformation, ΔG_C (see Ref. [4]), apply to *U* martensite, but cannot describe the kinetics of *S* martensite formation. For the latter case, a different approach appears necessary.

An alternative approach to describe the kinetics of isothermal martensite formation was presented by Borgenstam and Hillert [27]. They focused on the evolution of the transformation rate versus temperature and described the transformation rate in terms of chemical reaction rate theory as the product of the normalized chemical driving force,

$\Delta G/RT$, and the probability for growth, $exp(-E_A/RT)$, where ΔG is the chemical driving force for martensite formation, R is the gas constant, T is the temperature and E_A is the activation energy for the formation of isothermal martensite. This approach can be considered to describe the kinetics of *S* martensite formation because it (i) does not require a priori assumptions on the rate-determining mechanism and (ii) indicates that, provided that cooling is sufficiently fast, martensite formation can be suppressed. However, modelling of the kinetics of transformation requires information on E_A.

Following Borgenstam and Hillert, E_A is determined from the slope of the straight line obtained by plotting isothermal data in terms of $-1/T$ versus $\ln(t(M_{si}/T - 1))$, where T, t, and M_{si} are the temperature of isothermal holding, the time for obtaining a low fraction, say <0.05, of martensite, and the maximum temperature at which martensite formation can progress isothermally, respectively. Unfortunately, this quantification method can be applied only in a very limited number of cases, where marked isothermal behaviour is obtained, which makes it suitable only for the case of Fe-Ni-Mn and Fe-Ni-Cr alloys transforming isothermally at sub-zero Celsius temperatures.

In recent work [26], E_A was determined by applying a Kissinger-like approach (see Ref. [28]). In a Kissinger-like analysis, E_A is determined either from the time lapse to a fixed degree of transformation in a series of *isothermal* tests at various holding temperatures or from the evolution of the temperature at which a certain transformed fraction is reached in a series of *isochronal* experiments at various heating/cooling rates. Kissinger-like methods are less robust than the approach by Borgenstam and Hillert because they do not take ΔG into account. Practically, this would imply a systematic underestimation of E_A. The effect is more pronounced for data acquired close to equilibrium conditions (i.e., at temperatures close to M_{si}). An advantage of a Kissinger-like method is that it can be applied for isochronal conditions, thus allowing determination of E_A in all systems where martensite formation can be, at least partially, suppressed upon fast cooling to a sufficiently low temperature. Consistently, in Ref. [26], various Fe-based alloys and commercial steel grades, developing $\{557\}_\gamma$, $\{225\}_\gamma$, and $\{259\}_\gamma$ martensites, were cooled to 77 K by immersion in boiling nitrogen. The transformation was then followed during subsequent isochronal (re)heating and E_A was determined from the slope of the straight line obtained by plotting $\ln(T_{f'}^2/\phi)$ versus $1/T_{f'}$, where $T_{f'}$ is the temperature corresponding to a fixed stage of transformation, f', and ϕ is the heating rate.

In relation to the results of these analyses, Borgenstam and Hillert [27] suggested that small variations of chemical composition do not significantly affect the kinetics of the transformation. Under this assumption, isothermal data collected for Fe-Ni-Mn and Fe–Cr–Ni alloys with comparable total contents of substitutional atoms, but different levels of interstitial purity, were grouped together. For both series of alloys, their analysis yielded an approximate value of $E_A = 7$ kJ mol^{-1}. On the other hand, the Kissinger-like method applied to a broad range of alloys [26] indicated that E_A increased with the fraction of interstitial atoms and ranges from 8 to 27 kJ mol^{-1}. Extrapolation of this data set to low interstitial contents showed striking compatibility with the results obtained according to the approach of Borgenstam and Hillert. The present work aims to validate the two analyses and to obtain reliable information on E_A for future modelling of the transformation kinetics. The following two steps were taken to arrive at this validation.

Firstly, a new series of experiments was conducted on (wt%) 17Cr-7Ni-1Al-0.09C (17-7 PH) and 15Cr-7Ni-1Al-0.08C (15-7 PH) stainless steels, wherein $\{557\}_\gamma$ lath martensite (i.e., interpreted as pure S martensite) develops at sub-zero Celsius temperatures. The experiments included both isothermal and isochronal tests, and the two sets of data were used to assess E_A according to the approach presented by Borgenstam and Hillert and the Kissinger-like methods for isothermal and isochronal analysis (cf. Ref. [28]), respectively. This part of the work aimed at exploring the importance of the systematic underestimation of E_A by Kissinger-like methods.

Secondly, isothermal data considered in Ref. [27] were re-evaluated. As suggested in Ref. [26], E_A varies with the logarithm of the interstitial content. Consequently, small variations in low purity level can significantly affect the kinetics of the transformation. To verify this, each data set in Ref. [27] was re-evaluated independently. Additionally, it is noted that the analysis in Ref. [27] did not distinguish between $\{557\}_\gamma$, $\{225\}_\gamma$ and $\{112\}_\gamma$ martensites, which cannot be reconciled with our interpretation of S martensite. Data referring to $\{112\}_\gamma$ martensite were excluded in the present analysis.

Materials and Methods

The materials chosen for investigation were steels of types 17-7 PH (17Cr-7Ni-1.Al-0.08C) and 15-7 PH (15Cr-7Ni-2Al-1Al-009C). In these alloys, the kinetics of the transformation can be adjusted at convenience by varying the austenitization conditions, and martensite formation on cooling can be fully suppressed [9]. Samples were Ø 3 mm disks with a thickness of 0.15 μm (17-7 PH) and 0.25 μm (15-7 PH) thick, supplied by Goodfellow Inc. in as-rolled (17-7 PH) and annealed (15-7 PH) condition, respectively. Samples were electro-plated with a layer of pure Ni (approx.

0.5 μm thick) prior to austenitization in order to prevent preferential formation of martensite at the free surface. Austenitization was performed in a continuous Ar flow and consisted in heating at an average rate of 1 K s^{-1} to the austenitization temperature, followed by 180 s austenitization at temperature and cooling to room temperature at an average rate of 0.7 K s^{-1}. The austenitization temperatures chosen were 1253 K (980 °C) for PH 15-7 and 1283 K (1010 °C) for PH 17-7. In both cases, the microstructure of the material upon cooling to room temperature consisted of austenite and a minor (3–5%) presence of delta ferrite situated at the austenite grain boundaries.

The formation of martensite was followed applying magnetometry. Details on the experimental setup as well as on the quantification procedure were given elsewhere [26, 29, 30]. Two types of tests were performed: isothermal and isochronal. In order to attain identical starting conditions for the isochronal and isothermal data sets, the samples were first cooled to 77 K in the vibrating sample magnetometer before investigation.

Results and Discussion

From Isothermal and Isochronal Transformation Curves to Sub-zero Celsius Transformation Diagrams

Examples of the experimentally obtained transformation curves for 17-7 PH are presented in Fig. 1. Isothermal data in Fig. 1a indicates that martensite formation is time dependent in the investigated temperature interval, 120–270 K. The highest transformation rate was observed at 195 K. The lowest fraction of martensite forms at 270 K, where less than 0.2% martensite is obtained after 76 ks isothermal holding. Isochronal data in Fig. 1b shows that martensite forms during continuous heating starting from 77 K. For the highest heating rate of 0.833 K s^{-1} (i.e., 50 K min^{-1}), the transformation barely starts and the process stops at approx. 272 K with only 1.5% martensite formed. This temperature is interpreted as M_{si}. For the slowest applied heating rate of 0.00167 K s^{-1} (i.e., 0.1 K min^{-1}), the transformation starts at approx. 110 K and saturates at M_{si} at a maximum value of 74.5% fraction of martensite, f. In the case of PH 15-7 (not shown), the investigated temperature interval was 130–310 K, the highest transformation rate was observed at 230 K and M_{si} equalled 315 K.

All data from Fig. 1 for PH 17-7 and the data for PH 15-7 are presented in Fig. 2a, b, respectively. The isothermal data are represented by the data arranged along horizontal lines, while the isochronal data are arranged along the curves with increasing slope. Interconnecting points of equal transformed

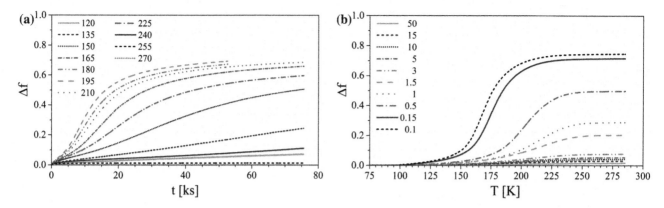

Fig. 1 Fraction of martensite Δf formed in 17-7 PH during **a** isothermal holding at various temperatures and **b** isochronal heating from 80 K at various heating rates. The legend refers to **a** the temperature of isothermal holding in K and **b** the rate of isochronal heating in K min^{-1}

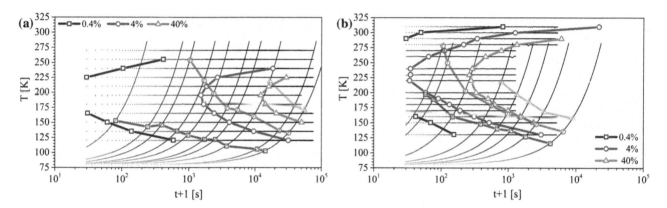

Fig. 2 Superposition of TTT (full colour) and CHT (faint colour) transformation diagrams for: **a** 17-7 PH stainless steel austenitized at 1283 K (1010 °C) and **b** 15-7 PH stainless steel austenitized at 1253 K (980 °C)

fraction provide Time-Temperature-Transformation (TTT) and Continuous-Heating-Transformation (CHT) diagrams. Analogous to Continuous-Cooling-Transformation, (CCT) diagrams, where the transformation lines are shifted to a lower temperature as compared to the corresponding TTT diagrams, the transformation lines for a CHT diagram are shifted to higher temperature as compared to the lines for the corresponding TTT diagrams. In the following, the experimental data were used to quantify E_A.

Assessment of Activation Energy for Martensite Formation

Following the analysis introduced by Borgenstam and Hillert [27], isothermal data in Fig. 2 are presented as $-1/T$ versus $\ln(t(M_{si}/T - 1))$ in Fig. 3. At a transformed fraction of 0.4%, there is good correspondence between data for PH 17-7 (open symbol) and PH 15-5 (closed symbols). For the higher transformed fractions, the slope of the low temperature asymptote, which is used to quantify E_A, is of comparable magnitude for the two steels and remains virtually

unchanged during transformation. This indicates that the rate-determining step for time-dependent martensite formation in the two materials is comparable and remains largely unaltered during the transformation. Differences in absolute kinetics are ascribed to the effect of the microstructure on the evolution of the phase fraction versus time. The activation energy, E_A, was evaluated from the slopes of the low temperature asymptotes. To secure a sufficiently robust analysis, only data acquired at temperatures equal or lower than the maximum transformation rate were considered. The analysis was performed for every increase in f by 0.001 and yielded E_A within the ranges 9.8–14 kJ mol^{-1} and 8.6–15.3 kJ mol^{-1} (mean values 11.3 kJ mol^{-1} and 12.9 kJ mol^{-1}) for PH 17-7 and PH 15-7, respectively.

The same data sets were used to estimate E_A according to Kissinger-like isothermal method [28], where E_A is evaluated from the slope of the straight line obtained by $1/T$ versus $\ln(t)$. Additionally, isochronal analysis was performed as previously reported [26] using the second data set. Isothermal analysis yielded 7.4–12 kJ mol^{-1} and 7.7–13 kJ mol^{-1} (mean values 9.1 kJ mol^{-1} and 10 kJ mol^{-1}) for PH 17-7 and PH 15-7, respectively; isochronal analysis

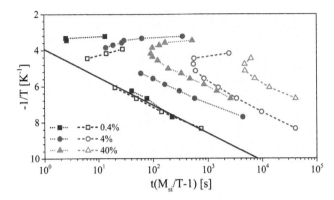

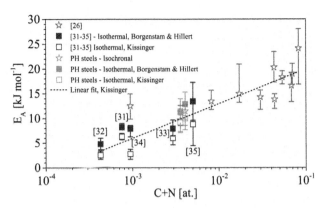

Fig. 3 Isothermal data presented in Fig. 2 according to the analysis proposed by Borgenstam and Hillert [27]. Open symbols connected by dashed lines and full symbols connected by dotted lines refer to PH 15-7 and PH 15-7 steels, respectively. In this present form, data are used to quantify E_A from the slope of the low temperature asymptote (schematically presented for 0.4% fraction transformed)

Fig. 4 Activation energy, E_A, for time-dependent martensite formation as a function of the total atomic fraction of interstitials in the alloy/steel, C + N. Full symbols refer to data obtained based on the method presented by Borgenstam and Hillert. Isothermal data: Refs. [32, 33] extracted at 1 and 5% Δf from the reported Figs.; Ref. [31], extracted at 1 and 5% from Fig. 1 present tabulated at 0.2%; Ref. [34], extracted at 20% f from Fig. 2; Ref. [35], data at 1% f extracted from the presented figure. Isochronal data were reported in previous work by the present authors [26]. The data is presented such that the error bars in E_A are given as the minimum, maximum and average values taking into account the standard error of the estimate for linear regression. Dashed line represent linear fit of data obtained by Kissinger-like methods

yielded E_A within the ranges 7.1–11.9 kJ mol^{-1} and 9.9–12.7 kJ mol^{-1} (mean values 9.8 kJ mol^{-1} and 11.5 kJ mol^{-1}) for PH 17-7 and PH 15-7, respectively. Evidently, isothermal and isochronal analyses yield consistent results. Moreover, there is a fair agreement between the results of the Kissinger-like methods and the Borgenstam-Hillert method. As anticipated, the self-consistent (and driving force omitting) Kissinger-like methods yield systematically lower values, albeit negligible within experimental accuracy. The trend is that the Kissinger analysis of the isochronal data set yields activation energy values in between those obtained with the Borgenstam-Hillert analysis and the Kissinger-like analysis of the isothermal data set.

In the following, isothermal data from the literature for Fe-Ni-Mn and Fe–Cr–Ni and previously used in Ref. [27] are revisited to determine E_A according to the Borgenstam-Hillert method and the Kissinger-like method. The analysis considered only those alloys developing $\{5\,5\,7\}_\gamma$ and $\{2\,2\,5\}_\gamma$ martensites and was performed for individual compositions to verify the dependence on interstitial content. Only data sets consisting of at least 3 data points at and below the maximum transformation temperature were taken into account, provided that a linear regression coefficient better than 0.8 was obtained. The results are presented in Fig. 4 along with the values obtained for PH 17-7 and PH 15-7 as described above and compared with the data in Ref. [26].

Again, a systematic underestimation of E_A for the Kissinger-like method is found as compared to the Borgenstam-Hillert analysis. Clearly, the data is consistent with those for the PH steels investigated in this work. A trend is observed that the activation energy increases with interstitial content for the data in Refs. [31–35], which remained unobserved in the evaluation in Ref. [27]. In

comparison with an assessment of the dependence of the activation energy on interstitial content obtained for a broad range of iron-based alloys and steels in Ref. [26] (and earlier in Ref. [7]), excellent correspondence is obtained and the trend of decreasing activation energy with logarithmic lowering of the interstitial content is confirmed (cf. Fig. 4). A rough quantitative relationship can be obtained by linear fit of data. Recognizing that the data in Ref. [26] rely on a Kissinger-like analysis of isochronal data, the linear fit in Fig. 4 was restricted to this type of analysis, yielding:

$$E_A = (27.0 \pm 1.9) + (7.0 \pm 0.8) \cdot Log(C + N) \quad (1)$$

where C + N represents the total content of C and N atoms in at. fraction. In line with the above-mentioned omission of the driving force in Kissinger-like analyses, this equation is likely to represent an underestimation.

The strong dependence of the activation energy on the interstitial content would be consistent with solid-solution strengthening of austenite and with the idea by Ghosh and Olson [36] that the interaction of solute atoms with the thermally assisted motion of the martensitic interface rate control the isothermal process. At present it is not clear whether E_A would eventually reach zero for a sufficiently low interstitial content, or whether the formation of S martensite is intrinsically time-dependent. In perspective, to fully understand the nature of martensitic transformation in steel, work should be initiated to address this fundamental question.

Conclusions

The time-dependent formation of martensite in (wt%) 17Cr-7Ni-1Al-0.09C and 15Cr-7Ni-2Mo-1Al-0.08C steels can be fully suppressed by immersion in boiling nitrogen. The transformation kinetics were studied in isothermal tests and isochronal heating experiments and yielded TTT and CHT (continuous heating transformation) diagrams for martensite formation in the sub-zero Celsius temperature regime.

The activation energy of time-dependent martensite formation was determined by applying the Borgenstam-Hillert and Kissinger-like methods for isothermal and isochronal analysis. The two approaches yield consistent results and indicated that the activation energy in these steels is 9–13 kJ mol^{-1}.

The present data were combined with a large number of partly re-interpreted literature data and establishes a logarithmic dependence on the total fraction of interstitials in the Fe-based alloys/steels, suggesting that solution strengthening determines the rate of isothermal martensite formation at sub zero Celsius temperature.

Acknowledgements This work was financially supported by the Danish Council for Independent Research [grant number: DFF-4005-00223]. The first author acknowledges Otto Mønsted fund for financially supporting the participation in ICOMAT 2017.

References

1. Koistinen DP, Marburger RE (1959) A general equation prescribing the extent of the austenite-martensite transformation. Acta Metall 7(1):59–60
2. Fei HY, Hedstrom P, Hoglund L, Borgenstam A (2016) A thermodynamic-based model to predict the fraction of martensite in steels. Metall Mater Trans 47A(9):4404–4410
3. Kurdyumov GV, Maksimova OP (1948) O kinetike prevrashcheniya austenitica v martensit pri nizikh temperaturah. Dokl Akad Nauk SSSR 61(1):83–86
4. Thadhani NN, Meyers MA (1986) Kinetics of isothermal martensitic transformation. Prog Mater Sci 30:1–37
5. Lobodyuk VA, Estrin EI (2005) Isothermal martensitic transformations. Physics-uspekhi 48(7):713–732
6. Zhao JC, Notis MR (1995) Continuous cooling transformation kinetics versus isothermal transformation kinetics of steels: a phenomenological rationalization of experimental observations. Mater Sci Eng R 15(4–5):135–207
7. Villa M, Christiansen T, Hansen MF, Somers MAJ (2015) Investigation of martensite formation in Fe based alloys during heating from boiling nitrogen temperature. Metall Ita 11–12:39–46
8. Cech RE, Hollomon JH (1953) Rate of formation of isothermal martensite in Fe-Ni-Mn alloy. Trans AIME J Metals 197(5):685–689
9. Gulyaev AP, Makarov VM (1960) Martensitic transformation, mechanical properties, and structure of austenitic-martensitic stainless steels. Metal Sci Heat Treat 2(8):419–423
10. Ullakko K, Nieminen M, Pietikäinen J (1990) Prevention of martensitic transformation during rapid cooling. Mater Sci Forum 56–58:225–228

11. Kulin S, Cohen M (1950) On the martensitic transformation at temperatures approaching absolute zero. Trans AIME 188 (9):1139–1143
12. Wiester HJ (1932) Die MartensitKristallisation in Filmbild. Ztsch Metallkunde 24(11):276–277
13. Foerster F, Scheil E (1940) Unterschung des zeitlichen Ablaufes von Umklappvorgängen in Metallen. Ztsch Metallkunde 32 (6):165–173
14. Bunshah RF, Mehl RF (1953) Rate of propagation of martensite. Trans AIME 197(9):1251–1258
15. Yu ZZ, Clapp PC (1989) Growth dynamics study of the martensitic transformation in Fe-30 Pct Ni alloys. I. Quantitative measurements of growth velocity. Metall Trans 20A(9):1601–1615
16. Nishiyama Z (1978) Martensitic transformations. Academic Press, New York, pp 236–238
17. Honma T (1958) Studies on lattice transformation of iron alloy and cobalt alloy (No.1). DENKI-SEIKO 29(4):261–276
18. Nemirovskiy VV (1968) A study of the martensitic transformation in iron-nickel near compositions corresponding to a change in the mechanisms of the transformation. Fiz Met Metalloved 25(5): 900–909
19. Ya Georgiyeva I, Maksimova OP (1971) Relation between kinetics and structure during martensitic transformations. Phys Metal Metallogr 32(2):135–146
20. Takashima K, Higo Y, Nunomura S (1980) Identification of acoustic emission during the martensite transformation of 304 stainless steel. Scripta Metall 14(5):489–491
21. Takashima K, Higo Y, Nunomura S (1984) The propagation velocity of the martensitic transformation in 304 stainless steel. Philos Mag A 49(2):231–241
22. Wilson EA (1994) The $\gamma \rightarrow \alpha$ transformation in low carbon irons. ISIJ Int 34(8):615–630
23. Mirzayev DA, Shteynberg MM, Ponomareva TN, Schastlivtsev VM (1979) Influence of cooling rate on the position of martensitic transformation points. 1. Carbon steels. Phys Metal Metallogr 47(1):102–111
24. Mirzayev DA, Shteynberg MM, Ponomareva TN, Schastlivtsev VM (1979) Influence of cooling rate on the position of martensitic transformation points. 2. Alloyed steels. Phys Metal Metallogr 47(5):73–79
25. Huizing R, Klostermann JA (1966) The martensite transformation in small (0.1-0.3 Mm) iron-nickel single crystals. Acta Metall 14 (12):1693–1702
26. Villa M, Somers MAJ (2017) Thermally activated martensite formation in ferrous alloys. Scr Mater 142(1):46–49
27. Borgenstam A, Hillert M (1997) Activation energy for isothermal martensite in ferrous alloys. Acta Mater 45(2):651–662
28. Mittemeijer EJ (1992) Analysis of the kinetics of phase transformation. J Mater Sci 27(15):3977–3987
29. Villa M, Hansen MF, Pantleon K, Somers MAJ (2015) Anomalous kinetics of lath martensite formation in stainless steel. Mater Sci Technol 31(11):1355–1361
30. Villa M, Hansen MF, Somers MAJ (2015) The sub-zero celsius treatment of precipitation hardenable semi-austenitic stainless steel. In: Proceedings of the 28th ASM HTS conference. Detroit, MI, pp 431–435
31. Shih CH, Averbach BL, Cohen M (1955) Some characteristics of the isothermal martensitic transformation. J Metals 7(1):183–187
32. Ghosh G, Raghavan V (1986) The kinetics of isothermal martensitic transformation in an Fe-23.2 wt.%Ni-2.8 wt.%Mn alloy. Mater Sci Eng 800(1):65–67
33. Cech RE, Hollomon JH (1953) Rate of formation of isothermal martensite in Fe-Ni-Mn alloy. Trans AIME J Metals 197(5): 685–689

34. Holmquist M, Nilsson JO, Stigenberg AH (1995) Isothermal formation of martensite in a 12Cr-9Ni-4Mo maraging stainless steel. Scr Metal Mater 33(9):1367–1373

35. Imai Y, Izumiyama M, Sasaki K (1966) Isothermal martensitic transformation in Fe-Ni-Cr alloy. Sci Report Research Inst, Tohoku Uni A 18(1):39–48

36. Ghosh G, Olson GB (1994) Kinetics of fcc → bcc heterogeneous martensitic nucleation-II. Thermal-activation. Acta Metall Mater 42(10):3371–3379

The Effects of Prior-γ Grain Boundary Segregation of Phosphorus, Manganese and Molybdenum on Intergranular Fracture Stress in Low Carbon Martensite Steels

Masahide Yoshimura, Manabu Hoshino, Masanori Minagawa, and Masaaki Fujioka

Abstract

Influence of prior-γ grain boundary segregation of alloy elements on intergranular fracture stress is important for the mechanism of temper embrittlement. There are a few efforts based on pure-iron [1], but no report on low carbon martensitic steels. In this study, the effect of segregation of phosphorus (P), manganese (Mn) and molybdenum (Mo) was investigated. The samples were melted by changing the amount of P, Mn and Mo based on the base Fe-0.1%C-3%Mn-90 ppmP. The martensitic steels with coarse prior-gamma (γ) were made by quenching and tempering. The segregation was measured by Auger electron spectroscopy, and the intergranular fracture stress was regarded as the yield strength at ductile brittle transition temperature of Charpy V-notch test. This study revealed that the segregation of P weakened the fracture stress mostly in the order of P and Mn, and that of Mo strengthened the fracture stress quantitatively. Mn-P co-segregation was not observed. The segregation of P was decreased by the addition of Mo.

Keywords

Temper-embrittlement • Segregation • Intergranular fracture • Fracture stress • Martensite Grain boundary • Phosphorus • Manganese Molybdenum • Steel

Introduction

For conventional low carbon steels, with the increase in tensile strength, Charpy toughness decreases. Low temperature fractures are normally transgranular. However, tempered martensite shows embrittlement in some cases with fracture surfaces along prior-gamma (γ) grain boundaries. Charpy toughness deteriorates significantly. According to conventional knowledge [2], firstly phosphorus (P) segregates to the grain boundaries during tempering, the minimum energy required for fractures, so-called grain boundary cohesive energy decrease, intergranular fracture stress decreases, and finally ductile-brittle transition temperature (DBTT$_{GB}$) shifts to a higher temperature.

On the other hand, manganese (Mn) and molybdenum (Mo) are empirically known as embrittlement and toughness elements [3], but the mechanisms are not clear especially for carbon containing martensite. It is required to divide the direct effect of segregation of itself and the indirect effect mediated by the segregation of P. That is to say, it is important to determine whether Mn weakens grain boundaries or Mo strengthens grain boundaries quantitatively. In addition, it is important to determine whether Mn assists the segregation of P and Mo suppresses the segregation of P [4, 5]. The purpose of this work is to clarify the influence of grain boundary segregations of P, Mn, Mo on intergranular fracture stress and the influence of Mn, Mo on grain boundary segregations of P. Therefore, grain boundary segregations and intergranular fracture stress are measured independently.

Experiments

In order to investigate the influence of alloy elements, the samples were melted by changing the amount of P, Mn, Mo based on Fe-0.1%C-3%Mn-90 ppmP (see Table 1). The martensitic steels with coarse prior-γ were made by

M. Yoshimura (✉) · M. Hoshino · M. Fujioka
Nippon Steel & Sumitomo Metal Corporation, Futtsu, Japan
e-mail: yoshimura.f8c.masahide@jp.nssmc.com

M. Minagawa
Nippon Steel & Sumikin Technology Corporation, Futtsu, Japan

A. P. Stebner and G. B. Olson (eds.), *Proceedings of the International Conference on Martensitic Transformations: Chicago*,
The Minerals, Metals & Materials Series, https://doi.org/10.1007/978-3-319-76968-4_3

Table 1 Chemical compositions

Ferrous alloys	C	P*	Mn	Mo	Others	mass%, *ppm
3Mn-P	0.095	50	3.0	–	Al = 0.025, Si ≤ 0.01, S* ≤ 20 N* ≤ 11, O* ≤ 10, Fe	
3Mn	0.094	90	3.0	–		
3Mn + P	0.094	450	3.0	–		
5Mn	0.099	100	5.0	–		
3Mn + 0.25Mo	0.097	90	3.0	0.25		

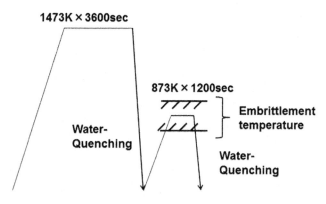

Fig. 1 Process

Fig. 2 Microstructure of 3Mn with nital eching

quenching from 1473 K and tempering at 873 K containing the embrittlement range below Ac1 and subsequently, quenching (Fig. 1). For the measurement of prior-gamma grain boundary segregation, Auger electron spectroscopy was used. Analysis conditions are beam voltage: 10 keV, beam current:10 nA. Segregation was determined by relative sensitivity methods. Auger peak values of analysis are P: 120 eV, C: 272 eV, Mn: 542 eV, Mo: 186 eV, Fe: 703 eV. Relative sensitivity coefficient value was used from JEOL data set. The samples were broken in a vacuum chamber at 77 K and measurements were performed except for precipitates with dispersion.

Experimental Results

Samples in this work are typical martensitic steels with lath structures (Fig. 2). Prior-gamma size is coarse 360–577 μm. All samples fracture along prior-γ grain boundaries in low temperature Charpy test. All samples have the almost same yield stress, between 615 and 688 MPa.

DBTT$_{GB}$ deteriorated with the increase in segregation of P (Fig. 3 line). By the addition of Mn, toughness remarkably deteriorates and the segregation of P was not changed (Fig. 3). Mn-P co-segregation was not observed. If the deterioration by added Mn was only caused by segregation of P, about 8% of segregation of P was required. On the other hand, by the addition of Mo, toughness improves and the results show suppression of the segregation of P and the

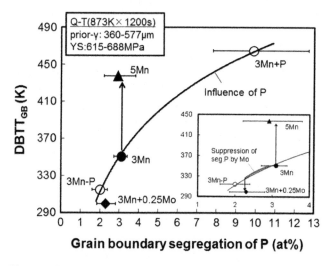

Fig. 3 Influence of P, Mn, Mo on the relationship between grain boundary segregation of P and DBTT$_{GB}$. The range of low concentration of P is shown in the inserted figure

direct effects of Mo itself (Fig. 3). The results show that Mn has only the direct effect of segregation of Mn itself, Mo has the direct effect and the indirect effect mediated by segregation of P.

Discussion

Intergranular fracture stress was estimated in this study as yield stress at DBTT$_{GB}$ temperature. When yield stress approaches intergranular fracture stress, fracture occurs. That

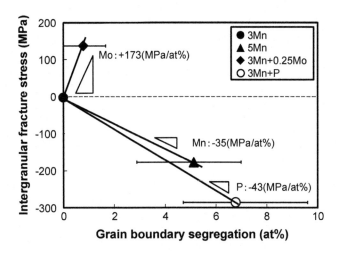

Fig. 4 Influence of grain boundary segregation on intergranular fracture stress

was calculated by yield stress in a tensile test at room temperature and conversion of the strain rate of the Charpy impact test from 10^{-3}(/s) to 10^{3}(/s) by using the strain rate temperature effect (Eq 1). Influence of temperature on yield stress was used by the reference [6] (Eq 2).

$$R = T \times \ln(10^8/\dot{e}) \qquad (1)$$

$$\sigma_{YS} = -2.5 \times T + Const. \qquad (2)$$

R is the strain rate temperature effect index. T is temperature (K). $\dot{e}$ is strain rate (/s). σ_{YS} is yield stress (MPa).

By calculating intergranular fracture stress without the indirect effect mediated by segregation of P, the influence of yield stress can be excluded, and intergranular fracture stress

corresponds to the segregation of P, Mn and Mo itself independently (Fig. 4). Influence of alloy elements on intergranular fracture stress was clarified quantitatively. This study revealed that the segregation of P weakened the fracture stress mostly in the order of P and Mn, and that of Mo strengthened the fracture stress. The degree of contribution per 1 at % in this work approximately corresponds to the grain boundary cohesive energy of sigma 3 ab initio calculations based on pure-Fe [7, 8].

Summary

This study revealed that the segregation of P weakened the fracture stress mostly in the order of P and Mn, and that of Mo strengthened the fracture stress quantitatively. The results of this work agree with ab initio calculations. As an indirect effect, the addition of Mo depresses the segregation of P. Mn-P co-segregation was not observed.

References

1. Lee DY, Barrera EV, Stark JP, Marcus HL (1984) Metall Trans A 15:1415
2. Mulford RA, McMahon CJ Jr, Pope DP, Feng HC (1976) Metall Trans A 7:1183
3. McMahon CJ Jr (1989) Mater Sci Forum 46:61
4. Yu J, McMahon CJ Jr (1980) Metall Trans 11A:277
5. Yu J, McMahon CJ Jr (1980) Metall Trans 11A:291
6. Petch NJ (1986) Acta Metall 34:1387
7. Geng WT, Freeman AJ, Olson GB (2001) Phys Rev B 63:165415
8. Yamaguchi M (2011) Metall Trans A 42:319

Effect of Molybdenum Content on Hardenability of Boron and Molybdenum Combined Added Steels

Kyohei Ishikawa, Hirofumi Nakamura, Ryuichi Homma, Masaaki Fujioka, and Manabu Hoshino

Abstract

The upper limit of the Mo-B combined effect on hardenability was investigated in 0.15%C-B added steels containing 0 to 1.5%Mo. The hardenability of Mo-B steels increases up to 0.75%Mo suppressing the precipitation of $Fe_{23}(C,B)_6$. In contrast, the effect decreases over 0.75%Mo where Mo_2FeB_2 precipitates instead of $Fe_{23}(C,B)_6$. By thermodynamic calculation, it is suggested that Mo_2FeB_2 precipitates during reheating and the precipitation of Mo_2FeB_2 decreases the solute B content in reheating, which determines the limit of the Mo-B combined effect.

Keywords

Hardenability • Boron • Segregation • Molybdenum Combined effect

Introduction

In making high tensile steels by low alloying cost, it is important to utilize Boron (B), because the addition of only several ppm B dramatically increases the hardenability of steel [1, 2]. The solute B atoms that are segregated on austenite grain boundaries increase the hardenability of steels [3–5], whereas, the precipitation of borides like BN, $Fe_{23}(C,B)_6$, or Fe_2B decrease hardenability [6–9]. To prevent the precipitation of $Fe_{23}(C,B)_6$ and Fe_2B [8, 10–12], Mo has an important role. Mo expands the optimal value of B content in increasing hardenability by suppressing $Fe_{23}(C, B)_6$, which is called the "Mo-B combined effect" [11].

However, the upper limit of Mo content in the combined effect is not clear. For instance, H. Asahi [1] reported that the combined effect decreases at more than 13 ppm of B content even in the 0.5%Mo steels. Therefore, in this study, to reveal the upper limit of the Mo-B combined effect, we investigated the effect of Mo and B content on the hardenability in a wider range of Mo contents than these previous studies up to 1.5%Mo. We also observed the borides to obtain the mechanism for determining the limit.

Experimental Procedure

The chemical compositions of the steels are given in Table 1. Each steel is named after the contents of Mo and B, for example, 0.50% Mo steel with 20 ppm B is named 05Mo20B. To prevent the precipitation of BN, Ti is added to all steels to fix N as TiN. Mo is added up to 1.5%. These steels are prepared in laboratory facilities, by being melted in a vacuum induction furnace, and cast into 50 kg ingots. The ingots were re-heated at 1250 °C for 60 min and hot-rolled into 35 mm thick plates, from which specimens were machined.

The ordinary Jiminy end-quench test was conducted to measure the hardenability of the steels. Figure 1a shows heat patterns of the specimens for the Jiminy test. The specimen dimensions are 25 mm in diameter and 100 mm in length. All specimens were water quenched at one end after being austenitized at 950 °C for 45 min. Hardness distribution from the quench end was measured on two pieces of flat ground 1 mm deep by a Rockwell hardness tester. In this study, the critical cooling rate V_{c-90} (°C/s) at which the hardness corresponds to 90% martensite structure was employed as the hardenability index [13, 14]. The reported data was used as the hardness corresponding to the 90% martensite structure [15].

Figure 1b shows heat patterns of the specimens for observations of B precipitation behavior. The specimen dimensions are 8 mm in diameter and 12 mm in length.

K. Ishikawa (✉) · H. Nakamura · M. Fujioka · M. Hoshino
Nippon Steel & Sumitomo Metal Corporation, Steel Research Laboratories, 20-1 Shintomi, Futtsu, Chiba 293-8511, Japan
e-mail: ishikawa.6mz.kyohhei@jp.nssmc.com

R. Homma
Nippon Steel & Sumitomo Metal Corporation, Kimitsu R&D Laboratories, 1-1 Kimitsu, Chiba, 299-1141, Japan

© The Minerals, Metals & Materials Society 2018
A. P. Stebner and G. B. Olson (eds.), *Proceedings of the International Conference on Martensitic Transformations: Chicago*,
The Minerals, Metals & Materials Series, https://doi.org/10.1007/978-3-319-76968-4_4

Table 1 Chemical compositions of sample steels (mass%, *ppm)

No.	C	Si	Mn	P*	S*	Ti	Al	Mo	B*	N*	O*
(0, 20)B	0.15	0.27	1.31	<20	19	0.020	0.020	0.00	(0, 20)	7	<10
05Mo(0, 20)B	0.14	0.27	1.29	<20	20	0.020	0.017	0.50	(0, 20)	8	<10
07Mo(0, 20)B	0.14	0.27	1.28	<20	20	0.020	0.017	0.76	(0, 20)	8	<10
10Mo(0, 20)B	0.14	0.28	1.27	<20	20	0.020	0.017	1.01	(0, 20)	13	<10
15Mo(0, 20)B	0.14	0.28	1.26	<20	20	0.020	0.017	1.52	(0, 20)	18	<10

Fig. 1 Heat patterns of the specimens for **a** Jiminy test and **b** TEM

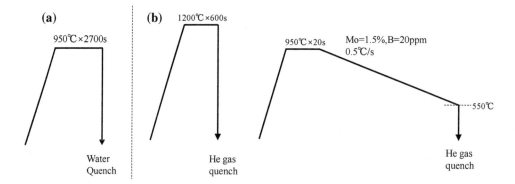

These specimens were reheated at 1200 °C for 600 s for solid solution treatment and quenched by helium gas blowing. Then they were austenitized at 950 °C for 20 s and helium(He)gas quenched after cooling down to 550 °C by He gas blowing to investigate the state of borides in the austenite phase. The cooling rate after being austenitized was 0.5 °C/s. After these heat treatments, the identification of B precipitates was conducted using a transmission electron microscope (TEM). The TEM sample used in this observation was prepared by the extraction replica method after electrolytic etching.

Results

Effect of Mo on the Hardenability

Figure 2 shows the effect of the Mo content on the V_{c-90} of Mo steels (B-free) and Mo-B steels (20 ppm B added). In B-free steels, the hardenability index V_{c-90} of each steel decreased with the increase in the Mo content. On the other hand, in Mo-B steels, the V_{c-90} decreased up to 0.75%Mo and saturated over 0.75%Mo. If we assume that the effect of Mo itself on hardenability in Mo-B steels is equal to that in B-free steels, we can recognize the hatched zone in Fig. 2 as the Mo-B combined effect. This suggests that the Mo-B combined effect has an optimum Mo content around 0.75%, and it decreases more than 0.75%.

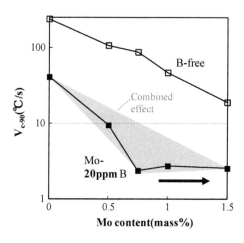

Fig. 2 Effect of Mo content on Hardenability of B-free steel and B-added steel

Precipitation Behavior

To understand the origin of the change in the Mo-B combined effect with Mo content, which decreases over 0.75% Mo, it is important to investigate borides mainly in the prior austenite grain boundary. In 15Mo20B steel, at the cooling rate of 0.5 °C/s, there was no $Fe_{23}(C, B)_6$ reported in previous studies regarding B added steel. On the other hand, a different type of boride, Mo_2FeB_2 was observed. Figure 3a shows the extraction replica TEM photograph of Mo_2FeB_2 in 15MoB20 steel. Figure 3b shows EDS spectra of

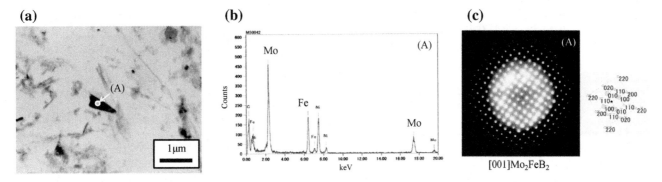

Fig. 3 **a** Extraction replica TEM photograph, **b** EDS spectra and **c** Electron diffraction pattern of boron precipitate along grain boundaries in 15Mo20B steel cooled at 0.5°C/s from austenitizing temperature

Mo_2FeB_2 and Fig. 3c shows the electron diffraction pattern. The precipitations shown in Fig. 3a were determined as Mo_2FeB_2 by the electron diffraction analysis. The intensity of Mo peaks in EDS spectra were almost twice as large as that of Fe, which is consistent with the diffraction analysis. Although Mo_2FeB_2 type bridges have already been reported in many studies, especially in the field of Maraging Steels [16] and cermets produced by the sintering method [17], it is the first time to report the precipitation in low-carbon steels and to have a role in determining the upper limit of the Mo-B combined effect on hardenability.

Discussion

In this study, we observed the increase of the Mo-B combined effect on the hardenability by 0.75%Mo and the decrease of the combined effect over 0.75%Mo where Mo_2FeB_2 precipitated instead of $Fe_{23}(C,B)_6$. It is clear that the hardenability property is affected by the precipitation of Mo_2FeB_2.

To estimate the effect of precipitation of Mo_2FeB_2, we conducted thermodynamic calculations by using Thermo-calc. In this calculation, considering the experimental results, we used a thermodynamic database [18] which contains M_3B_2 phases. To focus on the effect of Mo_2FeB_2, this calculation is composed of only M_3B_2 and FCC phases with 0.15%C, 1.3%Mn, 0.002%B and 0-1.5% Mo assuming bulk content.

Figure 4 shows the B contents in M_3B_2 and FCC phases calculated by Thermo-Calc. In the case of 1.5%Mo, Mo_2FeB_2 can precipitate below 1130 °C. Considering the reheating temperature of the experiment was 950 °C, the calculation means that Mo_2FeB_2 observed in 15Mo20B steel precipitated during reheating. Furthermore, the B content in FCC, which corresponds to the solute B content, decreases by the precipitation of Mo_2FeB_2. By this result, it is suggested that the decreases of the combined effect over 0.75% Mo were caused by the precipitation of Mo_2FeB_2 through the decrease of solute B content in reheating.

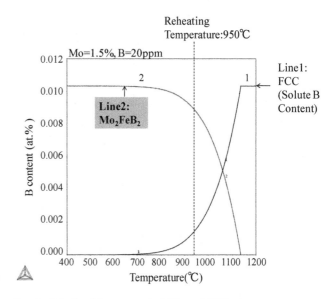

Fig. 4 Calculated B contents in M_3B_2 and FCC phases

Conclusions

- The hardenability of Mo-B steels increases up to 0.75% Mo. In contrast, the effect decreases over 0.75%Mo, and in 1.5%Mo-20 ppm B added steels, Mo_2FeB_2 was observed instead of $Fe_{23}(C,B)_6$.
- By experimental results and thermodynamic calculation, it is suggested that the decreases of the combined effect over 0.75%Mo were caused by the precipitation of Mo_2FeB_2 decreasing the solute B content in reheating.

References

1. Grossman MA (1942) Trans AIME 150:227
2. Ueno M, Itoh K (1988) Tetsu-to-Hagané 74:910
3. Simcoe CR, Elsea AR, Manning GK (1956) Trans AIME J Metals 8:984

4. Morral JE, Cameron JB (1977) Met Trans 8A:1817

5. Yoshida S, Ushioda K, Abe Y, Ågren J (2012) Tetsu-to-Hagané 98:482

6. Ueno M, Inoue T (1973) Trans Iron Steel Inst Jpn 13:210

7. Ohno Y, Okamura Y, Matsuda S, Yamamoto K, Mukai T (1987) Tetsu-to-Hagané 73:1010

8. Fujishiro T, Hara T, Shigesato G (2015) Tetsu-to-Hagané 101:300

9. Habu R, Miyata M, Sekino S, Goda S (1974) Tetsu-to-Hagané 60:1470

10. Tamehiro H, Murata M, Habu R, Nagumo M (1986) Tetsu-to-Hagané 74:458

11. Asahi H (2002) ISIJ Int 42:1150

12. Hara T, Asahi H, Uemori R, Tamehiro H (2004) ISIJ Int 44:1431

13. Ueno M, Itoh K (1988) Tetsu-to-Hagané 74:918

14. Ueno M, Itoh K (1988) Tetsu-to-Hagané 74:1073

15. Hodge JM, Orehoski MA (1946) Trans Metall Soc AIME 167:627

16. Kuribayashi K, Iloriuchi R (1988) Maraging steels; recent development and application. In: Wilson RK (ed) Proceedings of the symposium of TMS annual meeting. TMS, p 157

17. Kondo Y, Watanabe T, Ohira S, Takagi K, Ide T (1987) Bull Jpn Inst Met 26:302

18. Yamada K, Ohtani H, Hasebe M (2009) J Jpn Inst Met 73:180

Interaction of Martensitic Microstructures in Adjacent Grains

John M. Ball and Carsten Carstensen

Abstract

It is often observed that martensitic microstructures in adjacent polycrystal grains are related. For example, micrographs of Arlt (J Mat Sci 22:2655–2666, 1990) [1] (one reproduced in (Bhattacharya, Microstructure of martensite, 2003) [10, p 225]) exhibit propagation of layered structures across grain boundaries in the cubic-to-tetragonal phase transformation in BaTiO$_3$. Such observations are related to requirements of compatibility of the deformation at the grain boundary. Using a generalization of the Hadamard jump condition, this is explored in the nonlinear elasticity model of martensitic transformations for the case of a bicrystal with suitably oriented columnar geometry, in which the microstructure in both grains is assumed to involve just two martensitic variants, with a planar or non-planar interface between the grains.

Keywords

Bicrystal • Compatibility • Grain boundary
Hadamard jump condition

Description of Problem

Consider a bicrystal consisting of two columnar grains $\Omega_1 = \omega_1 \times (0, d)$ (grain 1), $\Omega_2 = \omega_2 \times (0, d)$ (grain 2), where $d > 0$ and $\omega_1, \omega_2 \subset \mathbb{R}^2$ are bounded Lipschitz domains whose boundaries $\partial \omega_1, \partial \omega_2$ intersect nontrivially, so that $\partial \omega_1 \cap \partial \omega_2$ contains points in the interior ω of $\overline{\omega}_1 \cup \overline{\omega}_2$ (see Fig. 1). Let $\Omega = \omega \times (0, d)$. The interface between the grains is the set $\partial \Omega_1 \cap \partial \Omega_2 \cap \Omega = (\partial \omega_1 \cap \partial \omega_2 \cap \omega) \times (0, d)$. Since by assumption the boundaries $\partial \omega_1, \partial \omega_2$ are locally the graphs of Lipschitz functions, and such functions are differentiable almost everywhere, the interface has at almost every point (with respect to area) a well-defined normal $\mathbf{n}(\theta) = (\cos \theta, \sin \theta, 0)$ in the (x_1, x_2) plane. We say that the interface is *planar* if it is contained in some plane $\{\mathbf{x} \cdot \mathbf{n} = k\}$ for a fixed normal $\mathbf{n}$ and constant k.

We use the nonlinear elasticity model of martensitic transformations from [6, 8], with corresponding free-energy density $\psi(\nabla \mathbf{y}, \theta)$ for a single crystal at temperature θ and deformation $\mathbf{y} = \mathbf{y}(\mathbf{x})$ with respect to undistorted austenite at the critical temperature θ_c at which the austenite and martensite have the same free energy. We denote by $\mathbb{R}^{n \times n}_+$ the set of real $n \times n$ matrices $\mathbf{A}$ with $\det \mathbf{A} > 0$, and by $SO(n)$ the set of rotations in $\mathbb{R}^n$. At a fixed temperature $\theta < \theta_c$, we suppose that

$$K = SO(3)\mathbf{U}_1 \cup SO(3)\mathbf{U}_2 \qquad (1)$$

is the set of $\mathbf{A} \in \mathbb{R}^{3 \times 3}_+$ minimizing $\psi(\mathbf{A}, \theta)$, where $\mathbf{U}_1 = \text{diag}(\eta_2, \eta_1, \eta_3)$, $\mathbf{U}_2 = \text{diag}(\eta_1, \eta_2, \eta_3)$ and $\eta_2 > \eta_1 > 0$, $\eta_3 > 0$. This corresponds to a tetragonal to orthorhombic phase transformation (see [10, Table 4.6]), or to an orthorhombic to monoclinic transformation in which the transformation strain involves stretches of magnitudes η_1, η_2 with respect to perpendicular directions lying in the plane of two of the orthorhombic axes and making an angle

J. M. Ball (✉)
Mathematical Institute, University of Oxford, Woodstock Road,
Oxford, OX2 6GG, UK
e-mail: ball@maths.ox.ac.uk

C. Carstensen
Department of Mathematics, Humboldt-Universität Zu Berlin,
Unter Den Linden 6, 10099 Berlin, Germany
e-mail: cc@math.hu-berlin.de

© The Minerals, Metals & Materials Society 2018
A. P. Stebner and G. B. Olson (eds.), *Proceedings of the International Conference on Martensitic Transformations: Chicago*,
The Minerals, Metals & Materials Series, https://doi.org/10.1007/978-3-319-76968-4_5

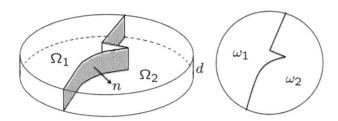

Fig. 1 Bicrystal consisting of two grains $\Omega_1 = \omega_1 \times (0, d), \Omega_2 = \omega_2 \times (0, d)$

of $\pi/4$ with respect to these axes.[1] Alternatively, for example, taking $\eta_3 = \eta_1$ the analysis of this paper can be viewed as applying to a cubic to tetragonal transformation under the a priori assumption that only two variants are involved in the microstructure.

We suppose that Ω_1 has cubic axes in the coordinate directions $\mathbf{e}_1, \mathbf{e}_2, \mathbf{e}_3$, while in Ω_2 the cubic axes are rotated through an angle α about $\mathbf{e}_3$. By adding a constant to ψ we may assume that $\psi(\mathbf{A}, \theta) = 0$ for $\mathbf{A} \in K$. Then a zero-energy microstructure corresponds to a gradient Young measure[2] $(\nu_\mathbf{x})_{\mathbf{x} \in \Omega}$ such that

$$\text{supp } \nu_\mathbf{x} \subset K \text{ for a.e. } \mathbf{x} \in \Omega_1, \text{ supp } \nu_\mathbf{x} \subset K\mathbf{R}_\alpha \text{ for a.e. } \mathbf{x} \in \Omega_2,$$
$$(2)$$

where

$$\mathbf{R}_\alpha = \begin{pmatrix} \cos\alpha & -\sin\alpha & 0 \\ \sin\alpha & \cos\alpha & 0 \\ 0 & 0 & 1 \end{pmatrix}.$$

It is easily shown that $K\mathbf{R}_\alpha = K$ if and only if $\alpha = n\pi/2$ for some integer n, and that $K\mathbf{R}_{\alpha+\pi/2} = K\mathbf{R}_\alpha$. We thus assume that $0 < \alpha < \frac{\pi}{2}$, since this covers all nontrivial cases.

As remarked in [5], by a result from [9] there always exists a zero-energy microstructure constructed using laminates, with gradient Young measure $\nu_\mathbf{x} = \nu$ satisfying (2) that is independent of x and has macroscopic deformation

---

[1]The general form of the transformation stretch for an orthorhombic to monoclinic transformation is given in [8, Sect. 2.10(4)]. In general one can make a linear transformation of variables in the reference configuration which turns the corresponding energy wells into the form (1). However, in [4, Sect. 4.1] and the announcement of the results of the present paper in [5] it was incorrectly implied that the analysis based on K as in (1) applies to a general orthorhombic to monoclinic transformation. This is not the case because the linear transformation in the reference configuration changes the deformation gradient corresponding to austenite in [4] and to the rotated grain in the present paper. A more general, but feasible, analysis would be needed to cover the case of general orthorhombic to monoclinic transformations.

[2]For an explanation of gradient Young measures and how they can be used to represent possibly infinitely fine microstructures see, for example, [2].

gradient $\bar{\nu} = \int_{\mathbb{R}^{3\times3}_+} \mathbf{A} \, d\nu(\mathbf{A}) = (\mathbf{U}_1)\mathbf{1}$. Our aim is to give conditions on the deformation parameters η_1, η_2, η_3, the rotation angle α and the grain geometry which ensure that *any* zero-energy microstructure has a degree of complexity in each grain, in the sense that it does not correspond to a pure variant with constant deformation gradient in either of the grains.

Rank-One Connections Between Energy Wells

Let $\mathbf{U} = \mathbf{U}^T > 0$, $\mathbf{V} = \mathbf{V}^T > 0$. We say that the energy wells $SO(3)\mathbf{U}$, $SO(3)\mathbf{V}$ are *rank-one connected* if there exist $\mathbf{R}, \mathbf{Q} \in SO(3)$, $\mathbf{a}, \mathbf{n} \in \mathbb{R}^3$, $|\mathbf{n}| = 1$ with $\mathbf{RU} = \mathbf{QV} + \mathbf{a} \otimes \mathbf{n}$, where without loss of generality we can take $\mathbf{Q} = \mathbf{1}$. By the Hadamard jump condition this is equivalent to the existence of a continuous piecewise affine map $\mathbf{y}$ whose gradient $\nabla \mathbf{y}$ takes constant values $\mathbf{A} \in SO(3)\mathbf{U}$ and $\mathbf{B} \in SO(3)\mathbf{V}$ on either side of a plane with normal $\mathbf{n}$. The following is an apparently new version of a well-known result (see, for example, [2, Theorem 2.1, 6, Prop. 4, 12]), giving necessary and sufficient conditions for two wells to be rank one connected. A similar statement was obtained by Mardare [13].

Lemma 1 *Let* $\mathbf{U} = \mathbf{U}^T > 0$, $\mathbf{V} = \mathbf{V}^T > 0$. *Then* $SO(3)\mathbf{U}$, $SO(3)\mathbf{V}$ *are rank-one connected if and only if*

$$\mathbf{U}^2 - \mathbf{V}^2 = \gamma(\mathbf{m} \otimes \mathbf{n} + \mathbf{n} \otimes \mathbf{m}) \tag{3}$$

for unit vectors $\mathbf{m}, \mathbf{n}$ *and some* $\gamma \neq 0$. *For suitable* $\mathbf{a}_1, \mathbf{a}_2 \in \mathbb{R}^3$ *and* $\mathbf{R}_1, \mathbf{R}_2 \in SO(3)$, *the rank-one connections between* $\mathbf{V}$ *and* $SO(3)\mathbf{U}$ *are given by*

$$\mathbf{R}_1\mathbf{U} = \mathbf{V} + \mathbf{a}_1 \otimes \mathbf{n}, \quad \mathbf{R}_2\mathbf{U} = \mathbf{V} + \mathbf{a}_2 \otimes \mathbf{m}. \tag{4}$$

We omit the proof, which is not difficult. The main point of the lemma is that the normals corresponding to the rank-one connections are the vectors appearing in (3). An interesting consequence is that if $\mathbf{U}$, $\mathbf{V}$ correspond to martensitic variants, so that $\mathbf{V} = \mathbf{Q}^T\mathbf{UQ}$ for some $\mathbf{Q} \in SO(3)$, then, taking the trace in (3) shows that the two possible normals are orthogonal (see [2, Theorem 2.1]).

Using Lemma 1 we can calculate the rank-one connections between K and $K\mathbf{R}_\alpha$. For example, for the rank-one connections between $SO(3)\mathbf{U}_1$ and $SO(3)\mathbf{U}_1\mathbf{R}_\alpha$ we find that

$$\mathbf{U}_1^2 - \mathbf{R}_\alpha^T\mathbf{U}_1^2\mathbf{R}_\alpha = (\eta_2^2 - \eta_1^2)\sin\alpha(\mathbf{m} \otimes \mathbf{n} + \mathbf{n} \otimes \mathbf{m}) \tag{5}$$

where $\mathbf{n} = (\sin(\alpha/2), \cos(\alpha/2), 0)$, $\mathbf{m} = (\cos(\alpha/2), -\sin(\alpha/2), 0)$, so that the two possible normals $\mathbf{n} = (n_1, n_2, 0)$ satisfy $\tan\alpha = 2n_1n_2/(n_2^2 - n_1^2)$. Swapping η_1 and η_2 we see that the possible normals for rank-one connections between $SO(3)\mathbf{U}_2$ and $SO(3)\mathbf{U}_2\mathbf{R}_\alpha$ are the same. Similarly, we find that the possible normals for rank-one connections between $SO(3)\mathbf{U}_1$ and $SO(3)\mathbf{U}_2\mathbf{R}_\alpha$ or between $SO(3)\mathbf{U}_2$ and $SO(3)\mathbf{U}_1\mathbf{R}_\alpha$ satisfy $\tan \alpha = (n_1^2 - n_2^2)/2n_1n_2$.

Main Results

Suppose there exists a gradient Young measure of the form (2) such that $v_{\mathbf{x}} = \delta_{\mathbf{F}}$ for a.e. $\mathbf{x} \in \Omega_1$ for some $\mathbf{F} \in K$, corresponding to a pure variant in grain 1. It follows that the corresponding macroscopic gradient $\nabla\mathbf{y}(\mathbf{x}) = \bar{v}_{\mathbf{x}} = \int_{M_+^{3\times 3}} \mathbf{A} dv_x(\mathbf{A})$ satisfies

$$\nabla\mathbf{y}(\mathbf{x}) = \mathbf{F} \text{ for a.e. } \mathbf{x} \in \Omega_1,$$
$$\nabla\mathbf{y}(\mathbf{x}) \in (K\mathbf{R}_\alpha)^{qc} \text{ for a.e. } \mathbf{x} \in \Omega_2, \qquad (6)$$

where E^{qc} denotes the quasiconvexification of a compact set $E \subset \mathbb{R}^{3\times 3}$, that is the set of possible macroscopic deformation gradients corresponding to microstructures using gradients in E. As determined in [6, Theorem 5.1] (and more conveniently in [10, p 155]) K^{qc} consists of those $\mathbf{A} \in \mathbb{R}_+^{3\times 3}$ with

$$\mathbf{A}^T\mathbf{A} = \begin{pmatrix} a & c & 0 \\ c & b & 0 \\ 0 & 0 & \eta_3^2 \end{pmatrix} \quad \text{and} \quad (a,b,c) \in P, \qquad (7)$$

$$P = \{(a,b,c) : a \geq 0, \ b \geq 0, \ ab - c^2 = \eta_1^2\eta_2^2, \\ a + b + |2c| \leq \eta_1^2 + \eta_2^2\}, \qquad (8)$$

and is equal to the polyconvexification K^{pc} of K. It follows that $(K\mathbf{R}_\alpha)^{qc} = K^{qc}\mathbf{R}_\alpha = K^{pc}\mathbf{R}_\alpha$. The proof of (7) shows also that the quasiconvexification $\tilde{K}^{qc}$ of $\tilde{K} = SO(2)\tilde{\mathbf{U}}_1 \cup SO(2)\tilde{\mathbf{U}}_2$, where $\tilde{\mathbf{U}}_1 = \begin{pmatrix} \eta_2 & 0 \\ 0 & \eta_1 \end{pmatrix}$, $\tilde{\mathbf{U}}_2 = \begin{pmatrix} \eta_1 & 0 \\ 0 & \eta_2 \end{pmatrix}$, is equal to $\tilde{K}^{pc}$ and is given by the set of $\tilde{\mathbf{A}} \in \mathbb{R}_+^{2\times 2}$ such that $\tilde{\mathbf{A}}^T\tilde{\mathbf{A}} = \begin{pmatrix} a & c \\ c & b \end{pmatrix}$ for $(a,b,c) \in P$.

Let $\mathbf{x}_0 = (\tilde{\mathbf{x}}_0, \delta)$, where $\tilde{\mathbf{x}}_0 \in \partial\omega_1 \cap \partial\omega_2 \cap \omega$, $0 < \delta < d$, be such that the interface has a well-defined normal $\mathbf{n} = (\cos \theta, \sin \theta, 0) = (\tilde{\mathbf{n}}, 0)$ at $\mathbf{x}_0$. By [7], there exists $\varepsilon > 0$ such that in $\mathcal{U} := B(\tilde{\mathbf{x}}_0, \varepsilon) \times (\delta - \varepsilon, \delta + \varepsilon)$, $0 < \varepsilon < \delta$, the map $\mathbf{y}$ is a plane strain, that is

$$\mathbf{y}(\mathbf{x}) = \mathbf{R}(z_1(x_1, x_2), z_2(x_1, x_2), \eta_3 x_3 + \gamma) \text{ for a.e. } \mathbf{x} \in \mathcal{U} \qquad (9)$$

for some $\mathbf{R} \in SO(3)$, $\mathbf{z} : B(\tilde{\mathbf{x}}_0, \varepsilon) \to \mathbb{R}^2$ and $\gamma \in \mathbb{R}$, where $B(\tilde{\mathbf{x}}_0, \varepsilon)$ denotes the open ball in $\mathbb{R}^2$ with centre $\tilde{\mathbf{x}}_0$ and

radius ε. Without loss of generality we can take $\mathbf{R} = \mathbf{1}$. Then, for $\tilde{\mathbf{x}} = (x_1, x_2) \in B(\tilde{\mathbf{x}}_0, \varepsilon)$ we have

$$\nabla\mathbf{z}(\tilde{\mathbf{x}}) = \tilde{\mathbf{F}} \in \tilde{K} \text{ for a.e. } \tilde{\mathbf{x}} \in \omega_1,$$
$$\nabla\mathbf{z}(\tilde{\mathbf{x}}) \in (\tilde{K}\tilde{\mathbf{R}}_\alpha)^{pc} = \tilde{K}^{pc}\tilde{\mathbf{R}}_\alpha \text{ for a.e. } \tilde{\mathbf{x}} \in \omega_2, \qquad (10)$$

where $\tilde{\mathbf{R}}_\alpha = \begin{pmatrix} \cos \alpha & -\sin \alpha \\ \sin \alpha & \cos \alpha \end{pmatrix}$. By a two-dimensional generalization of the Hadamard jump condition proved in [3] this implies that there exists $\tilde{\mathbf{A}} \in (\tilde{K}\tilde{\mathbf{R}}_\alpha)^{pc}$ such that

$$\tilde{\mathbf{F}} - \tilde{\mathbf{A}} = \tilde{\mathbf{a}} \otimes \tilde{\mathbf{n}} \qquad (11)$$

for some $\tilde{\mathbf{a}} \in \mathbb{R}^2$. Conversely, if the interface is planar with normal $\mathbf{n} = (\tilde{\mathbf{n}}, 0)$, then the existence of $\tilde{\mathbf{A}} \in (\tilde{K}\tilde{\mathbf{R}}_\alpha)^{pc}$ satisfying (11) implies the existence of a gradient Young measure $v = (v_{\mathbf{x}})_{\mathbf{x}\in\Omega}$ satisfying (6). Indeed there then exists a sequence of gradients $\nabla\mathbf{z}^{(j)}$ generating a gradient Young measure $(\mu_{\tilde{\mathbf{x}}})_{\tilde{\mathbf{x}}\in\omega}$ such that

$$\bar{\mu}_{\tilde{\mathbf{x}}} = \tilde{\mathbf{F}} \text{ for a.e. } \tilde{\mathbf{x}} \in \omega_1, \quad \bar{\mu}_{\tilde{\mathbf{x}}} = \tilde{\mathbf{A}} \text{ for a.e. } \tilde{\mathbf{x}} \in \omega_2, \qquad (12)$$

and then $\nabla\mathbf{y}^{(j)}(\mathbf{x}) = \begin{pmatrix} z_{1,1}^{(j)} & z_{1,2}^{(j)} & 0 \\ z_{2,1}^{(j)} & z_{2,2}^{(j)} & 0 \\ 0 & 0 & \eta_3 \end{pmatrix}$ generates such a gradient Young measure.

It turns out that we can say exactly when it is possible to solve (11). Set $\tau := \eta_2/\eta_1 > 1$, $s^* = (\tau^4 - 1)/(\tau^4 + 1)$ and define for $0 \leq s \leq 1$ the C^1 convex increasing function

$$f(s) := \begin{cases} (\tau^4 + 1 - 2\tau^2\sqrt{1 - s^2})/(\tau^4 - 1) & \text{if } s \leq s^*, \\ s & \text{if } s > s^*. \end{cases} \qquad (13)$$

Theorem 2 *There exist $\tilde{\mathbf{F}} \in \tilde{K}$ and $\tilde{\mathbf{A}} \in \tilde{K}^{pc}\tilde{\mathbf{R}}_\alpha$ with $\tilde{\mathbf{F}} - \tilde{\mathbf{A}} = \tilde{\mathbf{a}} \otimes \tilde{\mathbf{n}}$ for $\tilde{\mathbf{n}} = (\cos \theta, \sin \theta)$ and some $\tilde{\mathbf{a}} \in \mathbb{R}^2$ if and only if*

$$|\cos 2\theta| \leq f(|\cos 2(\alpha + \theta)|). \qquad (14)$$

Proof It is easily checked that the existence of $\tilde{\mathbf{F}} \in SO(3)\tilde{\mathbf{U}}_i$ and $\tilde{\mathbf{A}}$ is equivalent to the existence of $(a,b,c) \in P$ such that $|\tilde{\mathbf{U}}_i\tilde{\mathbf{n}}^\perp|^2 = aN_1^2 + bN_2^2 + 2cN_1N_2$, where $\tilde{\mathbf{n}}^\perp = (-n_2, n_1)$ and $\mathbf{N} = (N_1, N_2) = \tilde{\mathbf{R}}_\alpha\tilde{\mathbf{n}}^\perp$. That is

$$\text{either } n_1^2\eta_1^2 + n_2^2\eta_2^2 \text{ or } n_2^2\eta_1^2 + n_1^2\eta_2^2 \ \in [m_-(\mathbf{N}), m_+(\mathbf{N})], \qquad (15)$$

where $m_\pm(\mathbf{N}) = \max_{(a,b,c)\in P} \min (aN_1^2 + bN_2^2 + 2cN_1N_2)$. Changing variables to $x = a+b$ and $y = a-b$ we find that $m_+(\mathbf{N}) = \max_{(x,y)\in P_2}\psi_+(x,y)$, $m_-(\mathbf{N}) = \min_{(x,y)\in P_2}\psi_-(x,y)$, where

$$2\psi_\pm(x,y) = x + y(N_1^2 - N_2^2) \pm 2|N_1N_2|\sqrt{x^2 - y^2 - 4\eta_1^2\eta_2^2},$$

(16)

$$P_2 = \left\{ \begin{array}{l} (x,y) \in \mathbb{R}^2 : |y| \le \eta_2^2 - \eta_1^2 \text{ and} \\[2mm] \sqrt{y^2 + 4\eta_1^2\eta_2^2} \le x \le \dfrac{y^2 + 4\eta_1^2\eta_2^2 + \left(\eta_1^2 + \eta_2^2\right)^2}{2\left(\eta_1^2 + \eta_2^2\right)} \end{array} \right\}.$$

The region P_2 is bounded by the two arcs $C_1 := \{x = \sqrt{y^2 + 4\eta_1^2\eta_2^2}\}$, $C_2 := \left\{x = \left(y^2 + 4\eta_1^2\eta_2^2 + \left(\eta_1^2 + \eta_2^2\right)^2\right)/\left(2\left(\eta_1^2 + \eta_2^2\right)\right)\right\}$, defined for $|y| \le \eta_2^2 - \eta_1^2$, which intersect at the points $(x,y) = (\eta_1^2 + \eta_2^2, \pm(\eta_2^2 - \eta_1^2))$. Note that $\psi_\pm(x,y)$ have no critical points in the interior of P_2. In fact it is immediate that $\nabla\psi_+$ cannot vanish, while $\nabla\psi_-(x,y) = 0$ leads to $y = x(N_2^2 - N_1^2)$ and hence to the contradiction $0 = x^2 - y^2 - 4x^2N_1^2N_2^2 = 4\eta_1^2\eta_2^2 > 0$. Thus the maximum and minimum of $\psi_\pm(x,y)$ are attained on either C_1 or C_2. After some calculations we obtain

$$m_+(\mathbf{N}) = \frac{1}{2}(\eta_1^2 + \eta_2^2 + (\eta_2^2 - \eta_1^2)|N_1^2 - N_2^2|),$$

$$m_-(\mathbf{N}) = \begin{cases} \dfrac{\eta_1^2\eta_2^2}{\eta_1^2 + \eta_2^2}\left(1 + \sqrt{1 - |N_1^2 - N_2^2|^2}\right) & \text{if } |N_1^2 - N_2^2| \le s^*, \\[3mm] \frac{1}{2}\left(\eta_1^2 + \eta_2^2 - |N_1^2 - N_2^2|(\eta_2^2 - \eta_1^2)\right) & \text{if } |N_1^2 - N_2^2| \ge s^*. \end{cases}$$

Noting that $m_-(\mathbf{N}) \le \frac{1}{2}\left(\eta_1^2 + \eta_2^2\right) \le m_+(\mathbf{N})$, that $m_+(\mathbf{N}) + m_-(\mathbf{N}) \le \eta_1^2 + \eta_2^2$, and that $\frac{1}{2}\left(\eta_1^2\eta_1^2 + \eta_2^2\eta_2^2\right) + \frac{1}{2}\left(\eta_2^2\eta_1^2 + \eta_1^2\eta_2^2\right) = \frac{1}{2}\left(\eta_1^2 + \eta_2^2\right)$ it follows that (15) is equivalent to

$$m_-(\mathbf{N}) \le \min\{\eta_1^2\eta_1^2 + \eta_2^2\eta_2^2, \eta_2^2\eta_1^2 + \eta_1^2\eta_2^2\}$$
$$= \frac{1}{2}\left(\eta_1^2 + \eta_2^2\right) - \frac{1}{2}|\eta_2^2 - \eta_1^2|(\eta_2^2 - \eta_1^2). \quad (17)$$

With the relations $n_1^2 - n_2^2 = \cos 2\theta$ and $N_2^2 - N_1^2 = \cos 2(\theta + \alpha)$ this gives (14). □

Theorem 3 *If the interface between the grains is planar then there always exists a zero-energy microstructure which is a pure variant in one of the grains.*

Proof The case of a pure variant in grain 2 and a zero-energy microstructure in grain 1 corresponds to replacing θ by $\theta + \alpha$ and α by $-\alpha$ in the above. Hence, since $f(s) \ge s$, if the conclusion of the theorem were false, Theorem 2

would imply that $|\cos 2\theta| > f(|\cos 2(\alpha + \theta)|) \ge |\cos 2(\theta + \alpha)| > f(|\cos 2\theta|) \ge |\cos 2\theta|$, a contradiction. Thus to rule out having a pure variant in one grain the interface *cannot* be planar.

Theorem 4 *There is no zero-energy microstructure which is a pure variant in one of the grains if the interface between the grains has a normal $\mathbf{n}^{(1)} = (\cos\theta_1, \sin\theta_1, 0) \in E_1$ and a normal $\mathbf{n}^{(2)} = (\cos\theta_2, \sin\theta_2, 0) \in E_2$, for the disjoint open sets*

$$E_1 = \{\theta \in \mathbb{R} : f(|\cos 2(\theta + \alpha)|) < |\cos 2\theta|\},$$
$$E_2 = \{\theta \in \mathbb{R} : f(|\cos(2\theta)|) < |\cos 2(\theta + \alpha)|\}.$$

Proof This follows immediately from Theorem 2 and the preceding discussion. □

In the case $\alpha = \pi/4$ the sets E_1, E_2 take a simple form (note that the normals not in $E_1 \cup E_2$ correspond to the rank-one connections between the wells found in Sect. 2).

Theorem 5 *Let $\alpha = \pi/4$ and suppose[3] that $\eta_2^2/\eta_1^2 < 1 + \sqrt{2}$. Then*

$$E_1 = \bigcup_j ((4j-1)\pi/8, (4j+1)\pi/8),$$
$$E_2 = \bigcup_j ((4j+1)\pi/8, (4j+3)\pi/8).$$

(18)

Proof Note that $\eta_2^2/\eta_1^2 < 1 + \sqrt{2}$ if and only if $s^* < 1/\sqrt{2}$. Therefore if $|\sin 2\theta| < 1/\sqrt{2}$ then $f(|\sin 2\theta|) < f(1/\sqrt{2}) = 1/\sqrt{2} < |\cos 2\theta|$. Hence $f(|\sin 2\theta|) < |\cos 2\theta|$ if and only if $|\sin 2\theta| < 1/\sqrt{2}$. This holds if and only if $\theta \in (\pi/2)\mathbb{Z} + (-\pi/8, \pi/8) = E_1$. The case of E_2 is treated similarly. □

Discussion

Compatibility across grain boundaries in polycrystals using a linearized elastic theory is discussed in [10, Chapter 13, 11]. Whereas we use the nonlinear theory we are restricted to a very special assumed geometry and phase transformation. Nevertheless we are able to determine conditions allowing or excluding a pure variant in one of the grains without any a priori assumption on the microstructure (which could potentially, for example, have a fractal structure near the interface). This was possible using a generalized Hadamard jump condition from [3]. The restriction to a two-dimensional situation is due both to the current

---

[3] This extra condition, typically satisfied in practice, was accidentally omitted in the announcement in [5].

unavailability of a suitable three-dimensional generalization of such a jump condition, and because the quasiconvexification of the martensitic energy wells is only known for two wells.

Acknowledgements The research of JMB was supported by the EU (TMR contract FMRX - CT EU98-0229 and ERBSCI**CT000670), by EPSRC (GRlJ03466, EP/E035027/1, and EP/J014494/1), the ERC under the EU's Seventh Framework Programme (FP7/2007–2013) / ERC grant agreement no 291053 and by a Royal Society Wolfson Research Merit Award.

References

1. Arlt G (1990) Twinning in ferroelectric and ferroelastic ceramics: stress relief. J Mat Sci 22:2655–2666
2. Ball JM (2004) Mathematical models of martensitic microstructure. Mater Sci Eng A 78(1–2):61–69
3. Ball JM, Carstensen C Hadamard's compatibility condition for microstructures (In preparation)
4. Ball JM, Carstensen C (1999) Compatibility conditions for microstructures and the austenite-martensite transition. Mater Sci Eng A 273–275:231–236
5. Ball JM, Carstensen C (2015) Geometry of polycrystals and microstructure. MATEC Web Conf 33:02007
6. Ball JM, James RD (1987) Fine phase mixtures as minimizers of energy. Arch Ration Mech Anal 100:13–52
7. Ball JM, James RD (1991) A characterization of plane strain. Proc Roy Soc London A 432:93–99
8. Ball JM, James RD (1992) Proposed experimental tests of a theory of fine microstructure, and the two-well problem. Phil Trans Roy Soc London A 338:389–450
9. Bhattacharya K (1992) Self-accommodation in martensite. Arch Ration Mech Anal 120:201–244
10. Bhattacharya K (2003) Microstructure of martensite, Oxford University Press
11. Bhattacharya K, Kohn RV (1997) Elastic energy minimization and the recoverable strain of polycrystalline shape-memory materials. Arch Ration Mech Anal 139:99–180
12. Khachaturyan AG (1983) Theory of structural transformations in solids, Wiley
13. Mardare S Personal communication

Different Cooling Rates and Their Effect on Morphology and Transformation Kinetics of Martensite

Annika Eggbauer (Vieweg), Gerald Ressel, Marina Gruber,
Petri Prevedel, Stefan Marsoner, Andreas Stark, and Reinhold Ebner

Abstract

The characteristics of martensitic transformation is strongly dependent on the cooling rate applied to the material. For a quenched and tempered steel, the martensitic transformation occurs below 500 °C, but in industry, cooling rates are normally characterized for cooling in the temperature regime between 800 and 500 °C. The effects of different cooling rates in the lower temperature regime were thus, not intensively investigated in the past. To this end, a 50CrMo4 steel is quenched in a dilatometer applying varying cooling rates below 500 °C. The martensite microstructure is analyzed by APT, TEM and EBSD in regard to carbon distribution, lath width and block sizes. Additionally, hardness measurements are carried out and martensite start temperatures as well as the retained austenite phase fractions are evaluated. It can be shown, that lowering the cooling rate leads to increased carbon segregation within the martensitic matrix. The main effect is a decrease in martensite hardness. Also the block size increases with lower cooling rate.

A. Eggbauer (Vieweg) (✉) · G. Ressel · M. Gruber
P. Prevedel · S. Marsoner · R. Ebner
Materials Center, Leoben Forschung GmbH, Leoben, Austria
e-mail: Annika.Vieweg@mcl.at; Annika.eggbauer@mcl.at

G. Ressel
e-mail: Gerald.Ressel@mcl.at

M. Gruber
e-mail: Marina.Gruber@mcl.at

P. Prevedel
e-mail: Petri.Prevedel@mcl.at

S. Marsoner
e-mail: Stefan.Marsoner@mcl.at

R. Ebner
e-mail: Reinhold.Ebner@mcl.at

A. Stark
Helmholtz Zentrum Geesthacht, Geesthacht, Germany
e-mail: andreas.stark@hzg.de

Keywords

Cooling parameter • Cooling rate • Martensite
Auto tempering • Quenched and tempered steel

Introduction

Steel offers the possibility to achieve a diversity of microstructures and mechanical properties just by adapting the cooling program from the austenitic state. For a conventional quenched and tempered (QT) steel, fast cooling procedures lead to a hard martensitic microstructure, while slow air cooling causes a soft ferritic-pearlitic microstructure [1–3].

For industrial applications the cooling rate is often specified by means of the cooling parameter λ, which is defined as the time passing from 800 to 500 °C divided by 100 [1]. Looking into typical time temperature transformation diagrams (Fig. 1) it becomes obvious that the time from 800 to 500 °C is important for the transformation of austenite to phases such as ferrite or pearlite. For a martensitic microstructure, however, the cooling parameter is solely defining the cooling rate needed to avoid bainite or ferrite phase formation. However, below 500 °C there is still room for a diversity of cooling strategies to achieve a fully martensitic microstructure. Since for common quenched and tempered steels the martensite start temperature (M_s) is around 300 °C, cooling below 500 °C will continue to affect the formation of martensite.

Martensitic transformations are defined as diffusionless transformations, although local atomic diffusion might occur to some extent within the martensitic state at high temperatures during the cooling procedure. This behavior is described as auto-tempering [4]. Carbon can either segregate to lattice defects or, if cooling is sufficiently slow, carbides, e.g. cementite, are formed. Both segregation and precipitation can lead to different martensitic properties, such as hardness and toughness in the as-quenched state.

© The Minerals, Metals & Materials Society 2018
A. P. Stebner and G. B. Olson (eds.), *Proceedings of the International Conference on Martensitic Transformations: Chicago*,
The Minerals, Metals & Materials Series, https://doi.org/10.1007/978-3-319-76968-4_6

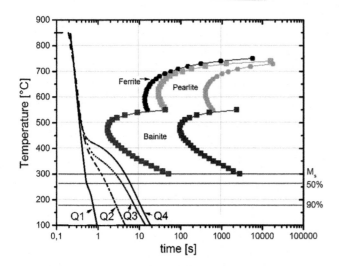

Fig. 1 Time temperature transformation (TTT-) diagram for the investigated steel grade calculated with JMAT Pro™. Black circles correspond to 0.1% ferrite, light green circles to 0.1% pearlite and dark green circles to 99.9% pearlite. Light blue squares refer to 0.1% bainite and dark blue squares to 99.9% bainite. The experimental cooling curves Q1 (solid line), Q2 (dashed line), Q3 (dotted line) and Q4 (dashed-dotted line) are depicted within the diagram

Furthermore, the tempering kinetics might change for different martensitic structures.

Therefore, this study discusses the differences within the martensitic microstructure as a consequence of different cooling strategies below 500 °C in regard to segregation and auto-tempering phenomena as well as transformation kinetics. To analyze the differences, a comprehensive set of high resolution methods such as dilatometry, 3D atom probe tomography (3D-APT), transmission electron microscopy (TEM), electron back scattered diffraction (EBSD) as well as high energy X-ray diffraction (HEXRD) is used.

Experimental

The investigated steel in this study was a 50CrMo4-steel with 0.49 wt% C, 0.71 wt% Mn, 1.05 wt% Cr, 0.18 wt% Mo, 0.27 wt% Si, 0.02 wt% P and 0.01 wt% S.

For dilatometer experiments, cylindrical samples with a diameter of 4 mm and a length of 10 mm were manufactured at half radius of a rolled material.

The heat treatments were conducted using a DIL 805L/A dilatometer from TA Instruments (formerly BAEHR).

Austenitization of all samples was done at 850 °C for 20 min. To obtain different quenching rates, the gas flow during quenching was either set to maximum (Q1) or was regulated according to the specific programs (Q2-Q4). The quenching rate from 850 to 500 °C was constant (2300 K/s, t = 0.15 s) for all experiments. All cooling programs were conducted twice and exhibited equal phase transformation behavior. The different cooling procedures are depicted in Fig. 1, which shows a time temperature transformation (TTT) diagram calculated with JMAT Pro™ for the analyzed steel grade. Table 1 lists the cooling times from 500 to 300 °C ($t_{500/300}$), from M_s to 100 °C ($t_{Ms/100}$) and the overall cooling time from 850 to 100 °C ($t_{850/100}$). The dilatometer data were analyzed to determine the volume fraction of martensite using the lever rule. To apply the lever rule, linear fits for austenite were done between 500 and 300 °C, linear fits for martensite between 100 and 50 °C for all data.

The TEM analysis was carried out on a FEI Tecnai F20, equipped with a field emission gun at an accelerating voltage of 200 kV. The samples were electrochemically etched on a Struers Tenupol 5, using a 7% solution of perchloric acid at temperatures around −10 °C. In order to achieve reasonable statistics, lath widths were determined by acquiring and evaluating 10 TEM images of each condition, hence around 150 laths were evaluated per condition.

The 3D-APT measurements were carried out at a LEAP™ System 3000XHR in voltage mode at a temperature of 60 K with a pulse frequency of 200 kHz and a pulse fraction of 20%. The tips were reconstructed using the IVAS 3.6.8. software tool provided by CAMECA.

Block sizes were analysed using electron backscattered diffraction (EBSD). Samples were ion-polished using a Hitachi IM4000plus ion milling device. EBSD measurements were carried out with a Zeiss Auriga cross beam workstation by investigating an area of 70 × 70 μm and a step size of 50 nm. The data were analysed using the Orientation Imaging Microscopy (OIM) data analysis software from EDAX and subsequently the data was cleaned using grain dilatation, with a dilation angle of 10.5° (as seen to be the best fit for martensitic materials in literature [5])

To obtain the amount of retained austenite, diffraction patterns of the samples were acquired by means of high energy X-ray diffraction (HEXRD) measurements at the HEMS beamline (P07) at Petra III [6]. For these measurements, the as-quenched martensitic microstructures were

Table 1 Nomenclature of the different cooling strategies. $t_{500/300}$ time gives the cooling time from 500 to 300 °C, $t_{Ms/100}$ from M_s to 100 °C and $t_{850/100}$ from 850 to 100 °C

Name	$t_{500/300}$ [s]	$t_{Ms/100}$ [s]	$t_{850/100}$ [s]
Q1	0.13	0.4	0.59
Q2	0.8	3	3.95
Q3	2.2	10.5	12.85
Q4	4.1	12.9	17.15

placed into the beam at room temperature with an exposure time of 0.2 s. In order to penetrate the 4 mm thick samples, high energy X-rays were used with a photon energy of 87.1 keV, corresponding to a wavelength of 0.14235 . The beam size was 0.7 × 0.7 mm. The resulting diffraction rings were recorded with a Perkin Elmer XRD1621 flat panel detector. The data were integrated using the fit2D software from ESRF [7]. Retained austenite amounts were analyzed using TOPAS 4-2 and the Rietveld Method [8]. The hardness measurements were performed on a Qness Q10A+ Vickers hardness tester.

Results

The dilatometer data were evaluated in regard to the transformation kinetics using the lever rule. The evolution of the martensite volume fraction over temperature is depicted in Fig. 2 for the four different cooling strategies. Differences in

"zero" level occur due to the evaluation method, since all data were fitted in the same temperature region, this is not a transformation effect. Only distinct bends are real transformation phenomena. Figure 2 shows that no significant difference in martensite start temperature (M_s) occurs for the different cooling strategies. Little variations are subject to deviation of the analysis method, hence M_s is at 267 ± 2 °C. Nevertheless, Q3 and Q4 show a slight increase in volume fraction of martensite prior to M_s (around 290 °C). This increase is more pronounced for Q4 than for Q3. Additionally, Q1 seems to exhibit the lowest slope at the beginning of the transformation.

Lath widths as well as block sizes are analyzed and depicted in Fig. 3a. The lath width (closed squares) remains constant around 0.2 μm for all samples. The measuring uncertainty of this evaluation method is large, so no trend can be stated. However, with increasing cooling time the block size (open circles) is significantly increasing from 2 to 2.5 μm. Due to the little differences in blocksize values, Q1

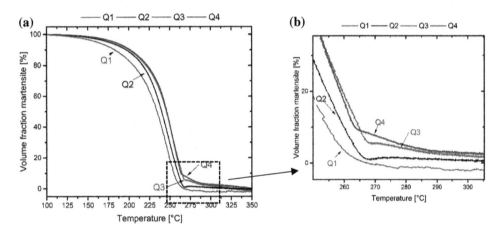

Fig. 2 Amount of martensite in vol% as a function of the temperature determined by means of the lever rule applied on the dilatometer data (a) and the region of interest around M_s in detail (b)

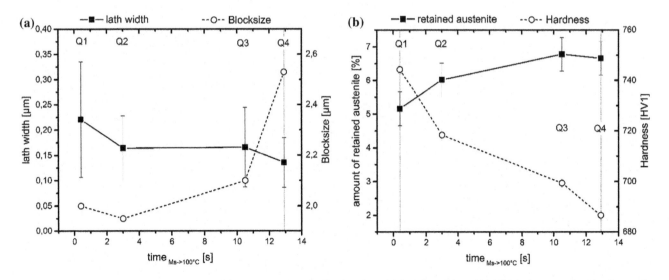

Fig. 3 Lath width (closed squares) and block size (open circles) depicted against the time from M_s to 100 °C (a); Amount of retained austenite (closed squares) and hardness (open circles) over time from Ms to 100 °C (b)

and Q2 are stated to be within the same range of 2 μm, and the decrease in block-size is due to deviations across the analyzed area.

Another important aspect for the as-quenched state is the amount of retained austenite, shown in Fig. 3b by means of closed squares. The amount of retained austenite is slightly increasing from 5.2 to 6.7% with increasing cooling time. In contrast to this trend, the hardness is significantly decreasing and drops from 744 HV1 for Q1 to 686 HV1 for Q4. An evaluation of the HEXRD patterns (not depicted) does not reveal indications for the occurrence of cementite in any of the as-quenched states. Additionally, the lattice constants of the austenite and ferrite were evaluated but no trend between the four states can be stated.

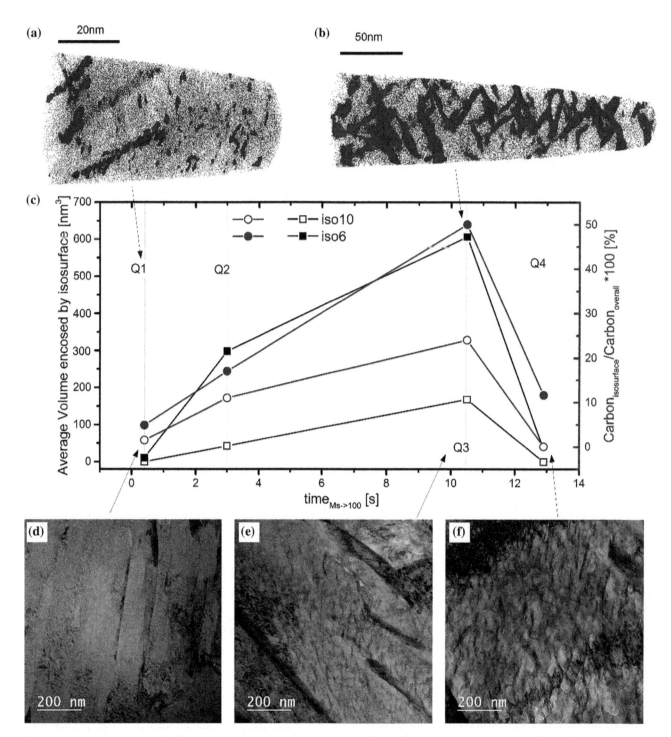

Fig. 4 Carbon enrichment indicated by means of isosurfaces of 6 (closed symbols) and 10 at.% carbon (open symbols) and by the ratio carbon within isosurface/overall carbon evaluated from the APT samples and selected 3D atom probe tips (**c**). Additionally, APT tips of Q1 (**a**) and Q3 (**b**) are depicted. Additionally, TEM bright field images of Q1 (**d**), Q3 (**e**) and Q4 (**f**) are depicted

To analyze the carbon segregation as a function of the cooling rate 3D-APT measurements were conducted. To this end, isosurfaces of 6 (closed symbols) and 10 at.% carbon (open symbols) were analyzed in regard to their volume (black squares) and the amount of carbon within these isosurfaces in relation to the overall amount of carbon ions (blue circles) within the tip. The results can be seen in Fig. 4c. For better imagination, the APT tips of Q1 (Fig. 4a) and Q3 (Fig. 4b) are depicted as well. The overall amount of carbon within the investigated martensite laths was 2.1 (Q1, Q3), 2.3 (Q2) and 2.6 at.% C (Q4). The differences are due to rolling segregation across the material, but are still in the range of the nominal composition of the steel (2.23 at.% C). The features of the isosurfaces are developed along bands, in a cylindrical manner. It is clearly visible, that the amount of carbon within the isosurfaces in relation to the overall carbon ions, as well as the average volume of the isosurfaces, is increasing from Q1 to Q3. For Q4 these attributes drastically drop to a value between Q1 and Q2, due to less isosurfaces within this condition. Furthermore, also by visual inspection an enrichment of carbon within the isosurfaces as well as a coarsening can be observed for Q1–Q3. In order to analyze carbon segregation, transmission electron microscopy was conducted and bright field (BF) images of Q1 (d) Q3 (e) and Q4 (f) are depicted in Fig. 4. While for Q1 nothing is visible inside the lath, Q3 and Q4 show increased precipitations within the laths. These precipitations did not show any scattering within selected area diffraction, hence no indexing could be done.

Discussion

The cooling strategies were chosen to avoid any other phase transformation than the martensitic one. The bainite phase nose visible in Fig. 1 was not passed by the cooling curves. Nevertheless, phase fraction analysis in Fig. 2 show the presence of a second transformation for the two slowest cooling strategies Q3 and Q4. It can be assumed, that the calculated TTT diagram differs from the real transformation behavior determined by dilatometry for the actual chemical composition of the investigated steel since a slight bainitic phase transformation is visible. Nevertheless, M_s was not quantifiably affected by the different cooling strategies below 500 °C.

The martensitic transformation itself is diffusionless [9], however, significant diffusion of carbon can occur in the formed martensite during subsequent quenching. Since all carbon enrichments within the APT tips, occur along lines and little round features are detected within the conditions, segregation of carbon is present [10]. Once a martensite lath is formed carbon can segregate to lattice defects which act as traps for carbon. The segregated areas can subsequently act as

nucleation sites for carbides, i.e. cementite [10]. As the segregation and formation of carbides is dependent on diffusion, their size and overall content depend on the martensite start temperature as well as on the cooling rate. As a result, at constant M_s temperature higher cooling rates lead to shorter diffusion paths of carbon and consequently to a lower amount of segregations. This is confirmed by the APT measurements, since the highest cooling rate shows a low amount of carbon within the segregations and also a smaller fraction of segregations compared to the conditions Q2 and Q3.

The sample of the condition Q4 shows less segregation as well as lower carbon isosurface value within the APT tips compared to the conditions Q3 and Q2. This trend was reproduced over several tips. Comparing the TEM images of Q1, Q3 and Q4 (Fig. 4) it is obvious, that segregation is increasing with increasing cooling time. Q4 shows segregations [10]. The amount of carbon trapped at dislocations is depending on the overall amount of carbon as well as the dislocation density [11]. Due to less dislocations within Q4, less segregated areas are present. Using APT the distances between the segregations are increasing so that APT shows less segregation due to the small size of the tips. However, due to the small volume measured in APT, segregations might obtain a longer distance than APT is able to detect. Therefore, it is possible, that none of the increased segregated areas was evaluated. Concluding, attention needs to be paid evaluating segregated areas using APT, due to the very local evaluation.

Despite the fact, that the prior austenite grain size is equal in all samples, since austenitization was always done equally, the block size is decreasing with decreasing cooling time (Fig. 3a). Blocks form within a packet which is built through a shift along the same habit plane [5, 12]. At higher temperatures austenite yield strength is lower [13] and hence it is assumed, that its resistance against lattice displacement (i.e. martensite formation) is decreased. As evaluated through Fig. 2, phase transformation obtains a higher slope around M_s for slower cooling rates, meaning a higher transformation rate. The martensite formed at higher temperatures is forming within an austenite with lower yield strength compared to lower temperatures. Consequently, larger martensite packets can be formed, since the resistance against crystallographic shift is lower at elevated temperatures (close to M_s) compared to lower temperatures. This leads to larger packets for slow cooling rates. At higher cooling rates, higher undercooling takes place (visible through the lower slope after M_s for Q1) and the dwell time at higher temperatures is thus limited. The crystallographic shift occurs at multiple habit planes simultaneously and due to the increasing yield strength of the austenite with decreasing temperature the shifted area is decreased. This leads to increased amounts of packets per austenite grain and, therefore, smaller packets. Smaller packets subsequently lead to a smaller block size [12, 14]. Consequently, high cooling rates lead to smaller block sizes.

From the experimental results of this work it can be derived that faster cooling of martensite results in higher hardness. This higher hardness is caused by a combination of three effects. The most dominant reason for the higher hardness can be related to minimized segregation of carbon to lattice defects upon fast cooling, as can be observed within the APT tips in Fig. 4. Consequently the distortion of ferrite crystal reaches a maximum. For Q4 the dislocation density is assumed to be lower than for Q1 and Q2, hence the matrix is less distorted through dislocation relieve during cooling and consequently softer. Secondly, the decreasing block size with decreasing cooling time can strengthen the matrix according to Hall-Petch [15, 16]. A correlation of mechanical properties of martensitic materials and the block size of lath martensite is an essential factor in the strength-structure relationship [17, 18]. Finally, the lower amount of retained austenite (Fig. 3b) causes a less decrease of the hardness of the material. However, this effect is assumed be the least dominant (only around 10 HV10), concluding from the rule of mixture (hardness of austenite ~ 100 HV10, hardness of martensite ~ 800 HV10).

Conclusions

This work shows that during the quenching process of a 50CrMo4 steel the cooling rate below 500 °C has essential influence on the resulting martensitic microstructure. Therefore, it should be chosen carefully to obtain the desired martensite properties, such as hardness, in the as-quenched material. Special focus needs also to be drawn to possible bainitic phase transformations prior martensite formation. To sum up, the following conclusions can be drawn:

- The hardness is increasing with decreasing cooling time. This is due to three main effects being the increased crystal lattice distortion due to less segregated carbon and higher dislocation density, increasing block size and decreasing amounts of retained austenite
- The martensite start temperature is not influenced by the cooling rate below 500 °C. A slight transformation before M_s is detected for the two slowest cooling procedures. This is supposed to be due to (beginning) bainite formation.
- Due to dislocation recovery during slow cooling, less dislocations are present for segregation, hence the distance between the segregated areas increases and APT measurements are no longer accurate.

- The block size is increasing with increasing cooling time below 500 °C, since less undercooling occurs and larger packets can shift from one habit plane due to lower austenite resistance at higher temperatures.

Acknowledgements Financial support by the Austrian Federal Government (in particular from Bundesministerium für Verkehr, Innovation und Technologie and Bundesministerium für Wissenschaft, Forschung und Wirtschaft) represented by Österreichische Forschungsförderungsgesellschaft mbH and the Styrian and the Tyrolean Provincial Government, represented by Steirische Wirtschaftsförderungsgesellschaft mbH and Standortagentur Tirol, within the framework of the COMET Funding Programme is gratefully acknowledged. The TEM investigations were carried out using facilities at the University Service Centre for Transmission Electron Microscopy, Vienna University of Technology, Austria. The authors want to thank Jozef Keckes for the help with the synchrotron measurements and Francisca Martin Mendez for the help with the 3D APT measurements.

References

1. Totten GE, Howes MA (1997) Steel heat treatment handbook. CRC Press
2. Bleck W (2004) Werkstoffkunde Stahl für Studium und Praxis, 2nd edn. Mainz, Aachen
3. Franz Wever (1961) Atlas zur Wärmebehandlung der Stähle, Teil 1&2. Max-Planck-Institut für Eisenforschung
4. Speich G, Leslie W (1972) Metall Trans 3(5):1043–1054
5. Morito S, Huang X, Furuhara T, Maki T, Hansen N (2006) Acta Mater 54(19):5323–5331
6. King A, Beckmann F, Müller M, Schreyer A, Schell N, Fischer T (2014) Mechanical stress evaluation by neutrons and synchrotron radiation VI. In: Volume 772 of Materials science forum. Trans Tech Publications, 2, pp 57–61
7. Hammersley AP (1997) Fit2d: an introduction and overview. ESRF internal report, ESRF97HA02T
8. Young RA (1993) The Rietveld method. Oxford University Press, New York
9. Bhadeshia H (2002) Martensite in steel
10. Vieweg A, Povoden-Karadeniz E, Ressel G, Prevedel P, Wojcik T, Martin-Mendez F, Stark A, Keckes J, Kozeschnik E (2017) Mater Des 136:214–222
11. Fischer FD, Svoboda J, Kozeschnik E (2013) Modell Simul Mater Sci Eng 21(2):025008
12. Morito S, Tanaka H, Konishi R, Furuhara T, Maki T (2003) Acta Mater 51(6):1789–1799
13. Byun TS, Hashimoto N, Farrell K (2004) Acta Mater 52(13):3889–3899
14. Hanamura T, Torizuka S, Tamura S, Enokida S, Takechi H (2013) ISIJ Int 53(12):2218–2225
15. Hall E (1951) Proc Phys Soc Sect B 64(9):747
16. Petch NJ (1953) Iron Steel Inst 174:25–28
17. Morito S, Yoshida H, Maki T, Huang X (2006) Mater Sci Eng: A 438–440(0):237–240. In: Proceedings of the international conference on martensitic transformations
18. Morris E (2001) The influence of grain size on the mechanical properties of steel

Part III

Interactions of Phase Transformations and Plasticity

TRIP Effect in a Constant Load Creep Test at Room Temperature

N. Tsuchida and S. Harjo

Abstract

In order to investigate TRIP (transformation induced plasticity) effect in different deformation style, a room temperature creep test under the constant load was conducted by using a TRIP-aided multi-microstructure steel. As a result, the volume fraction of deformation-induced martensite in the constant load creep test was larger than that in the tensile test. In situ neutron diffraction experiments during the constant load creep test were performed to discuss its reason. It is found from the in situ neutron diffraction experiments during the constant load creep tests that the phase strain of the austenite phase in the creep tests was larger than that in the tensile tests at the same applied stress.

Keywords

TRIP • Deformation-induced martensitic transformation Creep • Constant load

Introduction

TRIP-aided multi-microstructure steels can obtain high strength and better uniform elongation by the TRIP effect due to the deformation-induced martensitic transformation of retained austenite (γ_R) [1, 2]. It is important to clarify the possibility of TRIP effect by the effective use of γ_R. The deformation-induced martensitic transformation is affected by various factors such as temperature, strain rate, and so on, [1–3] and the effect of deformation style on deformation-induced transformation and the TRIP effect was

focused on in this study. As a way to accomplish this objective, constant load creep tests [4, 5] were conducted. The constant load creep test is a mechanical test to investigate tensile deformation behavior, and changes of load, strain rate, and work-hardening rate are different from those of tensile tests [4, 5]. In the constant load creep tests, the work-hardening rate ($d\sigma/d\varepsilon$) equals true stress (σ) and increase with true strain (ε) [6]. It is significant to study the relationship between the TRIP effect and deformation-induced martensitic transformation behavior during constant load creep tests and to discuss the difference of the TRIP effect between the creep test and the tensile test. In this study, the constant load creep tests were conducted using a low-carbon TRIP steel at room temperature.

Experimental

A TRIP-aided multi-microstructure steel obtained from a 0.2C-1.5Si-1.2Mn steel (0.2C TRIP) was used [2]. The microstructure of the 0.2C TRIP steel was observed using an optical microscope (OM). The OM observations were performed by a tint etching procedure [2, 7]. The volume fractions of γ_R and deformation-induced martensite (α') after the constant load creep tests and the carbon content of γ_R were estimated by X-ray diffraction experiments [2, 3]. The quantitative estimations of the γ_R and α' volume fractions by X-ray diffraction were based on the principle that the total integrated intensity of all diffraction peaks for each phase is proportional to the volume fraction of that phase.

The uniaxial creep test under a constant applied stress was conducted at room temperature using a uniaxial creep testing machine [4]. In the constant load creep tests, a test specimen with a gage width of 5 mm and gage length of 25 mm was prepared from the 0.2C TRIP steel. The applied stress conditions in the constant load creep test were determined by the nominal stress–strain curve, which was obtained by the static tensile test with an initial strain rate of 3.3×10^{-4} s^{-1} at 296 K [2], and were between 402 and

N. Tsuchida (✉)
Graduate School of Engineering, University of Hyogo, 2167 Shosha, Himeji, 671-2280, Japan
e-mail: tsuchida@eng.u-hyogo.ac.jp

S. Harjo
J-PARC Center, Japan Atomic Energy Agency, 2-4 Shirane Shirakata, Tokai-mura, Naka-gun, Ibaraki 319-1195, Japan

© The Minerals, Metals & Materials Society 2018
A. P. Stebner and G. B. Olson (eds.), *Proceedings of the International Conference on Martensitic Transformations: Chicago*,
The Minerals, Metals & Materials Series, https://doi.org/10.1007/978-3-319-76968-4_7

731 MPa. The constant load creep tests were interrupted at a holding time of about 1.08×10^6 s (300 h) and the volume fractions of γ_R and α' of the interrupted test specimen were calculated using X-ray diffraction [2, 3].

In in situ neutron diffraction experiments during the constant load creep tests at room temperature, test specimens with a gage width of 6 mm, gage length of 55 mm, and thickness of 1.8 mm were prepared [8]. The neutron diffraction experiments were conducted at TAKUMI on a high resolution and high intensity time-of-flight (TOF) neutron diffractometer for engineering sciences at MLF of J-PARC [8]. The applied stress condition in the in situ neutron diffraction experiment during the creep test was 719 MPa and the holding time was 3.6×10^4 s (10 h). Data analyses were performed by single and multi-peak fitting methods using Rietveld software (Z-Rietveld) [9]. In this study, the phase strains in the austenite (γ_R) and ferrite (ferrite + bainite) phases were summarized in addition to the deformation-induced martensitic transformation behavior during constant load creep deformation.

Results and Discussion

Figure 1 shows nominal strain versus log time plots obtained by the constant load creep tests at room temperature in the 0.2C TRIP steel. The nominal strain at the same time increased with an increase in the applied stress. The test specimen at the applied stress of 731 MPa fractured within 1 h after the test started. The slope of the change in nominal strain against time also increased with increasing applied stress. But the slope of nominal strain became smaller after

about 3.0×10^4 s (about 8 h) and then the change of nominal strain was almost saturated. Figure 2 shows log strain rate versus log time plots. The strain rate decreased with an increase in holding time and at a given time increased with increasing applied stress. The slopes of the strain rate in Fig. 2 are −1 independent of the applied stress conditions [4], but the strain rate decreased drastically by the order of 10^{-8} s^{-1} at holding times of about 3.0×10^4 s. The decrease in strain rate up to the 10^{-8} s^{-1} order in the present 0.2C TRIP steel seems to be associated with dynamic strain aging based on the past studies [4, 5, 10]. Figure 3 shows the

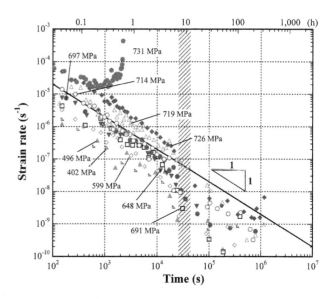

Fig. 2 Log strain rate versus log time plots in the 0.2C TRIP steel obtained by the constant load creep tests at room temperature with various applied stress conditions

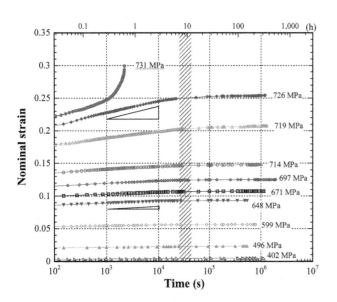

Fig. 1 Nominal strain versus log time plots in the 0.2C TRIP steel obtained by the constant load creep tests at room temperature with various applied stress conditions

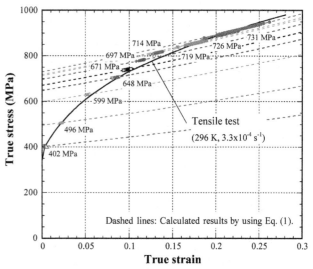

Fig. 3 True stress–true strain relationships obtained by the room-temperature creep tests with various applied stress conditions

true stress (σ)–true strain (ε) relationships at various applied stresses obtained by the constant load creep tests. Both σ and ε increased with an increase in the applied stress, and their changes during the creep tests for ~300 h also became larger with increasing applied stress. In the constant load creep tests, σ is identical to $d\sigma/d\varepsilon$ because the tensile deformation in the constant load creep test always maintains the plastic instability condition. The plastic instability condition in the constant load creep test can also be described by the following equation [4, 6]:

$$\frac{d\sigma}{d\varepsilon} = \sigma = s(1+e) = s \exp(\varepsilon) \qquad (1)$$

where s and e mean nominal stress and nominal strain. From Eq. (1), the σ–ε relationship in the constant load creep test can be estimated by the applied stress (s). The dashed lines in Fig. 3 show the calculated σ–ε relationships using Eq. (1), and the solid line the σ–ε relationship obtained by the static tensile test [2]. The σ at the same ε in the constant load creep tests were almost the same as those of the tensile tests, or a little larger. But the changes of the σ–ε relationship with ε and time are different between the creep and tensile tests.

Figure 4 shows the volume fraction of α' as a function of ε in the constant load creep and static tensile tests [2, 8]. Here, the results of creep tests were obtained using test specimens that were kept for about 300 h at each applied stress. The volume fractions of α' at the same ε obtained from the constant load creep tests are found to be larger than those from the static tensile tests. The deformation-induced martensitic transformation behaviors are different between the two tests despite obtaining almost the same σ–ε relationships. In order to investigate the difference of deformation-induced

transformation behavior between the constant load creep and tensile tests, in situ neutron diffraction experiments were conducted during the constant load creep test at room temperature. The creep test was conducted at the applied stress of 719 MPa and its holding time was 3.6×10^4 s (10 h). The reason for the holding time of 3.6×10^4 s is that the strain rate at 3.6×10^4 s was below the 10^{-8} s^{-1} order and the nominal strain changed little after the holding time of 3.6×10^4 s, as seen in Figs. 1 and 2.

Figure 5 shows the volume fraction of α' versus log time plots at the applied stress of 719 MPa in the in situ neutron diffraction experiments during the constant load creep test. The volume fraction of α' (0.04) obtained by the static tensile test at the same s or e [2, 8] is also shown in Fig. 5. The volume fraction of α' at immediately after the applied stress of 719 MPa was loaded in the constant load creep test was 3%. The volume fraction of α' increased to about 5% until the holding time of 30 s and showed an almost constant value after that. The volume fraction of α' in the constant load creep test became larger than that of tensile test for the first 30 s. Figure 6 shows phase strains of the austenite (γ) and ferrite (α) phases versus log time plots at the applied stress of 719 MPa. The phase strains of the γ and α phases at the same ε or the same applied stress obtained by the in situ neutron diffraction experiments during the tensile test [8] are also shown as red (γ) and blue (α) lines in Fig. 6. When the phase strains were compared between the constant load creep and tensile tests, the phase stain of γ phase in the constant load creep test was larger than that in the tensile test. The phase strain is associated with the phase stress [8, 11, 12]. The larger the phase strain, the larger phase stress

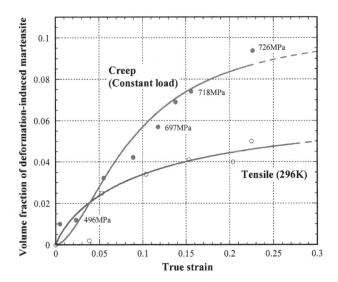

Fig. 4 Volume fraction of deformation-induced martensite as a function of true strain between the constant load creep and static tensile tests [2] in the 0.2C TRIP steel

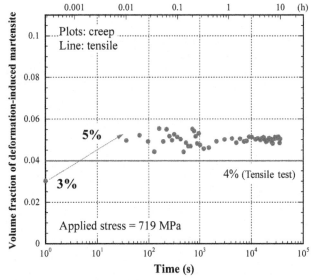

Fig. 5 Volume fraction of deformation-induced martensite versus log time plots between the constant load creep test (plots) and the static tensile test (blue line) at the applied load of 719 MPa

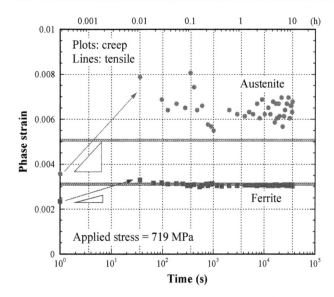

Fig. 6 Phase strains versus log time plots at the applied stress of 719 MPa in the in situ neutron diffraction experiment during the constant load creep test. Those in the tensile test at the applied stress of 719 MPa are also shown

becomes. The phase stress of γ in the constant load creep test is therefore larger than that in the tensile test at the same applied stress. The flow stress of γ phase (σ_γ) is associated with the descriptions of deformation-induced transformation behavior [13, 14]. The volume fraction of α' becomes larger with larger σ_γ. Judging from these factors, the larger volume fraction of α' in the creep test than the tensile test at the same σ or ε is ascribed a larger flow stress in the γ phase.

Summary

(1) Nominal strain and strain rate increased with an increase in applied stress in the constant load creep tests of the TRIP steel. The change of nominal strain was almost stagnated at holding times of about 3.0×10^4 s.
(2) The volume fractions of deformation-induced martensite (α') at a given true strain obtained from constant load creep tests were larger than those from tensile tests.
(3) From the in situ neutron diffraction experiments during the constant load creep tests, the phase strain of the austenite phase in the creep tests was larger than that in

the tensile tests at the same applied stress. This means that the true stress of the austenite phase in the TRIP steel is associated with the difference in the volume fraction of α' between the creep and the tensile tests.

References

1. Sugimoto K, Kobayashi J, Hojo T (2017) Microstructure and mechanical properties of ultrahigh-strength TRIP-aided steels. Tetsu-to-Hagane 103(1):1–11
2. Tsuchida N, Osaki K (2013) Effect of strain rate on TRIP effect in a 0.2C-1.5Si-1.2Mn steel. Tetsu-to-Hagane 99(8):527–531
3. Tsuchida N, Yamaguchi Y, Morimoto Y, Tonan T, Takagi Y, Ueji R (2013) Effects of temperature and strain rate on TRIP effect in SUS301L metastable austenitic stainless steel. ISIJ Int 53 (10):1886–1892
4. Tsuchida N, Baba E, Nagai K, Tomota Y (2005) Effects of interstitial solute atoms on the very low strain-rate deformations for an IF steel and an ultra-low carbon steel. Acta Mater 53:265–270
5. Tendo M, Takeshita T, Nakazawa T, Kimura H, Abo H (1993) Room temperature creep behavior of stainless steels. Tetsu-to-Hagane 79(1):98–104
6. Tsuchida N (2016) Basic of tensile tests. Bull Iron Steel Inst Jpn 21(7):389–396
7. Matsumura O, Sakuma Y, Takechi H (1991) Retained austenite in 0.4%C-Si-1.2%Mn steel sheet annealed in intercritical range and austempered. Tetsu-to-Hagane 77(8):1304–1311
8. Harjo S, Tsuchida N, Gong W, Abe J, Aizawa K (2013) TRIP steel deformation behavior by neutron diffraction. In: Materials Research Society symposium proceedings 1528. https://doi.org/10.1557/opl.2013.570
9. Oishi R, Yonemura M, Nishimaki Y, Torii S, Hoshikawa A, Ishigaki T, Morishima T, Mori K, Kamiyama T (2009) Rietveld analysis software for J-PARC. Nucl Instr Methods A 600:94–96
10. Tomota Y, Lukas P, Harjo S, Park JH, Tsuchida N, Neov D (2003) In situ neutron diffraction study of IF and ultra low carbon steels upon tensile deformation. Acta Mater 51:819–830
11. Asoo K, Tomota Y, Harjo S, Okitsu Y (2011) Tensile behavior of a TRIP-aided ultra-fine grained steel studied by neutron diffraction. ISIJ Int 51(1):145–150
12. Morooka S, Umezawa O, Harjo S, Hasegawa K, Toji Y (2012) Quantitative analysis of tensile deformation behavior by in-situ neutron diffraction for ferrite-martensite type dual-phase steels. Tetsu-to-Hagane 98(6):311–319
13. Perlade A, Bouaziz O, Furnemont Q (2003) A physically based model for TRIP-aided carbon steels behavior. Mater Sci Eng A 356:145–152
14. Patel JR, Cohen M (1956) Criterion for the action of applied stress in the martensitic transformation. Acta Metall 1:531–538

Modeling of Strain-Induced Phase Transformations Under High Pressure and Shear

Mehdi Kamrani, Biao Feng, and Valery I. Levitas

Abstract

The strain-induced $\alpha \rightarrow \omega$ phase transformation (PT) in a zirconium sample under compression and torsion under fixed load is investigated using the finite-element method (FEM), and results are compared to those for a sample in a diamond anvil cell (DAC) and a rotational diamond anvil cell (RDAC). Highly heterogeneous fields of stresses, strains, and concentration of the high-pressure phase are presented and analyzed. Some experimentally observed effects are analyzed and interpreted.

Keywords

Strain-induced phase transformations • Zirconium
High pressure • Diamond anvil cell • Rotational diamond anvil cell

Introduction

The diamond anvil cell (DAC) and the rotational diamond anvil cell (RDAC) are widely used in the field of high-pressure research. A very large plastic strain can be induced into a sample in DAC and RDAC compared to that from quasi-hydrostatic loading. Experiments show a drastic reduction in PT pressure by a factor of 2–5 [1–3] and even nearly a factor of 10 [4, 5] due to plastic strain induced into the sample. In some cases, the plastic straining results in formation of new phases that could not otherwise be achieved [3, 6]. Such phase transformations should be treated as strain-induced transformations under high pressure rather than pressure-induced transformations. Corresponding physical mechanisms and an underlying theory in which the plastic strain is a time-like parameter controlling the PT was proposed in [1, 2]. This model has been applied to study phase transformations in both DAC [7–9] and RDAC [10, 11]. In this paper, the strain-induced phase transformation in a Zr sample is studied using the finite element method (FEM), and results are compared for DAC and RDAC.

Problem Formulation

Geometry and Boundary Conditions

In this paper, both the diamond and the sample, with actual geometries as in experiments (e.g., in Ref. [12]), are considered to be deformable. Because of symmetries, just a quarter of the sample and the anvil are considered (see Fig. 1). A normal stress σ_n is applied at the top surface of the anvil. In RDAC, rotation is also applied to the top surface of the anvil while the applied stress σ_n is kept fixed. The radial displacement u_r and shear stresses τ_{rz} and $\tau_{z\varphi}$ are zero on the symmetry axis $r = 0$ (the lines BC and CD for the anvil and the sample, respectively). In the symmetry plane $z = 0$ (the plane DH), the radial shear stress and circumferential and axial displacement are all zero: $\tau_{rz} = 0$ $u_\varphi = u_z = 0$. Along the contact surface (the line CEF), the combined Coulomb and plastic friction model was implemented.

M. Kamrani (✉)
Department of Aerospace Engineering, Iowa State University, Ames, IA 50011, USA
e-mail: mkamrani@iastate.edu

B. Feng
Los Alamos National Laboratory, Los Alamos, NM 87545, USA
e-mail: fengbiao11@gmail.com

V. I. Levitas (✉)
Departments of Aerospace Engineering, Mechanical Engineering, and Material Science and Engineering, Iowa State University, Ames, IA 50011, USA
e-mail: vlevitas@iastate.edu

V. I. Levitas
Ames Laboratory, Division of Materials Science and Engineering, Ames, IA 50011, USA

© The Minerals, Metals & Materials Society 2018
A. P. Stebner and G. B. Olson (eds.), *Proceedings of the International Conference on Martensitic Transformations: Chicago*, The Minerals, Metals & Materials Series, https://doi.org/10.1007/978-3-319-76968-4_8

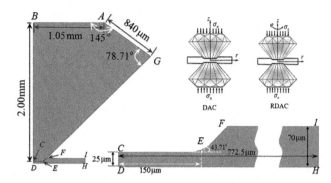

Fig. 1 The geometry and dimensions of a quarter of the sample and anvil in the DAC and RDAC

Complete System of Equations

Material Model

As shown in [13], for a wide range of materials, including metals, pressed powders, rocks, etc., above some level of plastic strain there is a saturation in strain hardening and a material behaves as isotropic and perfectly plastic. Besides, the elastic and plastic properties of the Zr sample are considered to be phase-dependent. The deformation gradient $F = \partial r/\partial r_0$ can be decomposed into elastic F_e, plastic F_p, and transformational F_t parts as in $F = F_e \cdot F_t \cdot F_p$, where r and r_0 are the position vectors in the deformed and un-deformed configurations. The symmetric part of the velocity gradient $\mathbf{d}$ is decomposed as:

$$\mathbf{d} = \left(\dot{\mathbf{F}} \cdot \mathbf{F}^{-1}\right)_s = \overset{\nabla}{\boldsymbol{\varepsilon}}_e + \dot{\varepsilon}_t \mathbf{I} + \mathbf{d}_p; \quad \varepsilon_t = \overline{\varepsilon}_t c \tag{1}$$

Here $\overset{\nabla}{\boldsymbol{\varepsilon}}_e$ is the Jaumann time derivative of the elastic strain, $\mathbf{I}$ is the unit tensor, $\overline{\varepsilon}_t$ is the volumetric transformation strain, and c is the concentration of high-pressure (ω) phase. The kinetics of a strain-induced PT is defined as:

$$\frac{dc}{dq} = 10k\frac{(1-c)\overline{p}_d H(\overline{p}_d)\frac{\sigma_{y2}}{\sigma_{y1}} - c\overline{p}_r H(\overline{p}_r)}{c + (1-c)\frac{\sigma_{y2}}{\sigma_{y1}}} \tag{2}$$

where k is the kinetic parameter, $\overline{p}_d = \frac{p - p_\varepsilon^d}{p_h^d - p_\varepsilon^d}$ and $\overline{p}_r = \frac{p - p_\varepsilon^r}{p_h^r - p_\varepsilon^r}$ are the dimensionless characteristic pressures for direct and reverse PTs, p_ε^d is the minimum pressure below which direct strain-induced PT (from a low-pressure to a high-pressure phase) cannot happen, p_ε^r is the maximum pressure above which reverse strain-induced PT is not possible, p_h^d and p_h^r are the pressures for direct and reverse PTs under hydrostatic condition, respectively, H is the Heaviside step function, and q is the accumulated plastic strain defined by the evolution equation $\dot{q} = (2/3\mathbf{d}_p : \mathbf{d}_p)^{1/2}$.

Friction Model

According to the Coulomb friction model, there can be no sliding between contact pairs unless the total shear stress reaches the critical shear stress $\tau_{crit} = \mu\sigma_c$, where μ is the friction coefficient and σ_c is the normal contact stress. If the shear stress reaches the yield strength in shear $\tau_y(c) = \sigma_y(c)/\sqrt{3}$, there will be plastic sliding along the contact surface regardless of the Coulomb friction condition being satisfied. Therefore, the critical friction stress is redefined [8, 10] as $\tau_{crit} = \min(\mu\sigma_c, \tau_y(c))$. To eliminate convergence problems in FEM due to this threshold-type change from a cohesive to a sliding condition, the cohesive condition is replaced with a small elastic sliding u_e, i.e., the total sliding along the contact surface is decomposed into [8, 10] u_e elastic and u_s plastic sliding portions, or, $u_c = u_e + u_s$. A simple linear relation [8, 10] $\tau = k_s u_e$ is assumed between the shear stress and the elastic sliding where the contact stiffness k_s is defined as $\tau_{crit} = k_s u_{crit}$. u_{crit} in which the maximum permissible elastic sliding along the contact surface is considered to be 0.5% of the average element size in accordance with ABAQUS documentation [14]. The elastic sliding vector can be defined as $u_e = (u_{crit}/\tau_{crit})\tau$ and the plastic sliding rule as:

$$\begin{aligned} |\dot{u}_s| &= 0 & if \quad \tau = |\tau| = <\tau_{crit} \\ \dot{u}_s &= \frac{|\dot{u}_s|}{\tau_{crit}}\tau & if \quad \tau = |\tau| \geq \tau_{crit} \end{aligned} \tag{3}$$

Material Parameters and Numerical Procedure

Isotropic elastic behavior is considered for diamond with Young's modulus E of 1048.5 GPa [15] and a Poisson's ratio v of 0.1055. At room temperature, the α phase of Zr can transform into the ω phase under a pressure of 1–7 GPa [16–18]. Here, the transformation pressure is assumed to be $p_\varepsilon^d = 1.7$ GPa and p_h^d is taken to be 7 GPa [17]. Since in experiments reverse PT does not occur during unloading, $p_\varepsilon^r = -2$ GPa and $p_h^r = -3.7$ GPa are considered. The following are the material properties for each of the two phases [19]:

for the α phase: $E_1 = 90.9$ GPa, $v_1 = 0.344$ and $\sigma_{y1} = 180$ MPa;
for the ω phase: $E_2 = 113.8$ GPa, $v_2 = 0.305$ and $\sigma_{y2} = 1180$ MPa

The kinetic parameter is taken as $k = 10$ and the transformational volumetric strain is $\overline{\varepsilon}_t = -0.014$ [20].

Evolution of Stress and Plastic Strain Fields During HPT

Figures 2 and 3 show the distributions and evolutions of the concentration of the high-pressure phase, pressure, and accumulated plastic strain in the sample in DAC and RDAC. In DAC, the maximum pressure first reaches the critical pressure p_ε^d at the center of the sample where there is also a concentration of plastic strain, so PT starts at the center of the sample. With an increase in the applied load and consequently in plastic strain, the PT propagates toward the contact surface and larger radii. Because the friction stress has reached yield strength in shear along the contact surface (Fig. 4), the maximum plastic strain is localized at the periphery of the contact surface due to the shear flow. Consequently, with an increase in applied load, the rate of

PT is higher at the contact surface compared to the symmetry plane for the same r.

With respect to loading in RDAC, the sample region on the left-hand side of the line $p = p_\varepsilon^d$ has already experienced a pressure larger than p_ε^d and is therefore prone to transform if there is a plastic flow. Similar to compression, PT initiates at the center of the sample. With increasing rotation angle, the isolines $p = p_\varepsilon^d$ move to the larger radii and the size of the transforming region increases. These isolines always remain at the diffuse border of the transforming region, which can be used for experimental determination of the critical pressure. For large rotation angles, transformed material is visible in the region where $p < p_\varepsilon^d$, in which transformation is impossible. High-pressure phase appears in this region due to reduction of the sample thickness during rotation of an anvil and radial plastic flow. If this is not

(a)

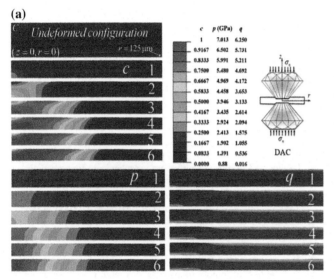

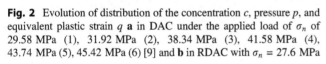

(b)

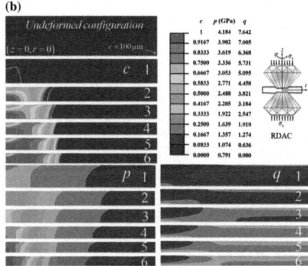

Fig. 2 Evolution of distribution of the concentration c, pressure p, and equivalent plastic strain q **a** in DAC under the applied load of σ_n of 29.58 MPa (1), 31.92 MPa (2), 38.34 MPa (3), 41.58 MPa (4), 43.74 MPa (5), 45.42 MPa (6) [9] and **b** in RDAC with $\sigma_n = 27.6$ MPa and rotation angle of $\varphi = 0.0$ (1), $\varphi = 0.16$ (2), $\varphi = 0.32$ (3), $\varphi = 0.48$ (4), $\varphi = 0.64$ (5), and $\varphi = 0.8$ (6) radian. White lines in the concentration distribution in **b** correspond to pressure $p = p_\varepsilon^d$

Fig. 3 Evolution of distributions of the concentration and pressure along the contact surface in **a** DAC and **b** RDAC

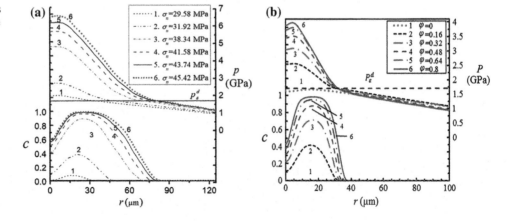

Fig. 4 Evolution of distribution of the radial component of shear stress τ_{rz} along the contact surface for **a** DAC [9] and **b** RDAC. The distribution of the yield strength in shear, $\tau_y(c)$, is also shown in **a**. Excluding central region, $\tau_{rz} = \tau_y(c)$ along the contact surface

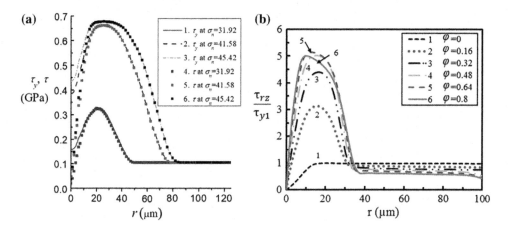

understood, an experimentalist may misinterpret measurements and report PT pressure below an actual value. Also, for high-pressure torsion, while averaged pressure (total force per unit area) is used to characterize PT pressure [18, 21], because of very strong pressure heterogeneity, maximum pressure can be more than three times that of this average. That is why we used $p_\varepsilon^d = 1.7$ GPa, much higher than the reported [21] value of 0.5 GPa based on averaged pressure.

For both DAC and RDAC, the radial friction shear stress τ_{rz} (Fig. 4) and consequent pressure gradient increase with increasing applied load or rotation angle due to an increase in the yield strength during phase transformation (see Figs. 2 and 3). For torsion in RDAC, increasing the circumferential component of shear stress $\tau_{\varphi z}$ leads to reduction in τ_{rz} because the magnitude of the total friction stress is limited by the yield strength in shear.

Concluding Remarks

In the paper, the main regularities of coupled plastic flow and strain-induced $\alpha \rightarrow \omega$ phase transformation in Zr were studied. Strong heterogeneity in pressure leads to significant error in reported value of transformation pressure based on averaged pressure (force/area), as reported in [18, 21], so we used $p_\varepsilon^d = 1.7$ GPa, more than three times higher than the reported value in [21] of 0.5 GPa based on averaged pressure. Note that pressure grows during compression up to $p \approx 3.5 p_\varepsilon^d$ and under torsion up to $p \approx 2.5 p_\varepsilon^d$, and the transformation is still not completed due to insufficient plastic strain. Such high pressure is not required for completing transformation but under such loadings appears as a result of a coupled phase transformation and plastic flow. While experimentalists who characterize transformation in terms of pressure without plastic strains would say that phase transformation is not completed even at 6.5 GPa, based on the model, this is a kinetic property of the phase

transformation that could occur until completion at $p = p_\varepsilon^d$ under large plastic straining. Thus, for economic synthesis of high-pressure phases, it is necessary to design a loading program that will allow the reaching of a complete phase transformation at the possible minimum pressure p_ε^d.

Acknowledgements The support of NSF (DMR-1434613), ARO (W911NF-12-1-0340), and Iowa State University (Schafer 2050 Challenge Professorship and Vance Coffman Faculty Chair Professorship) is gratefully acknowledged.

References

1. Levitas VI (2004) Continuum mechanical fundamentals of mechanochemistry. In: Gogotsi Y, Domnich V (eds) High-pressure surface science and engineering (chapter 3). Taylor & Francis. https://doi.org/10.1201/9781420034134.sec3
2. Levitas VI (2004) High-pressure mechanochemistry: conceptual multiscale theory and interpretation of experiments. Phys Rev B 70:184118
3. Novikov NV, Polotnyak SB, Shvedov LK, Levitas VI (1999) Regularities of phase transformations and plastic straining of materials in compression and shear in diamond anvils: experiments and theory. J Superhard Mater 3:39–51
4. Ji C, Levitas VI, Zhu HY, Chaudhuri J, Marathe A, Ma Y (2012) Shear-induced phase transition of nanocrystalline hexagonal boron nitride to wurtzitic structure at room temperature and lower pressure. P Natl Acad Sci 109:19108–19112
5. Levitas VI, Shvedov LK (2002) Low-pressure phase transformation from rhombohedral to cubic BN: experiment and theory. Phys Rev B 65:104109
6. Levitas VI, Ma Y, Selvi E, Wu JZ, Patten JA (2012) High-density amorphous phase of silicon carbide obtained under large plastic shear and High pressure. Phys Rev B 85:054114
7. Levitas VI, Zarechnyy OM (2010) Modeling and simulation of strain-induced phase transformations under compression in a diamond anvil cell. Phys Rev B 82:174123
8. Feng B, Levitas VI, Ma Y (2014) Strain-induced phase transformation under compression in a diamond anvil cell: simulations of a sample and gasket. J Appl Phys 115:163509
9. Feng B, Levitas VI (2017) Plastic flows and strain-induced alpha to omega phase transformation in zirconium during compression in a diamond anvil cell: finite element simulations. Mat Sci Eng A 680:130–140

10. Feng B, Levitas VI (2013) Coupled phase transformations and plastic flows under torsion at high pressure in rotational diamond anvil cell: effect of contact sliding. J Appl Phys 114:213514

11. Feng B, Levitas VI (2016) Effects of gasket on coupled plastic flow and strain-induced phase transformations under high pressure and large torsion in a rotational diamond anvil cell. J Appl Phys 119:015902

12. Hemley RJ, Mao HK, Shen GY, Badro J, Gillet P, Hanfland M, Hausermann D (1997) X-ray imaging of stress and strain of diamond, iron, and tungsten at megabar pressures. Science 276:1242

13. Levitas VI (1996) Large deformation of materials with complex rheological properties at normal and high pressure. Nova Science Publishers, New York

14. Abaqus V6.11 (2011) User subroutines. Abaqus INC, Providence RI, USA

15. Guler E, Guler M (2015) Elastic and mechanical properties of cubic diamond under pressure. Chin J Phys 53:040807

16. Zilbershtein VA, Chistotina NP, Zharov AA, Grishina NS, Estrin EI (1975) Alpha-omega transformation in titanium and zirconium during shear deformation under pressure. Fiz Met I Metalloved 39:445–447

17. Xia H, Duclos SJ, Ruoff AL, Vohra YK (1990) New high-pressure phase transition in zirconium metal. Phys Rev Lett 64:204–207

18. Srinivasarao B, Zhilyaev AP, Perez-Prado MT (2011) Orientation dependency of the alpha to omega plus beta transformation in commercially pure zirconium by high pressure torsion. Scr Mater 65:241–244

19. Zhao YS, Zhang JZ (2007) Enhancement of yield strength in zirconium metal through high-pressure induced structural phase transition. Appl Phys Lett 91:201907

20. Greeff CW (2005) Phase changes and the equation of state of Zr. Modell Simul Mater Sci Eng 13:1015

21. Zhilyaev AP, Sabirov I, Gonzalez-Doncel G, Molina-Aldareguia J, Srinivasarao B, Perez-Prado MT (2011) Effect of Nb additions on the microstructure, thermal stability and mechanical behavior of high pressure Zr phases under ambient conditions. Mat Sci Eng A 528:3496–3505

Modeling the Microstructure Evolutions of NiTi Thin Film During Tension

S. E. Esfahani, I. Ghamarian, V. I. Levitas, and P. C. Collins

Abstract

A microscale phase field model for the multivariant martensitic phase transformation is advanced and utilized for studying the pseudoelastic behavior of a thin film of equiatomic single crystal NiTi under tensile loading. The thermomechanical model includes the strain softening as a mechanism leading to strain (transformation) localization and discrete microstructure formation. To avoid a small scale limitation, gradient term is dropped. Numerical solutions have shown a negligible mesh sensitivity for different element shapes and densities, which is due to rate-dependent kinetic equations for phase transformation. Microstructure evolution and corresponding stress-strain curves are presented for several cases. Obtained stress-strain curves, band-like martensitic microstructure, a sudden drop in the stress at the beginning of the martensitic transformation, residual austenite, and multiple stress oscillations due to nucleation events are qualitatively similar to those in known experiments.

Keywords

Martensitic phase transition • NiTi • Localization Single crystal

Introduction

Studying the behavior of shape memory alloys has always been a demanding goal for the different experimental [1, 2] and theoretical [1, 3] groups. NiTi (nitinol), being an important shape memory alloy [4, 5], shows an unstable phase transformation behavior, which leads to a sudden drop in the global stress-strain curves and localized transformed regions during the loading and unloading of the sample. The thermomechanical phenomenological approaches were utilized in [6, 7] to study the influence of martensitic microstructure formation on the properties of shape memory alloys. The superelastic behavior and the localization during the loading of NiTi were observed experimentally in [3, 8]. The main focus of the current paper is to advance the phase field model presented in [9, 10] in order to investigate pseudoelastic behavior of the thin film of a single crystal NiTi. Finite element (FE) simulations based on the model result in the formation of martensitic microstructures were produced, which are in qualitative agreement with experimental observation. The mesh sensitivity and morphology as well as the effects of the athermal threshold and a pre-existing nucleus on the global stress-strain curves and the microstructure evolution are investigated (Figs. 1 and 2).

Model Description

A scale-independent thermomechanical model for studying multivariant martensitic phase transformation proposed in [9, 10] is advanced and utilized in the current study. In contrast to traditional phase field approaches (e.g., [11–13]), the current model is scale-free because the gradient term is

S. E. Esfahani (✉)
Department of Aerospace Engineering, Iowa State University, Ames, IA 50011, USA
e-mail: ehsan@iastate.edu

I. Ghamarian
Department of Materials Science & Engineering, University of Michigan, Ann Arbor, MI 48109, USA
e-mail: ghamaria@umich.edu

V. I. Levitas
Departments of Aerospace Engineering, Mechanical Engineering, and Material Science & Engineering, Iowa State University, Ames Laboratory, Division of Materials Science & Engineering, Ames, IA 50011, USA
e-mail: vlevitas@iastate.edu

P. C. Collins
Department of Materials Science & Engineering, Iowa State University, Ames, IA 50011, USA
e-mail: pcollins@iastate.edu

© The Minerals, Metals & Materials Society 2018
A. P. Stebner and G. B. Olson (eds.), *Proceedings of the International Conference on Martensitic Transformations: Chicago*, The Minerals, Metals & Materials Series, https://doi.org/10.1007/978-3-319-76968-4_9

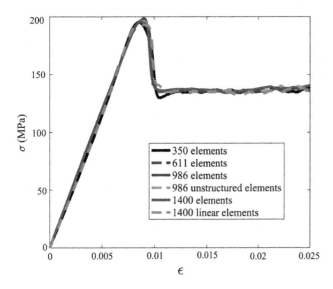

Fig. 1 The macroscopic stress-strain response of a thin NiTi film for a zero athermal threshold and for various mesh densities and regularities. The macroscopic strain rate is $\dot{\varepsilon} = 5.5 \times 10^{-2}$ 1/s.

only, and martensitic microstructure is determined in an averaged way in terms of volume fraction of the martensitic variants without explicit variant-variant interfaces. The model [9, 10] included isotropic elasticity and was implemented for cubic to tetragonal transformation with three martensitic variants. In the current work, cubic to monoclinic transformation in NiTi was studied with 12 martensitic variants, and anisotropic single crystal elasticity was implemented for both austenite and martensite. Transformation strains for cubic to monoclinic phase transition are taken from [14]. Components of the anisotropic elastic tensor can be learned from [15, 16]. Kinetic coefficients were accepted as $\lambda_{ij} = 10$ MPa; the difference in thermal energy of martensite and austenite was taken as $(\psi_i - \psi_0) = 18$ MPa, and the magnitude of the energy term $Ac_0(1 - c_0)$, which characterizes internal stresses and interfacial energy and determined strain softening, was considered as $A = 13$ MPa.

The plane stress condition was imposed to model a thin film. Finite element analysis results are presented for uniaxial tensile tests of a thin film of equiatomic NiTi. For this purpose, 8-node biquadratic quadrilateral finite elements are employed for all simulations. Temperature distribution in the thin film is considered homogeneous and constant. The rectangular sample shown in Fig. 3 is clamped at the left edge; at the right edge increasing normal displacement and zero shear stress are prescribed. The Bunge Euler angle set is $(\phi_1, \phi, \phi_2) = (86°, 17°, 0°)$ in all simulations.

dropped. This is done to consider micro- and macroscale samples, which is impossible with traditional phase field approaches due to the necessity to numerically resolve nanometer-wide interfaces. In the current model, the volume fraction of austenite, c_0, is the only order parameter, and the volume fraction of each martensitic variant is just the internal variable. Thus, austenite-martensite interfaces are resolved

Fig. 2 Austenite distribution within a thin NiTi film for various mesh densities and regularities for two different macrostrains. Zero athermal threshold is assumed, and the macroscopic strain rate is $\dot{\varepsilon} = 5.5 \times 10^{-2}$ 1/s.

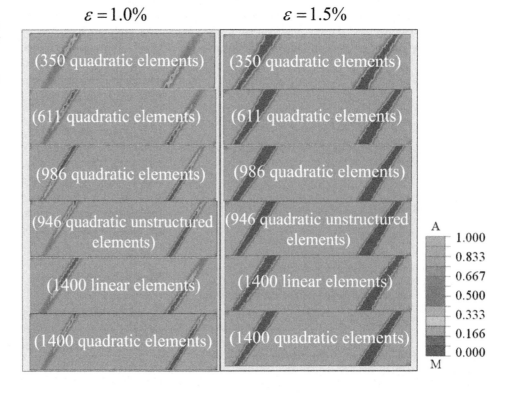

Fig. 3 The macroscopic stress-strain response, and the microstructure evolution within a thin NiTi film under a cycle of the uniaxial load. Macroscopic strain rate is $\dot{\varepsilon} = 1.65 \times 10^{-2}\ 1/s$ and zero athermal threshold is considered. 981 eight-node biquadratic FE are utilized

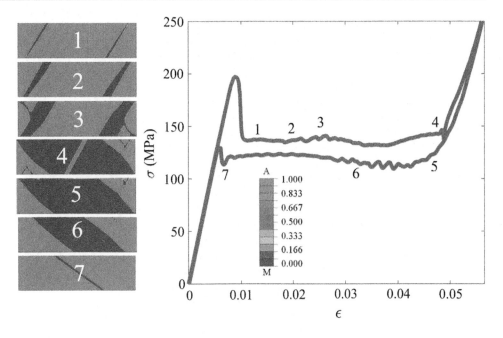

Simulation Results

It is well-known that the strain softening causes the strain localization, which without regularizing the gradient energy term may lead to strong mesh sensitivity of FE simulations; hence, we need to study the mesh sensitivity of the solution. In Fig. 1, the macroscopic stress-strain responses for various mesh densities and different mesh regularities (structured and unstructured meshes) are presented for the same conditions. The results demonstrate that despite the local and global strain softening, stress-strain curves are practically mesh-insensitive. In Fig. 2, the distribution of the austenitic phase for different mesh densities and regularities for two different macrostrains is presented. Solutions are weakly dependent on the mesh, especially for a larger strain because of a broader martensitic region, and for more than 611 FE. In particular, phase transformation starts and occurs in the same two bands, and band width and concentration distribution within bands are close for different meshes. Formally, the problem is well posed, even without the gradient energy term. It is regularized due to rate dependence of the kinetic equations for phase transformation, similar to viscoplastic regularization in plasticity. This prevents a theoretical solution with zero thickness of the interface between austenite and martensite. However, for a slow loading or stationary solution, the interface width tends to go to zero. In the FE solution, interfaces localize within one finite element, and the interface width is mesh-dependent. However, this does not necessarily affect the solution, especially when martensitic and austenitic bands are much larger than the element

(interface) size. Small differences observed in Fig. 2 for $\varepsilon = 0.01$ for different meshes are because of the interface width is comparable with the band width. For $\varepsilon = 0.02$ this difference practically disappears. The obtained mesh independence makes such a model efficient for the treatment of macroscale real-world problems on phase transformations in a large sample.

The microstructure evolution for the single crystal of NiTi and corresponding macroscopic stress-strain response for a cycle of tensile loading are presented in Fig. 3. From this example, the influence of the martensite plate's nucleation and propagation on the global stress-strain curve can be observed. As soon as phase transformation starts, a sudden drop in the stress can be seen. When two band-like martensitic regions start propagation, a plateau in the stress-strain response appears. Several oscillations can be obviously seen in this figure, which are due to the appearance and disappearance of the martensitic zones in the specimen. In point 4, there is one more drop in the stress-strain curve since two major bands coalesce and two interfaces are getting eliminated. Residual austenite at the corners of a sample is observed at high stresses. Thus, the "elastic" branch of martensite in the stress-strain curve is affected by residual austenite and a minor transformation in small regions, which causes narrow hysteresis under unloading. While direct transformation occurs within two major bands, the reverse transformation progresses within a single shrinking band. The stress pick after point 7 is due to the coalescence of two interfaces.

In Fig. 4, the effect of a nucleation site at the middle of the sample is investigated to simulate a pre-existing defect.

Fig. 4 The macroscopic stress-strain response and the microstructure morphology evolution within a thin NiTi film with a single nucleating site at the middle of a sample, under a cycle of the uniaxial load. Macroscopic strain rate $\dot{\varepsilon} = 1.65 \times 10^{-2}$ 1/s is applied and zero athermal threshold is assumed. 981 eight-node biquadratic FE are utilized

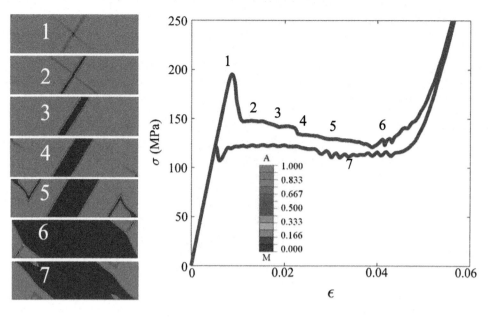

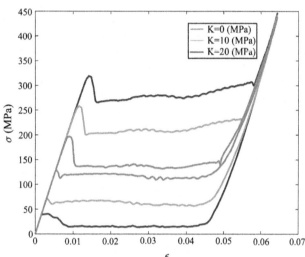

Fig. 5 Effect of the athermal threshold K on the global stress-strain curves of a thin NiTi film for the applied strain rate $\dot{\varepsilon} = 1.65 \times 10^{-2}$ 1/s

The concentration of the first variant was set in this FE as 1 and did not change during simulation. Due to stress concentration around the defect, two martensitic crossed bands nucleate. Further loading causes the reverse phase transformation in one of them, which leads to a single dominant band propagation. Afterward, the phase transformation starts near the stress concentrations at the corners. The third favorable regions for triggering the phase transformation are the locations at the surface due to the bending of a sample. Two martensitic bands, similar to the initial bands in Fig. 2, are produced at these locations. These bands intersect with the bands propagating from the corners. At point 6, all bands coalesce and two new bands from two other corners appear. Also, a single nucleation site increases the oscillations and

affects the plateau part of the stress-strain curve. In Fig. 5, the magnitude of the athermal threshold K is varied. An increase in K increases nucleation stresses for martensite and reduces stresses for initiation of the reverse transformation, thus increasing the hysteresis in the stress-strain curves. The athermal threshold produces some strain hardening for direct transformation while keeping a horizontal plateau for the reverse transformation. Also, hysteresis loops for $K > 0$ are shifted up with respect to the loop at $K = 0$.

To summarize, a scale-independent model proposed in [9, 10] is extended to study the phase transformation in a single crystalline thin film of nitinol. Microstructure evolution and corresponding stress-strain curves are presented for several cases. Obtained stress-strain curves, band-like martensitic microstructure, a sudden drop in the stress at the beginning of the martensitic transformation, residual austenite, and multiple stress oscillations due to nucleation events are qualitatively similar to those in known experiments, e.g., in [2].

Acknowledgements The support of NSF (CMMI-1536925 and DMR-1434613), ARO (W911NF-17-1-0225), XSEDE (TG-MSS140033), and ISU (Schafer 2050 Challenge Professorship and Vance Coffman Faculty Chair Professorship) is gratefully acknowledged.

References

1. Lagoudas DC (2008) Shape memory alloys. Science and Business Media LLC
2. Shaw JA, Kyriakides S (1997) Int J Plast 3:837–871
3. He YJ, Sun QP (2010) Int J Mech Sci 52:1198–1211
4. Duval A, Haboussi M, Zineb TB (2011) Int J Solids Struct 48:1879–1893

5. Iadicola MA, Shaw JA (2004) Int J Plast 20:577–605
6. Arghavani J, Auricchio F, Naghdabadi R, Reali A, Sohrabpour S (2010) Int J Plast 26:976–991
7. Panico M, Brinson LC (2007) J Mech Phys Solids 55:2491–2511
8. Zhang X, Feng P, He Y, Yu T, Sun Q (2010) Int J Mech Sci 52:1660–1670
9. Idesman AV, Levitas VI, Preston DL, Cho J-Y (2005) J Mech Phys Solids 53:495–523
10. Levitas VI, Idesman AV, Preston DL (2004) Phys Rev Lett 93:105701
11. Levitas VI, Lee DW (2007) Phys Rev Lett 99:245701
12. Idesman AV, Cho J-Y, Levitas VI (2008) Appl Phys Lett 93:0431028
13. Levitas VI (2013) Int J Plast 49:85–118
14. Bhattacharyya K (2003) Microstructure of martensite. Oxford Series on Materials Modeling OUP, Oxford
15. Ford DS, White SR (1996) Acta Mater 44:2295–2307
16. Thomasova M, Seiner H, Sedlak P, Frost M, Sevcik M, Szurman I, Kocich R, Drahokoupil J, Sittner P, Landa M (2017) Acta Mater 123:146–156

Residual Stress and Texture Evolution on Surface of 304L TRIP Steel Sheet Subjected to FLC Test

I. S. Oliveira, J. M. Alves, R. A. Botelho, A. S. Paula, L. P. M. Brandão, M. C. Cardoso, L. P. Moreira, and M. C. S. Freitas

Abstract

The martensite formation in the outer and inner blank sheet surfaces of the metastable 304L austenitic stainless steel subjected to the Forming Limit Curve (FLC) testing, performed according to the Marciniak method, was analysed in regions located near and away from the failure site. X-ray Diffraction (XRD) was also used for residual stress and texture quantification related to strain-induced martensitic phase transformation resulting from distinct stress and strain trajectories associated to FLC specimens. From the results were observed some aspects as greater diversity in the martensite formation and austenite hardening at outer sheet surface that are correlated to residual stress and austenite textural evolution.

Keywords

304L austenitic stainless steel • FLC curve
Phase transformation • Residual stress • Texture

Introduction

The austenitic stainless steel 304L, which will be studied in this work, comprises the AISI 304 series and has been intensively used in equipment for chemical, pressure vessels in oil refineries [1].

It can present a particular feature: the phase transformation induced by deformation phenomenon called TRIP effect (Transformation Induced Plasticity) which has basis in martensitic transformation. This effect is referred because it gives to steel favorable mechanical properties as high toughness [2].

In cold working conditions where the material is plasticly deformed ensuing in the TRIP effect, the homogeneous deformation belonging to austenitic phase, that in case of the austenitic stainless steels support this transformation, results in the martensitic transformation. This transformation occurs in the solid state without diffusion, resulting of the coordinated movement among atoms the parent phase, and that keep a narrow crystallographic correspondence between the parent phase (austenite, in steels) and product phase (martensite) resulting of the transformation, however, with new crystallographic structure without chemical composition change.

The martensitic transformation that entails in the TRIP effect is directly affected by the temperature and strain rate. The alloying elements relative effects on start martensitic transformation temperature (M_s) can be estimated by empiric relations [3]:

I. S. Oliveira · L. P. Moreira · M. C. S. Freitas
Department of Mechanical Engineering, Universidade Federal Fluminense, Avenida dos Trabalhadores, 420, Vila Santa Cecília, Volta Redonda, Rio de Janeiro, Brazil
e-mail: isabela.santana@yahoo.com.br

L. P. Moreira
e-mail: luciano.moreira@pq.cnpq.br

M. C. S. Freitas
e-mail: mariacarolinauff@hotmail.com

J. M. Alves · R. A. Botelho · A. S. Paula (✉) · L. P. M. Brandão
Department of Mechanical and Materials Engineering, Instituto Militar de Engenharia (IME), Praça General Tibúrcio, 80, Urca, Rio de Janeiro, Brazil
e-mail: andersan@ime.eb.br

J. M. Alves
e-mail: juciane_alves_rj@yahoo.com.br

R. A. Botelho
e-mail: ramonbotelho@gmail.com

L. P. M. Brandão
e-mail: brandao@ime.eb.br

M. C. Cardoso
Department of Chemical and Materials Engineering, Universidade Federal do Rio de Janeiro, Campus Macaé, Macaé, Rio de Janeiro, Brazil
e-mail: marcelocardoso@macae.ufrj.br

© The Minerals, Metals & Materials Society 2018
A. P. Stebner and G. B. Olson (eds.), *Proceedings of the International Conference on Martensitic Transformations: Chicago*, The Minerals, Metals & Materials Series, https://doi.org/10.1007/978-3-319-76968-4_10

$$Ms(°C) = 1305 - 1665(\%C + \%N) - 28(\%Si) \\ - 33(\%Mn) - 42(\%Cr) - 61(\%Ni) \quad (1)$$

The highest temperature to which the M_s temperature can be raised by applied stresses is defined as the M_d temperature [4, 5]. When this temperature lies above room temperature and the M_s is below room temperature, it is possible to retain the austenite at room temperature and then, some martensite can be formed by working the metastable austenite at room temperature. This can be important in highly alloyed steels such as stainless steels.

The direct effect of the composition on the formation of strain-induced martensite on alloy 304 was first investigated by Angel [6–8] who correlated elemental compositions with the temperature at which 50% of martensite formed are related to the application of 0.3 true strain in tension, denoted by M_{d30}, i.e.

$$M_{d30}(°C) = 413 - 462(\%C + \%N) - 9.2(\%Si) \\ - 8.1(\%Mn) - 13.7(\%Cr) - 9.5(\%Ni) \\ - 18.5(\%Mo) \quad (2)$$

The residual stresses in the material can arise as a result of interactions between deformation and microstructure, so can also result from the phase transformation in a material susceptible to austenite-martensitic transformation [9].

In most of the polycrystalline materials, crystal orientations are present in a definite pattern and a propensity for this occurrence is caused initially during crystallization from the melt or amorphous solid state and subsequently by thermomechanical processes This tendency is known as preferred orientation or, more concisely, the texture [10]. Many material properties (Young's modulus, Poisson's ratio, strength, ductility, toughness, electrical conductivity etc.) depend on the average texture of a material. For austenitic stainless steel in recrystallized state (from the hot-rolled or annealed after cold-rolling) revealed a weak preferential orientation on cfc grains structure—that is considered a random texture. However, provided only cold-rolled, with austenite-martensitic microstructure resulted to TRIP effect, develops a stronger texture for both phases due to this particular phase transformation [11].

The forming tests allow evaluate the material capability to be plastically deform in distinct modes before the crack occurs (proceed or by necking). These tests replace the normal manufacturing conditions that would be costly since it would be mandatory to use the same equipment of the manufacturing process of the final product.

According to Bresciani et al. [12], plastic formability is defined as the metal or metallic alloy capability to can be processing by plastic deformation without exhibit defects or crack on the processed pieces. The various trajectories that can be established in the possible forming processes in the

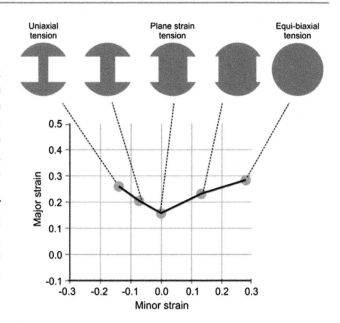

Fig. 1 FLC curve with relationship between deformation trajectory and specimen's width, according to Marciniak method

plane of the sheet use specimens with distinct width and constant length and thickness (as shown in Fig. 1): pure shear ($\alpha = -1$), uniaxial tension ($\alpha = 0$), pure strain ($\alpha = 1/2$), and biaxial stretching ($\alpha = 1$) [13]. Thus, as in stress plane state, exists strain trajectories on sheet plane (β) that correspond to the plastic deformation ratio that occur in sheet plane, i.e., $\beta = \varepsilon2/\varepsilon1$, pure shear ($\beta = -1$), uniaxial tension ($\beta = -1/2$), pure strain ($\beta = 0$) and biaxial stretching ($\beta = 1$).

Tourki et al. [14], study the influence of the plastically induced martensite on the 304 and 316 austenitic stainless steels drawing through the FLC curves (Forming Limit Curve) obtained by Nakazima test in conjunction with Marciniak-Kuczynski analytic model in order to evaluate if this model consisted of a good approximation of the experimental method. The metallographic analyze of the samples used in this study, whose width ranged from 20 to 240 mm and fixed length of 240 mm, shown that at room temperature the higher deformation imposed on the material results in greater number of shear bands intersections favorable to α' martensite nucleation.

The Tourki et al. [14] results are in agreement with Cardoso [15], that claim the occurrence of martensite formation more meaningful for the case of biaxial stretching if compared with a uniaxial tension.

In previous works Oliveira et al. [16, 17], studied a 304L austenitic stainless steel sheet submitted to the FLC test using a Marciniak methodology at room temperature. They observed that the Ferritoscopy, X-ray Diffraction and Vickers Hardness results exhibited a relationship with microstructural evolution, with mechanical behavior and deformation mode. In these studies, was possible to conclude that the FLC specimens that presented a uniaxial tension deformation trajectory associated ($\beta = -1/2$) the hardness was proportional to specimen width. However, the specimens associated to plane strain deformation trajectory ($\beta = 0$) exhibited a opposite behavior, i.e. the hardness decreased as width increase. On the other hand, the specimens with deformation trajectory close to biaxial stretching ($\beta = 1$) not shown any significant modification on hardness with specimen width increased.

In this paper, X-ray diffraction was used to correlate the stress and texture with the microstructural and mechanical evolution of the 304L austenitic stainless steel sheet submitted to FLC tests by Marciniak method at room temperature. This sheet was submitted to FLC test following by Marciniak methods associated to specimens with distinct widths (20, 40, 60, 80, 100, 120, 140, 150, 180, 200 and 220) and same length 220 mm (parallel to the rolling direction—RD). According to previous work [17], specimens with 20, 40 and 60 mm width exhibit a deformation trajectory close to uniaxial tension, i.e., $\varepsilon_1 = -2\varepsilon_2$ ($\beta = -1/2$). While, the specimen with 80 mm width revealed characteristics between uniaxial tension ($\beta = -1/2$) and plane strain ($\beta = 0$). However, the specimens with 100, 120, 130 and 140, near to crack, exhibit a behavior close to plane strain ($\beta = 0$). On the other hand, the specimen with 150 mm width shown a deformation trajectory between plane strain ($\beta = 0$) and biaxial stretching ($\beta = 1$). As long as, the specimens with 180 and 220 mm width were those closer to biaxial stretching ($\beta = 1$).

The graph and table showed in Fig. 1 summarize the available deformation from the test that represent the FLC curve and α' martensite fraction measured by Ferritoscopy on front surface of the specimens tested and as-received sample (AR1—without metallographic preparation; and AR —after grinding and mechanical polishing with diamond suspension). According to Oliveira et al. [18], the smaller

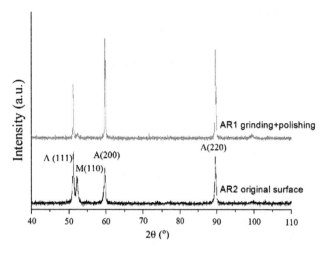

Fig. 3 XRD profile from 304L steel in as-received condition Adapted of Oliveira et al. [18]

martensite fraction in as received sample original surface and its reduction after metallographic preparation (Figs. 2 and 3 —Ferritoscopy and XRD results) are due to processing route in Aperam: after annealing the sheet was process in skin pass mill in order to promote a specular aspect on sheet surface, resulting in a significant superficial deformation without significant thickness reduction.

For this present work only, specimens which deformation trajectories in the range of the uniaxial tension to biaxial stretching (80, 100, 130, 150 and 200 mm) were selected.

Methodology

The material for this study is an austenitic stainless steel sheet, classified as 304L, supplied by Aperam South America. The chemical composition is shown in the Table 1, associated with M_s [3] and M_{d30} [7] equals to -40.89 and 61.72 °C, respectively, calculated by Eqs. 1 and 2.

Fig. 2 FLC curves on crack (line that connect the superior points) and marginal to safe zone (necking and adjacent regions) obtained for the 304L austenitic stainless steel in study in previous work [17], and α' martensite measured by ferritoscopy on front surface of the FLC specimens [16]

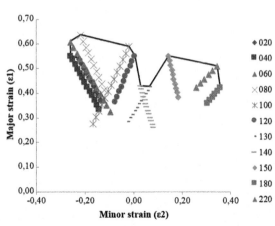

Specimen Width	%martensite (Ferritoscopy)	
	Average	Standard Deviation
AR1	0.91	0.17
AR2	0.68	0.18
020	25.6	0.47
040	26.2	0.25
060	29.7	0.53
080	26.4	0.94
100	22.9	0.85
120	21.5	0.42
130	20.8	0.73
140	25.0	0.16
150	24.9	0.33
180	23.7	0.78
220	31.7	0.76

Table 1 Chemical composition of the 304L austenitic stainless steel sheet in study

C	Mn	Si	P	S
0.018	1.2693	0.4786	0.0303	0.0015
Cr	Ni	Mo	Al	Ti
18.3639	8.0221	0.0261	0.0032	0.0018
Co	V	Nb	B	–
0.1015	0.0418	0.0071	0.006	–

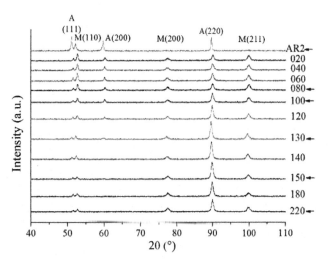

Fig. 4 Comparison of as-received and FLC specimens XRD profile (original surface)

According to the supplier data, the material was hot rolled to 2.85 mm thickness, cold rolled to 1.0 mm thickness and 500 mm width and submitted to box annealing at 1,060 °C (soaking temperature). After annealing the sheet was processed in skin pass mill in order to promote a specular aspect on sheet surface, which resulting only a superficial deformation without significant thickness reduction.

The present phases (austenite and martensite) in as-received steel samples (with and without electrolytic polishing) and FLC specimens (on front surface without metallographic preparation) were determined by XRD beyond ferritoscope measurements in previous study [17], in order to observe the phase transformation kinetic associated with modifications on profile of the consumed austenitic peaks and formed martensitic peaks. The diffractometer used is XRD-6000-Shimadzu model, Co Kα radiation, 30 kV and 30 mA on Bragg-Brentano geometry. The variations of the peaks intensity were measured with θ/2θ coupled, between 40 and 110° of 2θ with 0.02° step on continuous scanning.

In present work is discussed the analysis of the peak profile related to area, as for austenite and martensite XRD peaks were verified with "Spectroscopy/Baseline and Peaks" functions on the OriginPro 8 Software.

On the other hand, the conditions and equipment for residual stress and texture measurement were distincts. The incident optic consisted in focus-point, Co tube, iron filter, collimator policapillary (X-ray lenses) in order to optimize the beam parallelism; divergent optical collimator of parallel plates to ensure that X-ray beam reaches the detector in parallel, and PIXcel detector (PANalytical), voltage and current in diffractometer (PANalytical—X'Pert PRO MPD): 40 kV and 45 mA, respectively. The residual stress and crystallographic texture measurements were performed, respectively, at 2θ = 89° and in (111), (200) and (220) austenite (γ) crystallographic planes.

Results and Discussion

In Fig. 4 is shown the XRD profile that revealed the austenite and α′ martensite peak presents associated to the original surface from the FLC specimens compared with

Fig. 5 **a** Austenite (A) and **b** α′ Martensite (M) peak areas calculated from as-received and FLC specimens XRD profile (original surface)

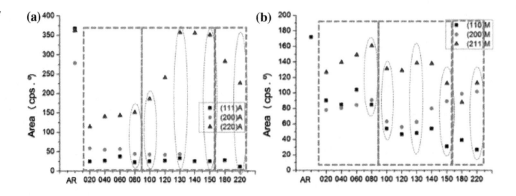

Table 2 Residual stress on surface of the FLC selected and as-received samples (original and electrolytical polishing surface)

Specimen width	Residual stress on surface (MPa)		
	Average	Standard deviation	Nature
AR1	−193.5	11.4	Compression
AR2	−30.7	19.0	Compression
080	−269.9	311.2	Compression
100	−247.3	209.7	Compression
130	−114.3	84.0	Compression
150	130.2	160.9	Tension
220	−101.7	99.0	Compression

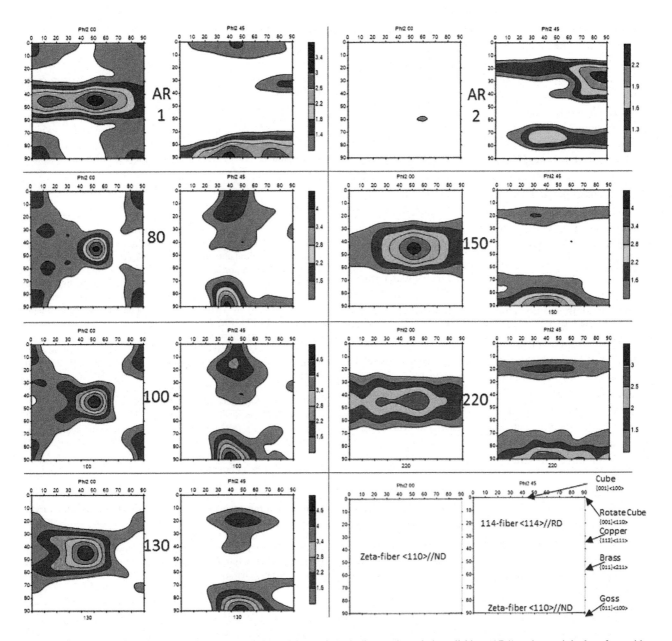

Fig. 6 FDOC: As-received sample after removed deformed layers by grinding + electrolytic polishing (AR1) and on original surface without metallographic preparation (AR2). FLC specimens selected (without metallographic preparation)

as-received sample. While is summarized in Fig. 5 the peak areas calculation for each austenite and α′ martensite peak that appear in Fig. 4.

The XRD results exposed that the (200) austenite peak was intensely consumed, and disappeared entirely for FLC specimens with width equal or higher than 130 mm. While the (111) austenite peak consume was less intense with the specimen width increased. However, the (220) austenite peak was kept almost intact in terms of intensity, independent of specimen width analyzed. On the other hand, the consumed austenite transformed into martensite, with a (111) martensite orientation intensification parallel with two others martensite orientations ((200) and (211)), only for specimen width up to 80 mm, which are closer to uniaxial tension. Opposite behavior was found for specimen higher than 100 mm width, included, which are between uniaxial tension and plain strain up to approach to biaxial stretching, where the highest martensite orientation intensity is linked to (200) and (211) the expense of (111) martensite orientation.

The residual stress and texture results for selected samples (AR and FLC specimens—80, 100, 130, 150 and 220 mm) are presented in Table 2 and Fig. 6, respectively.

There are few quantitative studies concerning with residual stress by XRD involving the martensitic transformation in austenitic stainless steel with TRIP effect due to the complexity of the stress measured, because it is generated superposed by phase transformation and stress accompanying the plastic deformation on material.

To understand a face centered cubic material that only plasticly deform when is submitted to mechanical request, i.e., without any phase transformation. The residual stress was measured in a fully annealed copper sample, which results −114.85 MPa and low standard deviation value (±4.4 MPa) with compression nature.

On the other hand, the residual stress results for TRIP steel in study, in as-received condition with low martensite fraction (AR1—Fig. 2 and Table 2), shows scattering that indicated heterogenous distribution of the residual stress at material microstructure, but describe clearly a compressive stress state. This value decrease when the as-received sample surface is submitted to grinding following by electrolytical polishing.

The residual stress results for FLC specimens selected exhibited a complex behavior of residual stress revealing heterogenous of these stresses on deformed microstructure with significant martensite fraction.

This residual stress behavior probably associated heterogenous microstructure due deformation reflected on crystallographic results by development of zeta-fiber on austenite grain orientation on sheet surface when deformation mode tends to biaxial stretching and related to the skin pass process.

Conclusion

According to the XRD results explored in this work, it can be concluded that: distinct strain states affect the martensitic transformation associated to TRIP effect on 304L austenitic stainless steel sheet subject to FLC according to Marcianick Method as was expected. These strain states result in modifications on the residual stress and crystallographic orientation in present phases, that inferred in austenitic reinforcement on FLC specimens surface.

Acknowledgements The authors acknowledge the Aperam South America (steel sheets supplier) and USIMINAS (conducted the FLC tests). JMA acknowledge the funding of Faperj (PhD scholarship related to "Aluno nota 10"). LPM acknowledge the funding of CNPq (research productivity scholarship PQ and Universal Program 2012). RD and ASP acknowledge the funding of CNPq (PhD scholarship and research productivity scholarship PQ-2—Process 307798/2015-1).

References

1. Souza JFP (2006) Propriedades mecânicas de aços inoxidáveis austeníticos nitretados submetidos à hidrogenação catódica. Master thesis, UFPR, Curitiba/PR, Brazil
2. Iawamoto T, Tsuta T, Tomita Y (1998) Investigation on deformation mode dependence of strain-induced martensitic transformation in TRIP steels and modelling of transformation kinetics. 40:173–182, Elsevier Science
3. Cina B (1954) J Iron Steel Inst 791
4. Lichtenfeld JA, Mataya MC, van Tyne CJ (2006) Effect of strain rate on stress-strain behavior of alloy 309 and 304L austenitic stainless steel. Metall Trans A 37A:147–161
5. Olsen GB, Cohen M (1975) Kinetics of strain-induced martensitic nucleation. Metall Trans A 6A:791–795
6. Andrade-Campos A, Teixeira-Dias F, Krupp U, Barlat F, Rauch EF, Grácio JJ (2010) Effect of strain rate, adiabatic heating and phase transformation phenomena on the mechanical behaviour of stainless steel. Strain 46:283–297
7. Angel T (1954) Formation of martensite in austenitic steels. J Iron Steel Inst 5:165–174
8. Krupp U, West C, Duan HP, Christ HJ (2002) Strain-induced martensite formation in metastable austenitic steels with varying carbon content. Z Metallkd 7:706–711
9. Noyan IC, Cohen JB (1987) Residual stress measurement by diffraction and interpretation. Springer, New York, p 265
10. Hosford WF (1993) The mechanics of crystals and textured polycrystals. Oxford

11. Raabe D (1997) Texture and microstructure evolution during cold rolling of a strip cast and of a hot rolled austenitic stainless steel. Acta Mater 45(3):1137–1151
12. Bresciani Filho E, Zavaglia CAC, Button ST, Gomes E, Nery FAC (1991) Conformação plástica dos metais, 4ª edição, Editora da Unicamp
13. Hosford WF, Caddell RM (1993) Metal forming—mechanics and metallurgy
14. Tourki Z, Bargui H, Sidhom H (2005) The kinetic of induced martensitic formation and its effect on forming limit curves in the AISI 304 stainless steel. J Mater Process Technol 166:330–336
15. Cardoso MC (2012) Avaliação do comportamento plástico de uma chapa de aço inoxidável austenítico 304L por meio de ensaios de tração uniaxial e curva limite de conformação. Master thesis, PPGEM-UFF, Volta Redonda/RJ, Brazil
16. Oliveira IS, Paula AS, Cardoso MC, Moreira LP, Freitas MCS (2014) Estudo comparativo do comportamento mecânico e evolução estrutural de chapas de aços 304L submetidos a ensaio CLC. In: 69° congresso anual da, ABM, São Paulo, Brazil
17. Oliveira IS, Paula AS, Cardoso MC, Moreira LP, Freitas MCS (2015) Estudo comparativo do comportamento mecânico e evolução estrutural de chapas de aço inoxidável austenítico 304L submetidas a ensaio CLC". In: 8° congresso brasileiro de engenharia de fabricação, Salvador/Bahia, Brazil
18. Oliveira IS, Paula AS, Cardoso MC, Andrade JG, Vieira TF, Almeida GM, Moreira LP, Freitas MCS (2013) Avaliação da fração de martensita pré-existente em chapa de aço inoxidável 304L laminado a frio e recozido. In: Proceedings of international 50th rolling seminar—processes, rolled and coated products, ABM, Ouro Preto/MG, Brazil

An Investigation on the Microstructure and Mechanical Properties of Hot Rolled Medium Manganese TRIP Steel

Yu Zhang and Hua Ding

Abstract

Microstructure and mechanical properties of hot rolled medium manganese TRIP steel were investigated in the present study. The heat treatment schedule was intercritical annealing at various temperature (640, 665 and 690 ° C) for 1 h and water quenched. The results showed that an excellent combination of high strength and adequate ductility was obtained. As the intercritical annealing temperature increased, the ultimate tensile strength and strain hardening rate increased, while the total elongation increased firstly and then decreased. A large amount of austenite could be retained at room temperature by adopting the method of intercritical annealing. Austenite stability and initial microstructure before deformation played an important role in the mechanical properties and strain hardening rate.

Keywords

TRIP steel • Intercritical annealing • Austenite stability

Introduction

In order to accommodate the requirement of improvement in passenger safety and fuel efficiency as well as reduction of CO_2 emission, advanced high strength steels (AHSS), with both high strength and good ductility, have received extensive interests in automobile industry in recent years. At present, three generations of AHSS have been developed. The first generation AHSS, including dual phase, complex phase and transformation induced plasticity (TRIP) steels, showed high strength but an inadequate ductility of below 20% [1–3], which restricted the applications in automobile industry. Twinning induced plasticity (TWIP) steels, regarded as the second generation AHSS, demonstrated high strength and high work hardening rate as well as superior ductility of exceeding 50% [4–6]. However, a large amount of alloying elements became a drawback in mass production and material cost. Thus, medium Mn steels, as the third generation AHSS, have attracted much attention due to their excellent mechanical properties. Medium Mn steels presented a combination of high tensile strength and adequate elongation about 30–40 GPa%, which was between the first generation AHSS and the second generation AHSS. Since a large amount of metastable retained austenite was obtained, the excellent properties of medium Mn steels resulted from the phase transformation from austenite to martensite during the deformation process. To obtain retained austenite, strategies of austenite reversion from full or partial martensitic microstructure during intercritical annealing, named as austenite reverted transformation (ART), were often adopted in previous studies [7, 8]. C and Mn, known as austenite stabilized elements, partitioned from martensite to austenite during the intercritical annealing process, leading to retain the intercritical austenite to room temperature. It was reported that ARTed medium Mn steels showed an excellent mechanical properties with high strength and good ductility [8].

In present work, an alternative method of intercritical annealing on the as hot rolled samples was proposed to obtain retained austenite in medium Mn steels. The aim of this paper was to report the excellent mechanical properties of the investigated steels and discuss the effect of austenite stability on mechanical properties and strain hardening ability of the steels.

Y. Zhang (✉) · H. Ding (✉)
School of Materials Science and Engineering, Northeastern University, Shenyang, 110819, China
e-mail: humanbeing1989@126.com

H. Ding
e-mail: hding2013@163.com

Experiment Methods

The chemical compositions of the medium Mn steel in the present study were Fe–9.2Mn–1.6Al–0.15C (in wt%). The experimental steels were cast in a vacuum furnace. The cast materials were forged to ingots with a thickness of 70 mm and then air cooled to room temperature. The forged materials were soaked at 1200 °C for 1 h, hot rolled from 70 mm to 4 mm with a hot rolling reduction of 94%, and then air cooled to room temperature. In order to obtain retained austenite, a schedule of intercritical annealing was adopted. The as hot rolled samples were intercritically annealed for 1 h at 640 °C (abbreviated as IA640), 665 °C (abbreviated as IA665) and 690 °C (abbreviated as IA690), respectively, and then water quenched. The dog-bone tensile specimens with a gauge length of 50 mm were machined from the intercritical annealing samples with the tensile axis parallel to the hot rolling direction. Tensile tests were carried out at room temperature and at an engineering strain rate of 1×10^{-3} s^{-1}. The microstructure were characterized by scanning electron microscope (SEM). For the sake of SEM observation, specimens were grinded, electro polished and then etched with 25% sodium bisulfite solution. X-ray diffraction (XRD) analysis was carried out to determine the volume fraction of retained austenite before and after the tensile test.

The Experimental Results

Mechanical Behavior

The engineering stress-strain curves of the investigated steels intercritically annealed at various temperatures were shown in Fig. 1, and the mechanical properties were summarized in Table 1. It could be found that an excellent combination of high tensile strength and adequate ductility was obtained. When the intercritical temperature was elevated from 640 to 690 °C, the ultimate tensile strength increased. The ductility of IA640 sample and IA665 sample presented to be similar, while a decrease in ductility was observed for IA690 sample.

The strain hardening rate curves of the investigated steels were illustrated in Fig. 2. It could be found that the samples in the present study showed a good ability in strain hardening. As the intercritical annealing temperature increased, the strain hardening rate increased. Compared with IA640 sample, IA665 sample presented a little increase in strain hardening rate, and both IA640 sample and IA665 sample demonstrated a durable strain hardening behavior during the deformation process. IA690 samples showed a much higher strain hardening rate than IA640 sample and IA665 sample but insufficient durability in strain hardening behavior.

Microstructure

The SEM images of the investigated steels intercritically annealed at various temperatures were given in Fig. 3. Two types of morphology could be observed in the intercritical annealing samples. Lath-typed ultrafine austenite and ferrite predominated in the microstructure. In addition, austenite strips were found to be along the hot rolling direction, and it

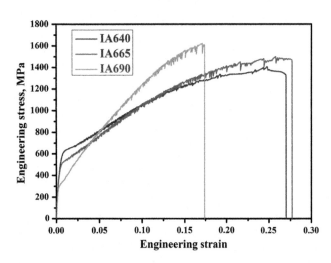

Fig. 1 Engineering stress-strain curves of the investigated steels intercritically annealed at different temperatures

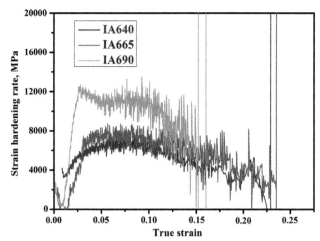

Fig. 2 Strain hardening rate curves of the investigated steels intercritically annealed at different temperatures

Table 1 Mechanical properties of the investigated steels intercritically annealed at different temperatures

Sample no.	Ultimate tensile strength (MPa)	Total elongation (%)	Product of strength and elongation (MPa%)
IA640	1406	27.00	37962
IA665	1499	27.70	41522
IA690	1623	17.38	28208

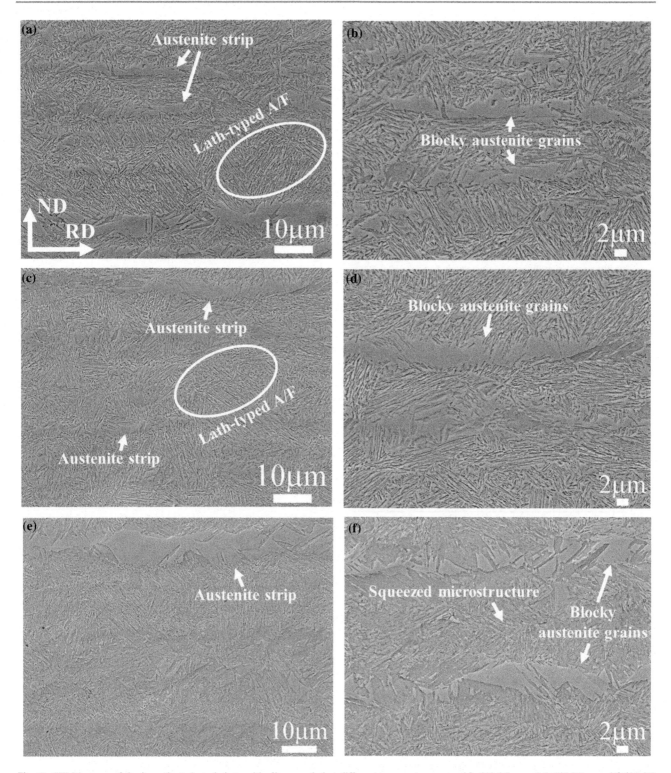

Fig. 3 SEM images of the investigated steels intercritically annealed at different temperatures: **a** and **b** 640 °C, **c** and **d** 665 °C, **e** and **f** 690 °C

could be found that the austenite strips were composed of several blocky austenite grains. In IA690 sample, the microstructure was found to be squeezed as a result of volume expansion of martensite transformation. It should be noted that as the intercritical annealing temperature increased, the austenite stability decreased, thus some austenite with lower stability might transform into martensite during the quenching process.

The volume fraction of retained austenite before and after tensile tests and austenite transformation ratio obtained from

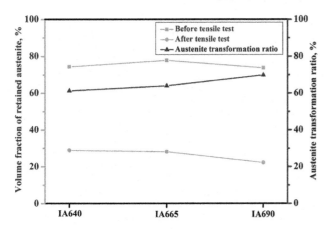

Fig. 4 Volume fraction of retained austenite before and after tensile test and austenite transformation ratio of the investigated steels intercritically annealed at different temperatures

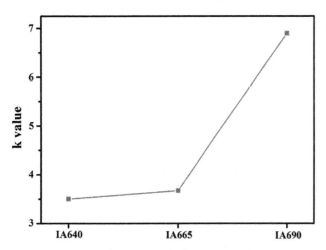

Fig. 5 k value of the investigated steels intercritically annealed at different temperatures

XRD results were demonstrated in Fig. 4. A large amount of austenite was retained at room temperature after intercritical annealing (all above 70%) due to the supplement of medium amount of Mn element since Mn was a strong austenite stabilized element. As the temperature increased, the amount of retained austenite before the tensile tests presented to firstly increase and then decrease, which was in accordance with the SEM results that some austenite with lower stability transformed into martensite when IA690 sample was quenched. Furthermore, an increase in austenite transformation ratio could be found as the intercritical temperature was elevated.

Discussion

To clarify the austenite stability further, k value was calculated from the equation as below,

$$f_\gamma = f_{\gamma 0} \exp(-k\varepsilon)$$

where, $f_{\gamma 0}$ represented the volume fraction of austenite before the tensile test and f_γ represented the volume fraction of austenite at the fracture strain of ε. Thus, a higher k value implied lower austenite stability. The results of k value were shown in Fig. 5. Compared with IA640 sample, a slight increase in k value was found for IA665 sample, suggesting a slight decrease in austenite stability, which led to an increase in tensile strength. Moreover, the total elongation presented to be similar for both IA640 sample and IA665 sample. It could be deduced that the samples intercritically annealed at 640 and 665°C possessed an appropriate austenite stability for high strength, high strain hardening rate and good ductility. For IA690 sample, a significant increase in k value meant much lower austenite stability, leading to a much higher strain hardening rate. A large amount of retained austenite transformed into martensite in the early stage of deformation due to the lower austenite stability, resulting in a decrease in ductility. Considering that some austenite transformed into martensite during the quenching process in the initial microstructure of IA690 sample before deformation, the lower austenite stability and initial martensite resulted in a higher strength but a lower ductility.

Conclusion

A large amount of austenite could be retained at room temperature and an excellent combination of high strength and adequate ductility could be obtained by adopting the method of intercritical annealing. As the intercritical annealing temperature increased, the ultimate tensile strength and strain hardening rate increased, while the total elongation increased firstly and then decreased. Austenite stability and initial microstructure before deformation played an important role in the mechanical properties and strain hardening rate.

References

1. Chung JH, Jeon JB, Chang YW (2010) Work-hardening and ductility enhancement mechanism of cold rolled multiphase TRIP steels. Met Mater Int 16:533–541
2. Zaefferer S, Ohlert J, Bleck W (2004) A study of microstructure, transformation mechanisms and correlation between microstructure and mechanical properties of a low alloyed TRIP steel. Acta Mater 52:2765–2778
3. De Cooman BC (2004) Structure–properties relationship in TRIP steels containing carbide-free bainite. Curr Opin Solid State Mater Sci 8:285–303
4. Lee YK (2012) Microstructural evolution during plastic deformation of twinning-induced plasticity steels. Scr Mater 66:1002–1006

5. Jin JE, Lee YK (2012) Effects of Al on microstructure and tensile properties of C–bearing high Mn TWIP steel. Acta Mater 60:1680–1688
6. Jung JE, Park J, Kim JS, Jeon JB, Kim SK, Chang YW (2014) Temperature effect on twin formation kinetics and deformation behavior of Fe–18Mn–0.6C TWIP steel. Met Mater Int 20:27–34
7. Shi J, Sun XJ, Wang MQ, Hui WJ, Dong H, Cao WQ (2010) Enhanced work-hardening behavior and mechanical properties in ultrafine-grained steels with large-fractioned metastable austenite. Scr Mater 63:815–818
8. Sun RM, Xu WH, Wang CY, Shi J, Dong H, Cao WQ (2012) Work hardening behavior of ultrafine grained duplex medium-Mn steels processed by ART-annealing. Steel Res Int 83:316–321

Correlation Between Deformation Texture and Martensitic Transformation in TWIP/TRIP Steels on Multiscale

Marton Benke, Erzsebet Nagy, Mate Sepsi, Peter Pekker, and Valeria Mertinger

Abstract

In the present manuscript, the texture variation of γ austenite, ε martensite and α' martensite phases are investigated in FeMn (Cr) steels exhibiting both TWIP and TRIP behaviour during uniaxial tensile tests. Samples of three steels with varying Cr content were subjected to tensile tests till fracture on different temperatures ranging from room temperature, at which ε martensite and γ austenite are stable to 453 K, where only γ austenite was present prior to tensile stressing. The developed texture of ε martensite, α' martensite and γ austenite was examined on multiscale, by TEM orientation mapping, Electron Backscattered Diffraction (EBSD) orientation mapping and X-ray-diffraction pole figure measurements. Correlations between the developed textures and the formation of the martensitic phases are discussed.

Keywords

TRIP/TWIP steels • Texture • Thermomechanical treatment

Introduction

TWIP/TRIP steels are favoured because of their combined high strength and elongation. In case of TWIP steels, the unique mechanical behaviour is due to Twinning Induced Plasticity (TWIP) which is based on the $\gamma \rightarrow \varepsilon$ martensitic transformation, while Transformation Induced Plasticity is characteristic to TRIP steels which is based on the $\gamma \rightarrow \varepsilon$

α' and $\gamma \rightarrow \alpha'$ martensitic transformations [1–4]. The unique mechanical behaviour of TWIP/TRIP steels can be utilised in energy absorbing components and/or during the manufacturing process of such steels.

Principles of the transformations occurring in TWIP and TRIP steels such as crystallographic correlation during fcc-hcp transformations has been described for long [1–6]. However, because of the components' mechanical behaviour during deformation must be well known and controlled for proper process design, detailed examinations on the behaviour of TWIP/TRIP steel with specific compositions are still being investigated. For instance, Kwon et al. have shown that reverse $\varepsilon \rightarrow \gamma$ transformation can occur during deformation in high Mn-content steels [6]. Besides the deformation induced transformations, crystallographic texture has also effect on the components' formability. Barbier et al. examined the texture evolution during room temperature tensile tests on macroscale and microscale by means of X-ray diffraction (XRD) and electron backscattered diffraction (EBSD). They showed that a pronounced <111> fiber texture formed during tensile tests [7]. Co-authors of the present paper investigated the macrotexture evolution of TRIP steels at sub-zero temperature levels using XRD [8–10].

However, the description of texture formed at elevated temperatures is also useful for practical applications if the thermomechanical process of TWIP/TRIP steels occurs at elevated temperatures.

The aim of the present paper is to characterize the orientation relationships of γ austenite, ε martensite and α' martensite within the same former austenite grain in samples subjected to tensile tests at different temperatures both above and below the martensite start (M_s) temperature of the $\gamma \rightarrow \varepsilon$ transformation.

M. Benke (✉) · M. Sepsi · V. Mertinger
Institute of Physical Metallurgy, Metalforming and Nanotechnology, University of Miskolc, Miskolc-Egyetemvaros, H3515, Hungary
e-mail: fembenke@uni-miskolc.hu

E. Nagy · P. Pekker
MTA-ME Materials Science Research Group, Miskolc-Egyetemvaros, H3515, Hungary

© The Minerals, Metals & Materials Society 2018
A. P. Stebner and G. B. Olson (eds.), *Proceedings of the International Conference on Martensitic Transformations: Chicago*, The Minerals, Metals & Materials Series, https://doi.org/10.1007/978-3-319-76968-4_12

Experimental

TWIP/TRIP steels with Cr content of 0.07, 2.26 and 6.12 were produced at TU Bergakademie Freiberg. The compositions of the examined steels are given in Table 1. Tensile test specimens with diameter of 5 mm and differential scanning calorimeter (DSC) samples were machined from hot rolled rods. The specimens were austenized at 1273 K in argon atmosphere for 30 min and subsequently quenched in room temperature water. XRD phase analysis confirmed that the microstructure of the as-quenched (AQ) samples consisted of thermally induced ε martensite and γ austenite of all three steel types (Table 2). The ε ↔ γ transformation temperatures were obtained using a Netzsch 204 heat flux DSC. The ε ↔ γ transformation temperatures are summarized in Table 1. Since the formation of α′ martensite could be induced only by mechanical stress in the examined alloys, only the ε ↔ γ transformation temperatures were measured. Accordingly, martensite start temperature of the ε ↔ γ transformation is referred to as 'M$_s$'. Tensile tests were performed at different test temperatures with an Instron 5982 universal mechanical tester machine equipped with a climate chamber. The test temperature range was set to reach A$_f$ temperatures and go below M$_f$ temperatures. Prior to tensile tests, the specimens were heated up to 573 K and held for 30 min. After that, the specimens were cooled down to the test temperature being: 473, 453, 433, 413, 398, 383 and 313 K and loaded until rupture. The detailed anisotropy investigation of tested samples at 453, 398 K are introduced in this paper.

After tensile tests, samples for XRD and transmission electron microscope (TEM) examinations were prepared from the cross-gauge section of the tensile specimens. Specimens stressed above M$_s$ consisted of γ austenite and strain induced ε martensite (ε$_{STR}$) on the test temperature. Specimens tensile tested below M$_s$ contained thermally induced ε martensite (ε$_{TH}$) in addition to γ and ε$_{STR}$. α′ martensite also formed in samples stressed at 413 K or lower temperatures. The volume fraction of phases was determined by full intensity fitting X ray diffraction method. Orientation relationships between γ austenite and ε martensite, adjacent ε plates and ε martensite and α′ martensite crystals in nano scale were characterised by TEM orientation mapping using a FEI Tecnai G^2 TEM and ASTAR system. The orientation relationships were described through pole figure generations. For every orientation relation, the examined crystal pairs were located within a common former austenite grain. For each examined sample, examinations were carried out on 5 different fields of view. For the colour version of the TEM map images the Reader is asked to see the online version. The colours correspond to different orientations of crystals, thus, plates of the same phase with different orientations can be distinguished. The orientations related to different colours are not given since pole figures were used for the orientation investigations. The microscale orientation mapping was described by EBSD using Zeiss EVO MA 10 scanning electron microscopy equipped with EBSD system. The macroscale orientation was characterised by Bruker D8 Advance X ray diffractometer and Stresstech G3R centreless diffractometer using inverse pole figure and pole figure descriptions respectively.

Table 1 Composition and transformation (ε ↔ γ) temperatures of the examined steels

Steel type	Composition						Transformation temperatures			
	C	Mn	Cr	Si	S	P	M$_s$	M$_f$	A$_s$	A$_f$,
	wt%						K			
Steel 2	0.028	18	0.07	0.03	0.025	0.0081	411	319	455	499
Steel 1	0.026	17.7	2.26	0.1	0.029	0.0051	404	319	453	494
Steel 3	0.08	17.7	6.12	0.06	0.025	<0.003	378	310	444	481

Table 2 Volume fraction of phases, %

Steel type	Test temperature								
	As quenched			398 K			453 K		
	γ	α	ε	γ	α	ε	γ	α	ε
	%			%			%		
Steel 2	8	0	92	13	31	56	45	0	55
Steel 1	14	0	86	6	68	26	51	0	49
Steel 3	14	0	85	36	14	50	47	0	53

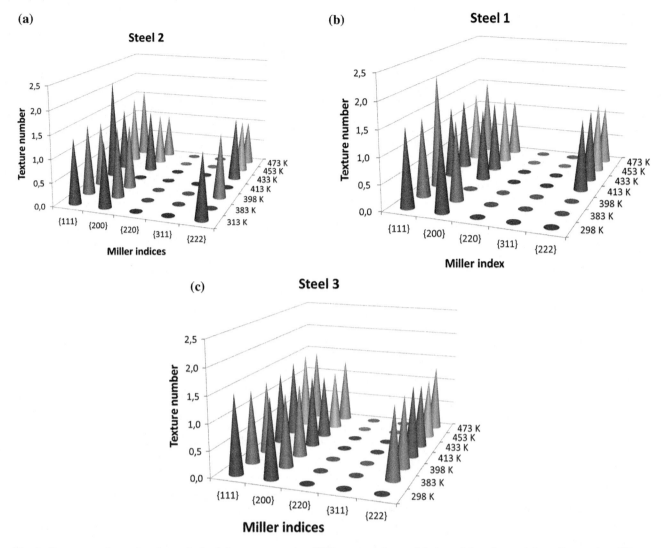

Fig. 1 Texture numbers of γ phase obtained from cross section XRD spectra of **a** steel 2, **b** steel 1 and **c** steel 3 tensile tested at different temperatures [11]

Results and Discussion

The Table 2 shows the calculated volume fraction for the fractured samples at 398 and 453 K. Austenite and ε martensite were determined at the higher temperature while the tree phases mixture (austenite, ε and α′ martensites) were found at the lower temperature. Formation of the strain induced martensite is evident at the 453 K in case of all steel types. In case of Steel 2 and Steel 1 the M_s temperature is higher than the lower test temperature so a mixture of thermally and stress induced ε martensite can be expected after fracture at 398 K. In case of Steel 3 this test temperature is slightly above to the M_s temperature (378 K) so the thermally induced ε martensite is unexpected. The loaded at lower temperature resulted strain induced α′ martensites in case of each steel.

The orientation relationships are performed in the rank of scale as it were determined.

Figure 1 and 2 show the texture numbers of γ and ε phases, respectively, calculated from the relative intensity ratios of Bragg reflections on cross section XRD spectra of the steel samples after tensile tests at different temperatures.

As seen in Fig. 1, the γ phase remained closely isotropic after all performed tensile tests regardless of test temperature or steel composition. However, according to Fig. 2,

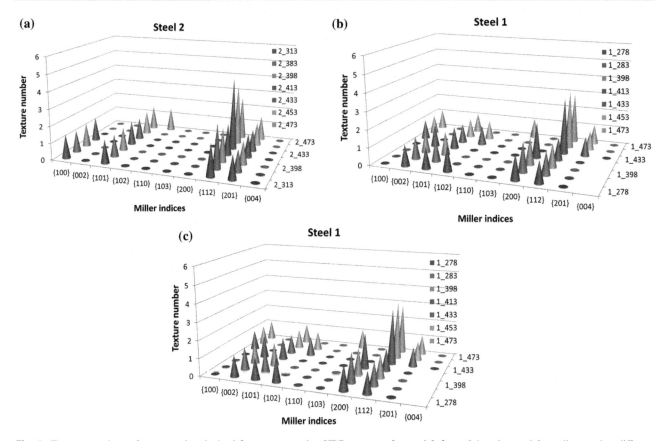

Fig. 2 Texture numbers of ε martensite obtained from cross section XRD spectra of **a** steel 2, **b** steel 1 and **c** steel 3 tensile tested at different temperatures [11]

ε martensite was isotropic at lower test temperatures, but a notable {112} texture developed at higher tensile test temperatures for all three steel types. We have already showed in our earlier paper [11] that during the DSC test the decomposition kinetic of martensites is related to this {112} texture gradient in function of test temperature. So, this temperature dependence of the {112} texture refers the different origin of ε martensites. According to the {112} pole figures of ε martensites (Fig. 3), fiber texture was developed at each case but it can be stated that the samples exhibit stronger texture at 453 K in case of all steel types.

The texture evolution of ε martensite with increasing test temperature was examined on microscale through EBSD orientation mapping and pole figures of Steel 1 tested at 398 and 453 K. Figures 4 and 5 show the EBSD orientation and phase maps and pole figures of γ phase and ε martensite. The γ {111} ∥ ε {001} relationship is evident in both cases.

The Figs. 6, 7, 8, 9, 10 show the results of nanoscale (TEM) orientation investigations. A γ-ε orientations is presented of Steel 3 tested at 398 K. Note that for Steel 3 the

test temperature is above the M_s temperature (see Table 1). The microstructure is dominated by ε plates in which some regions of γ phase are also present (all areas of γ phase are marked). No high angle austenite grain boundary can be seen; thus, all the examined area would be one austenite grain. The ε plates have two main orientations. To examine the orientation relationship between γ and ε phases, γ [111] and ε [001] pole figures were taken and are shown in Fig. 6. Both γ [111] and ε [001] pole figures have a pole in the same orientation meaning that there is a parallel γ [111] and ε [001] direction. This agrees with the Kurdjumov-Sachs (K-S) orientation relationship: γ {111} ∥ ε {001} ∥ α' {110} [4, 7, 9, 10, 12].

Figure 3 XRD {112} pole figures of ε martensites of Steel 2, 1 and 3 tested at 398 and 453 K. CHI range of pole figures is 12°–60°

Figure 7 shows the TEM orientation map and γ [111] and ε [001] pole figures of Steel 1 tensile tested at 398 K. For the composition of Steel 1, the applied test temperature is below the M_s temperature (see Table 1). The microstructure

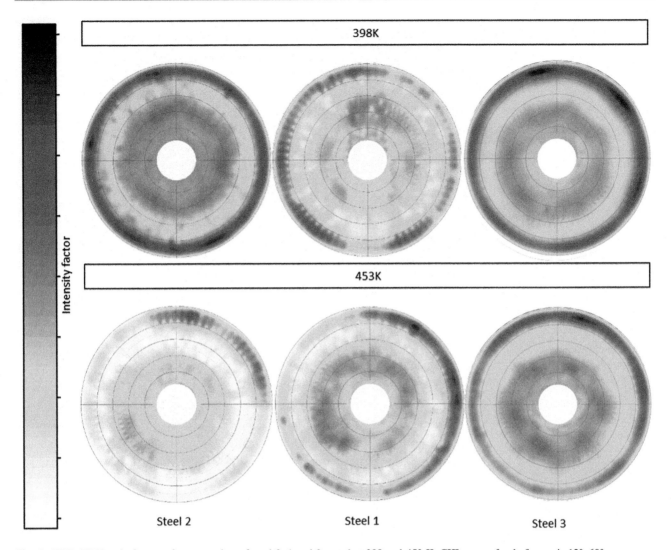

Fig. 3 XRD {112} pole figures of ε martensites of steel 2, 1 and 3 tested at 398 and 453 K. CHI range of pole figures is 12°–60°

contains large regions of ε and α′ martensite phases with some areas of γ austenite. A high angle boundary that was a grain boundary in the former austenitic structure can be seen in the upper part of the image. The locations of the γ-ε orientation examination are marked. The γ [111] and ε [001] pole figures have a common orientation meaning that there is a parallel γ [111] and ε [001] direction about which the γ and ε phases are rotated. The same correlation was found between γ and ε phases of a common former austenite grain during the examination of other areas as well. Furthermore, the same orientation relationship was observed between γ and ε phases in all the examined samples regardless of its state (as-quenched or tensile tested), test temperature or composition. Thus, thermomechanical treatment, temperature of tensile test and variation of Cr concentration of a few wt% have no effect on the orientation relationship between γ and ε phases originated from the same austenite grain.

As seen in Fig. 6 all the ε plates found within one former austenite grain have two main orientations. Figure 8 shows ε [100] pole figures of two neighbouring ε plates having the two main orientations. The adjacent ε plates have a common direction in the [100] pole figure meaning that one [100] direction of the ε plates are parallel to each other, and the rotation axis about the examined ε plates are rotated relative to each other is in the direction of the common [100] direction. The angle of rotation was obtained by measuring

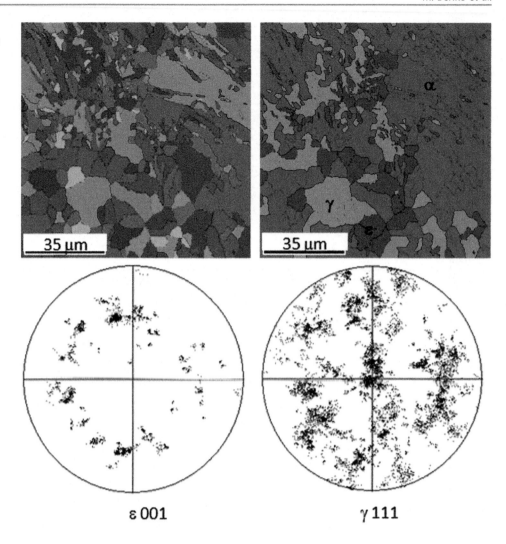

ε 001   γ 111

the misorientation angle along a line crossing the two adjacent ε plates. The line is shown in Fig. 6. The misorientation variation of the neighbouring ε plates is shown in Fig. 8. The measured misorientation angle is highlighted. The measured misorientation angle between the neighbouring ε plates is ∼71°. Values of ∼70°, 71° misorientation angle were found to be characteristic to ε-ε relations of the same austenite grain in all the examined samples, regardless of state, tensile test temperature and composition. The measured misorientation angle of 70°-71° which is very close to the theoretical inclination angle of the <111> directions in the fcc structures (70.53°). This is in agreement with the K-S relation according to which the γ ↔ ε transformation occurs along the γ {111}/ε {001} planes. This means that the application of thermomechanical treatment, the temperature of the thermomechanical treatment and variation of Cr concentration have no effect on the

orientation relationship between adjacent ε phases within the same austenite grain.

Due to the strain induce effect at lower testing temperature the α′ martensites also formed. The Fig. 9 shows the orientation map of Steel 2 tensile tested at 398 K. The microstructure contains ε and α′ martensites. The α′ martensite does not have a plate-like morphology, its appearance is granular. The orientation relationship between ε martensite and α′ martensite is shown by ε [001] and α′ [110] pole figures (Fig. 9). It can be seen that ε [001] and α′ [110] pole figures have a common pole orientation. Accordingly, there is a parallel ε [001] and α′ [110] directions. This relation was found to be characteristic to ε and α′ martensites formed from a common austenite grain. Similarly to the previous orientation examinations, the K-S relation was found to be valid for the ε martensite and stress induced α′ martensite as well.

Fig. 5 EBSD orientation map and pole figures of steel 1 tensile tested at 453 K

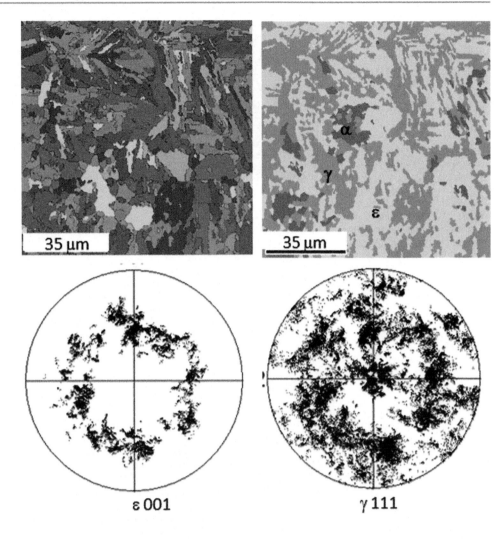

ε 001   γ 111

Fig. 6 TEM orientation map and pole figures of steel 3 tensile tested at 398 K

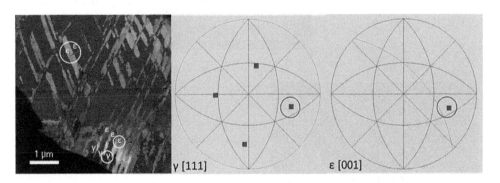

γ [111]   ε [001]

The γ-α′ orientations were also investigated. Figure 10 shows the microstructure of the Steel 1 tested at 398 K. Plates of α′ martensite and some regions of γ austenite are present in the ε matrix. A high angle grain boundary can be seen passing through horizontally in the middle that was an austenite grain boundary. The examination of the γ-α′ orientation was carried out by measuring pole figures in γ and α′ phases within the same former austenite grain (marked). As seen in Fig. 10, γ [111] and α′ [110] poles are in the same direction meaning that the γ and α′ phases have a parallel γ [111] and α′ [110] direction. Thus, the K-S relation was confirmed by the γ-α′ relationship as well.

Fig. 7 TEM orientation map and pole figures of steel 1 tensile tested at 398 K

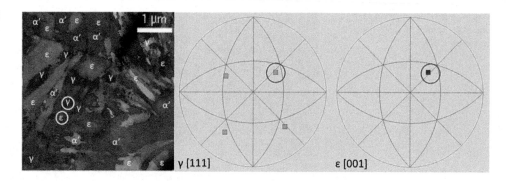

Fig. 8 ε [001] pole figures and misorientation angle between two adjacent ε plats of steel 3 tensile tested at 398 K

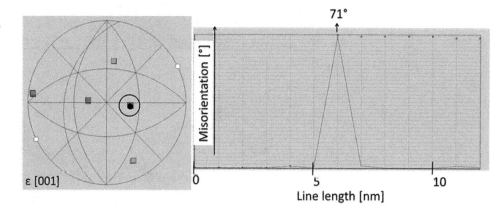

Fig. 9 Orientation map of steel 2 tensile tested at 398 K and the ε [001] and α′ [011] pole figures

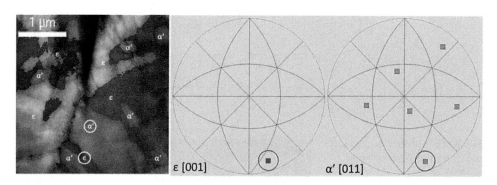

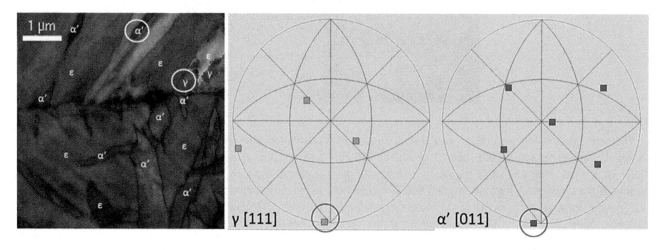

Fig. 10 Orientation map of steel 1 tensile tested at 398 K and the γ [111] and α′ [011] pole figures with the locations of γ-α′ orientation examination

Conclusions

Orientation relationships were investigated between γ austenite, ε martensite and α' martensite in TWIP/TRIP steel samples having different Cr content after tensile tests carried out at different temperatures using XRD, EBSD and TEM orientation mapping. Within the same former austenite grain, the Kurdjumov-Sachs (K-S) orientation relationship (γ $\{111\}$ $\|$ ε $\{001\}$ $\|$ α' $\{110\}$) was confirmed between adjacent γ-ε, ε-α' and γ-α' phases in all three examined steels (Cr content: 0, 2.26, 6.12 wt%), in as-quenched and tensile tested samples regardless of whether tensile test was carried out above or below the M_s temperature. Previous XRD texture investigations revealed that the texture number related to the ε $\{112\}$ planes increase with tensile test temperature. Now, it was shown that the ε $\{112\}$ has a fiber texture which is stronger at higher test temperatures. The $\{112\}$ fiber texture could only be observed on the macroscopic scale.

Acknowledgements The research work presented in this paper is based on the results achieved within the NKFI 119566 K project. One of the authors, Mate Sepsi, is grateful for the support of the new national excellence program of the ministry of human capacities.

References

1. Nishiyama Z (1978) Martensitic transformations. Academic Press
2. Porter DA, Easterling KE (1992) Phase transformations in metals and alloys. Chapman & Hall
3. Olson GB, Owen WS (1992) Martensite. ASM International
4. Bhadeshia HKDH, Honeycombe RWK (2006) Steels microstructure and properties. Elsevier, BH
5. Pandey D, Lele S (1986) Acta Metall 34(3):405–413
6. Kwon KH, Suh BC, Baik SI, Kim YW, Choi JK, Kim NJ (2013) Sci Technol Adv Mater 14(1):014204
7. Barbier D, Gey N, Allain S, Bozzolo N, Humbert M (2009) Mat Sci Eng A 500:196–206
8. Nagy E, Mertinger V, Tranta F, Solyom J (2004) Mat Sci Eng A 378:308–313
9. Mertinger V, Nagy E, Tranta F, Solyom J (2008) Mat Sci Eng A 481–482:718–722
10. Mertinger V, Nagy E, Benke M, Tranta F (2015) Mat Sci Forum 812:161–166
11. Mertinger V, Benke M, Nagy E (2015) Mater Today Proc 2S: S673–S676
12. Lu F, Yang P, Meng L, Cui F, Ding H (2011) J Mater Sci Technol 27(3):257–265

Phase Transitions and Their Interaction with Dislocations in Silicon

Valery I. Levitas, Hao Chen, and Liming Xiong

Abstract

In this paper, phase transformations (PTs) in silicon were investigated through molecular dynamics (MD) using Tersoff potential. In the first step, simulations of PTs in single crystal silicon under various stress-controlled loading were carried out. Results shows that all instability points under various stress states are described by criteria, which are linear in the space of normal stresses. There is a region in the stress space in which conditions for direct and reverse PTs coincide and a unique homogeneous phase transition (without nucleation) can be realized. Finally, phase transition in bi-crystalline silicon with a dislocation pileup along the grain boundary (GB) was carried out. Results showed that the phase transition pressure first decreases linearly with the number of dislocation pileups and then reaches a plateau with the accumulation of dislocations in the pileup. The maximum reduction of phase transition pressure is 30% compared to that for perfect single crystalline silicon.

Keywords

Molecular dynamics • Phase transition criteria
Homogeneous phase transition • Triaxial loading
Phase transition pressure • Grain boundary
Dislocation pileup

V. I. Levitas (✉)
Departments of Aerospace Engineering, Mechanical Engineering, and Material Science and Engineering, Iowa State University, Ames, IA 50011, USA
e-mail: vlevitas@iastate.edu

V. I. Levitas
Division of Materials Science and Engineering, Ames Laboratory, Ames, IA 50011, USA

H. Chen (✉) · L. Xiong
Department of Aerospace Engineering, Iowa State University, Ames, IA 50011, USA
e-mail: haochen@iastate.edu

Introduction

It is known that nonhydrostatic stresses and plastic deformation drastically reduce phase transformation pressure for various materials [1–5]. However, the reasons and mechanisms are still not completely clear. There is an analytical model [2–4, 6, 7] and phase field solutions for nucleation of a high-pressure phase at the tip of strain-induced dislocation pileup that suggest that this may be a possible mechanism for strong reduction in transformation pressure. In this paper, we will report results of some our atomistic studies. First, we will review the lattice instability criteria under six dimensional nonhydrostatic loadings [8, 9]. Then we introduce silicon bi-crystal and dislocation pileup along the GB to investigate the role of dislocation activities in promoting phase transformation.

Simulation Method

In this work, classical MD simulations were performed using the LAMMPS package [10]. Tersoff interatomic potential is employed as the interatomic force field for the interactions between Si atoms [11]. This potential has been demonstrated to be successful in describing the transition from the diamond-cubic to beta-tin in single crystal silicon (Si I to Si II) under a uniaxial stress of 12 GPa (see [12] and current results), which is close to the experimental value [13]. The majority of simulations have been performed for a Si sample containing 64,000 atoms. To prove a size-independence of the results, simulations under uniaxial loading were performed for varying sample sizes ranging from 5 nm to 40 nm, which contain 8,000–4,096,000 atoms, respectively. A time step of 1 fs was used in all simulations. The system temperature is set as T = 1 K to eliminate the possibility of the occurrence of thermally activated phase transitions (PTs). Effects of the free surfaces on the PTs were excluded by employing periodic boundary conditions along all three

© The Minerals, Metals & Materials Society 2018
A. P. Stebner and G. B. Olson (eds.), *Proceedings of the International Conference on Martensitic Transformations: Chicago*,
The Minerals, Metals & Materials Series, https://doi.org/10.1007/978-3-319-76968-4_13

Fig. 1 a Plane in stress space corresponding to the analytical instability criterion from [8, 9] for direct Si I → Si II PT and the lattice instability points from MD simulations. **b** The plot in (**a**) is rotated until theoretical plane is visible as a line. **c** and **d** are the same plots as in (**a**) and (**b**) but for reverse Si II → Si I PT [9]

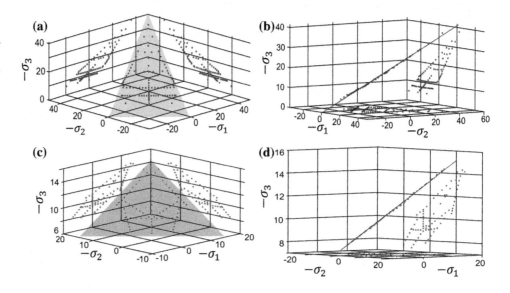

directions. For uniaxial loading, simulations were conducted under (a) a specified first Pila-Kirchhoff stress P; (b) a specified Cauchy stress σ; and (c) a strain-controlled loading. Here the first Piola-Kirchhoff stress P was applied to the system by enforcing constant forces on the top and bottom layers of the atomistic system along the directions of compression. The Cauchy stress σ was applied to the system using the Berendsen algorithm [14], in which the instantaneous stress of the system was calculated using the virial formula and controlled in two steps. First, a Cauchy stress increment of 0.01 GPa was applied to the simulation cell; this was then followed by an equilibration of the entire specimen for 10 ps. In order to ensure that a desired Cauchy stress has been achieved, the system virial stress at the end of each loading increment was calculated and was checked against the prescribed stress, assuming that the averaged Cauchy stress coincides with the virial stress [15]. It should be noted that such a weak-coupling stress-controlling strategy is different from that of using the Parinello-Rahman algorithm [16], which, in contrast, approximately controls the deviatoric component of the second Piola-Kirchhoff stress T [17]. In the strain-controlled loading, the fix deform method in LAMMPS was employed. That is, each time after the simulation box size along the main loading direction was changed at a value of 0.2 Å, the system was equilibrated for 100 ps. This atomistic system was equilibrated with a fixed box size along the loading direction and zero stress along the other direction. Multiaxial loading was applied to the simulation cell through controlling the normal components of the Cauchy stress utilizing the Berendsen algorithm [14]. However, shear stresses in LAMMPS cannot be applied through the Berendsen algorithm. They were applied with the Parinello-Rahman algorithm [15], which controls the deviatoric part of the second Piola-Kirchhoff stress T [16].

At the instability point, the Cauchy stress was calculated and substituted into the instability criterion.

Simulation Results

Instability Criteria Calibrated by Molecular Dynamics

Using a phase field approach, the lattice instability (phase transformation) criteria for cubic-tetragonal PTs, Si I ↔ Si II, was derived as a linear function of three normal prescribed Cauchy stresses σ_i along cubic axes [8, 9]. They are shown as the planes in Fig. 1. The negative stresses are compressive, and compressive stress σ_3 has the largest magnitude. In order to validate these lattice instability criteria, corresponding MD simulations were performed. Microstructure evolution during PTs Si I ↔ Si II and typical uniaxial stress-strain curves for σ, P, and the second Piola-Kirchhoff stress T for direct and reverse PTs are shown in Fig. 2 under prescribed σ, P, and displacements (strains). Under a prescribed σ, instability for the PT of Si I → Si II starts at maximum Cauchy stress (point I, Lagrangian strain $E = 0.2293$), i.e., at a zero elastic modulus, which is typical for a sample under a multiaxial loading as well; P and T continue growing beyond the instability point I. However, reverse PT starts at a minimum stress but nonzero value of any elastic moduli, i.e., it cannot be described by a tradiional zero-moduli approach. Instability is easily detected by the impossibility of equilibrating the system under fixed σ until it transforms to an alternative phase. After instability point I, the microstructure initially evolves homogeneously, then heterogeneously with stochastic fluctuations, then with bands consisting of some intermediate phases (Fig. 2). At

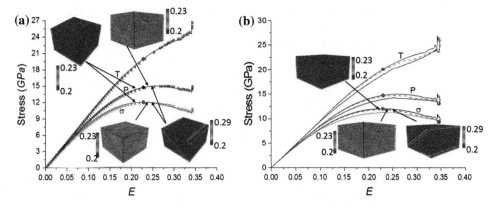

Fig. 2 Stress-Lagrangian strain E curves for uniaxial compression ($\sigma_1 = \sigma_2 = 0$) for the Cauchy σ, the first Piola-Kirchhoff P, and the second Piola-Kirchhoff stress T for direct ((**a**) and upper curves in (**b**)) and reverse (lower curves in (**b**)) PTs Si I $\leftrightarrow$ Si II. Dots mark instability points that correspond to stresses above (or below for reverse PT) the stress at which crystal cannot be at equilibrium at prescribed σ, or multiple (homogeneous and heterogeneous) microstructures exist. After loss of stability, the microstructure initially evolves homogeneously, then heterogeneously with stochastic fluctuations, then with bands consisting of some intermediate phases and, at larger strains, bands with fully formed Si II [9]

larger strains, bands with fully formed Si II appear and grow. However, if starting with band structure, the stress increases (i.e., strain reduces) toward instability point I, and the heterogeneous fluctuating structure is observed even in the vicinity of instability point I (Fig. 2). Thus, multiple solutions—including homogenous and various heterogeneous ones—are observed after instability.

The main result is that instability stresses for both direct and reverse PTs in silicon under a broad variation of all three stresses fall within a plane (see Fig. 1). Thus, it is sufficient to find just two material parameters for two different stress states in order to describe instability at any other stress state.

Homogeneous Hysteresis-Free Phase Transformation and Continuum of Intermediate Phases

For $\sigma_2 = \sigma_1$, lattice instability and initiation of PT in silicon can be described by equations $\sigma_3^d = 11.8286 + 0.6240\sigma_1$ and $\sigma_3^r = 9.3888 + 0.3840\sigma_1$, for direct and reverse PTs, respectively. Because instability lines possess different slopes in $\sigma_3 - \sigma_1$ plane (Fig. 3), they should intersect at the point $\sigma_1 = 10.1658$ and $\sigma_3 = 5.4851$. Instead, the instability line for Si I $\rightarrow$ Si II PT bends and merges with the line for Si II $\rightarrow$ Si II PT within a broad stress range. The phase equilibrium line (corresponding to the equality of the Gibbs energy of phases (Fig. 3)) is between instability lines, and consequently, it should also coincide with the merged lines. The stress hysteresis, defined as the difference in values of σ_3 between instability stresses for direct and reverse PTs for the same σ_1, decreases down to zero when σ_1 increases toward the merged region. Within the merged region, the energy barrier between phases disappears and Gibbs energy

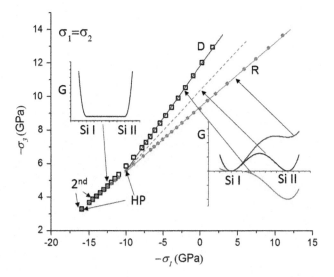

Fig. 3 Relationships between stresses σ_3 and $\sigma_1 = \sigma_2$ for the crystal lattice instability for direct and reverse Si I $\leftrightarrow$ Si II PTs and existence of the continuum of homogenous intermediate phases. Each instability line is related to the disappearance of the minimum in the Gibbs energy G plot for the corresponding phase. The dashed line is the tentative phase equilibrium line corresponding to equality of the Gibbs energy G of phases. For the stress states at the merge of two instability lines, Gibbs energy has a plateau with constant value leading to unique homogeneous and hysteresis-free first-order Si I $\leftrightarrow$ Si II PT, with a continuum of intermediate homogeneous phases (HP), which are in indifferent thermodynamic equilibrium. With a further increase in σ_1, the first-order transformation changes to the second-order transition (designated as 2nd) and then (not shown) to a disordered phase [8]

possess the flat portion with constant energy between strains corresponding to each of the phases. Consequently, each intermediate phase along the transformation path has the same Gibbs energy as both phases and is in an indifferent (i.e., intermediate between stable and unstable) thermodynamic equilibrium state. If one of the strains

Fig. 4 Nanostructure evolution in silicon during phase transformation. **a** Transformation of two-phase Si I-Si II mixture into intermediate homogeneous phase at prescribed compressive strain $E_3 = -0.31$ and increasing tensile stresses $\sigma_1 = \sigma_2$. **b** Homogeneous transformation process from Si I to Si II through continuum of homogeneous phases with increasing strain E3 at fixed stresses $\sigma_1 = \sigma_2 = 11$ GPa. Colors characterize the local von-Mises shear strain [8]

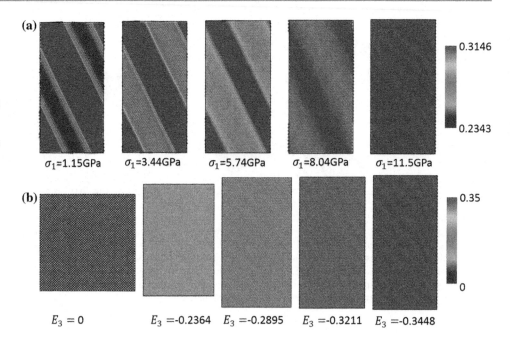

(a)

σ_1=1.15GPa σ_1=3.44GPa σ_1=5.74GPa σ_1=8.04GPa σ_1=11.5GPa

0.3146
0.2343

(b)

$E_3 = 0$ E_3 =-0.2364 E_3 =-0.2895 E_3 =-0.3211 E_3 =-0.3448

0.35
0

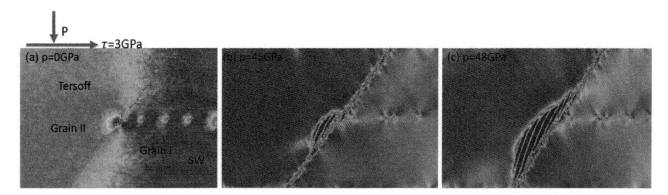

Fig. 5 Nucleation of Si II at perfect 60° dislocation pileup. **a** along the GB. **a** Dislocations were generated in Grain I under constant shear stress $\tau = 3$ GPa. **b** Hydrostatic pressure $p = 45$ GPa was applied to the sample while keeping the constant shear stress τ. The high-pressure phase Si II nucleated along the GB around the stress concentration due to dislocation pileup; **c** Martensitic phase Si II grows along the GB

(i.e., displacement at the boundary) is prescribed, then any intermediate crystal structure can be arrested (see Fig. 4b).

Away from the merged region, when Si I becomes unstable, transformation occurs through nucleation of Si II followed by formation of multiple bands of Si II (Fig. 4a) and their growth until the completeness of PTs. This happens in a material sample under both prescribed stresses and prescribed or changing strains. Interestingly, homogeneous intermediate structures are not observed and cannot be stabilized and studied. In contrast, within and in the close vicinity of the merged region, the transformation process is homogeneous (Fig. 4b) and each intermediate homogeneous crystal structure can be arrested and studied.

Phase Transformation Induced by a Dislocation Pileup at the GB

Here, dislocations and phase transitions simultaneously occur within one computer model. In the literature, the best potential to describe dislocation behavior in silicon is the Stillinger Weber (SW) [18] potential while the best potential to describe phase transition in silicon is the Tersoff potential [11]. We failed to find an interatomic potential in literature that can accurately describe a simultaneous occurrence of dislocations and phase transitions. One way to escape the limitation of the existing potentials is to use different potentials for different parts of the simulations [19]. In this paper, we apply the strategy similar to that in [19]. In Grain I, where dislocations will be generated, the SW potential is used, while in Grain II, where phase transition happens, the Tersoff potential is used. We also used the Tersoff potential to commute the forces between Grain I and Grain II. In this case, dislocations are generated in Grain I and pile up along the GB. The stress concentration in Grain II is obvious. Thereafter, a hydrostatic pressure was applied to the sample; martensitic phase nucleates around the stress concentration and propagates along the GB. Notice that now the critical stress to nucleate the Si II phase is 45 GPa while for perfect crystal it is 80 GPa. The nucleation pressure has been reduced, which demonstrated that the dislocation pileup plays a critical role in the nucleation of a new phase at the GB (Fig. 5).

Concluding Remarks

In this paper, phase instability criterion is calibrated by MD simulations. Through MD simulations, a homogeneous phase path is found by applying tension stress along the two transverse directions. Furthermore, dislocation pileup along the GB was generated in the simulation. It is shown that dislocation pileup can induce phase nucleation and greatly reduce the nucleation pressures.

Acknowledgements The support of NSF (DMR-1434613 and CMMI-1536925), ARO (W911NF-12-1-0340), and Iowa State University (Schafer 2050 Challenge Professorship) is gratefully acknowledged.

References

1. Blank, VD, Estrin, EI (2014) Phase transitions in solids under high pressure, Y Gogotsi, V Domnich (eds). CRC Press, New York Science and Engineering, Inst. of Physics, Bristol, Sect 3, pp 159–292
2. Levitas VI (2004) Continuum mechanical fundamentals of mechanochemistry. In: High pressure surface
3. Levitas VI (2004) High-pressure mechanochemistry: conceptual multiscale theory and interpretation of experiments. Phys Rev B 70:184118
4. Ji C, Levitas VI, Zhu H, Chaudhuri J, Marathe A, Ma Y (2012) Shear-induced phase transition of nanocrystalline hexagonal boron nitride to wurtzitic structure at room temperature and low pressure. Proc Natl Acad Sci USA 109:19108–19112
5. Levitas VI, Shvedov LK (2002) Low pressure phase transformation from rhombohedral to cubic BN: experiment and theory. Phys Rev B 65:104109
6. Levitas VI, Javanbakht M (2014) Phase transformations in nanograin materials under high pressure and plastic shear: nanoscale mechanisms. Nanoscale 6:162–166
7. Javanbakht M, Levitas VI (2016) Phase field simulations of plastic strain-induced phase transformations under high pressure and large shear. Phys Rev B 94:214104
8. Levitas VI, Chen H, Xiong L (2017) Triaxial-stress-induced homogeneous hysteresis-free first-order phase transformations with stable intermediate phases. Phys Rev Lett 118:025701
9. Levitas VI, Chen H, Xiong L (2017) Lattice instability during phase transformations under multiaxial stress: modified transformation work criterion. Phys Rev B 96:054118
10. Plimpton S (1995) Fast parallel algorithms for short-range molecular dynamics. J Comput Phys 117(1):1–19
11. Dodson BW (1987) Development of a many-body Tersoff-type potential for silicon. Phys Rev B 35(6):2795
12. Mizushima K, Yip S, Kaxiras E (1994) Ideal crystal stability and pressure-induced phase transition in silicon. Phys Rev B 50 (20):14952
13. Golovin YI (2008) Nanoindentation and mechanical properties of solids in submicrovolumes, thin near-surface layers, and films: a review. Phys Solid State 50(12):2205–2236
14. Ryckaert JP, Ciccotti G, Berendsen HJ (1977) Numerical integration of the cartesian equations of motion of a system with constraints: molecular dynamics of n-alkanes. J Comput Phys 23 (3):327–341
15. Subramaniyan AK, Sun CT (2008) Continuum interpretation of virial stress in molecular simulations. Int J Solids Struct 45 (14):4340–4346
16. Martoňák R, Laio A, Parrinello M (2003) Predicting crystal structures: the Parrinello-Rahman method revisited. Phys Rev Lett 90(7):075503
17. Miller RE, Tadmor EB, Gibson JS, Bernstein N, Pavia F (2016) Molecular dynamics at constant cauchy stress. J Chem Phys 144 (18):184107
18. Stillinger FH, Weber TA (1985) Computer simulation of local order in condensed phases of silicon. Phys Rev B 31(8):5262
19. Buehler MJ, Tang H, van Duin AC, Goddard WA III (2007) Threshold crack speed controls dynamical fracture of silicon single crystals. Phys Rev Lett 99(16):165502

Thermal Cycling Induced Instability of Martensitic Transformation and the Micro-Mechanism in Solution-Treated Ni$_{51}$Ti$_{49}$ Alloy

Cai-You Zeng, Zhong-Xun Zhao, Yuan-Yuan Li, Shanshan Cao, Xiao Ma, and Xin-Ping Zhang

Abstract

The effect of the maximum temperature of thermal cycling (T_{max}) on the instability of martensitic transformation (MT) in the solution-treated Ni$_{51}$Ti$_{49}$ alloy was investigated by differential scanning calorimetry (DSC). Results manifest that the peak temperature of martensitic transformation (M_p) decreases linearly with the increase of cycle number, while the transformation hysteresis (H) increases linearly. The instability of MT is promoted by increasing T_{max} from 20 to 100 °C, with variation of M_p increasing from 0.9 to 12.3 °C and variation of H increasing from 0.4 to 4.3 °C after 10 thermal cycles. Transmission electron microscopy (TEM) study demonstrates the appearance of transformation-induced dislocations in the NiTi matrix, which are responsible for the instability of MT. Moreover, the dislocation multiplication is obviously enhanced with the increase of T_{max} during thermal cycling, as a result of the interaction between dislocations and quenched-in point defects (QIDs) in the solution-treated Ni-rich Ni$_{51}$Ti$_{49}$ alloy, which consequently leads to the temperature dependence of MT instability during thermal cycling.

Keywords

Thermal cycling • Martensitic transformation instability Solution-treated NiTi alloy • Quenched-in defect Transformation-induced dislocation

C. Y. Zeng · Z. X. Zhao · Y. Y. Li · S. Cao · X. Ma
X. P. Zhang (✉)
School of Materials Science and Engineering, South China University of Technology, Guangzhou, 510640, China
e-mail: mexzhang@scut.edu.cn

Introduction

NiTi shape memory alloys (SMAs) have attracted increasing attention in widespread engineering fields owing to the combination of fantastic functional properties, including shape memory effect (SME) and superelasticity (SE), and excellent mechanical properties [1]. Such unique functional properties originate from the intrinsic thermoelastic martensitic transformation (MT) between the cubic B2 austenite and the monoclinic B19′ martensite. In practical applications, the components made of NiTi SMAs are commonly subjected to repetitive loading-unloading cycles, which may induce prolonged MT cycling, i.e., unstable martensitic transformation. Unfortunately, instability of MT is hardly avoidable under MT cycling condition, which can be described by the variation of the characteristic parameters of MT, e.g., characteristic temperatures, transformation hysteresis and enthalpy, associated with MT cycling [2, 3]. Such instability of MT generally leads to degradation of MT-induced properties in functional level, which is defined as functional fatigue [3], and has become one of the major limitations hindering the further application of NiTi SMAs.

Solution treatment is usually necessary to ensure the homogeneity in composition and microstructure of the NiTi alloy, and also to make the alloy possess one-step B2 ⇌ B19′ MT and take advantage of excellent SME and SE, in particular for some applications requiring high shape memory strain or recovery stress (e.g., one-way bias spring or orthodontic arch wire). Normally, thermal cycling was often employed to explore the instability of MT so as to characterize the functional fatigue of NiTi SMAs [4–6]. Many studies did observe a considerable decrease in characteristic temperatures of MT in solution-treated NiTi SMAs just after several thermal cycles. Recently, a few studies reported that quenched-in point defects (QIDs) existed in solution-treated Ni-rich NiTi SMAs have important influence on the instability of MT [7–9]. Notably, a decrease in martensitic transformation temperature of solution-treated Ni-rich NiTi

© The Minerals, Metals & Materials Society 2018
A. P. Stebner and G. B. Olson (eds.), *Proceedings of the International Conference on Martensitic Transformations: Chicago*, The Minerals, Metals & Materials Series, https://doi.org/10.1007/978-3-319-76968-4_14

SMAs was observed after prolonged exposure at room temperature, which was called the room temperature aging effect [7]. Further, a recent study indicated that such room temperature aging effect is more remarkable if the solution-treated Ni-rich NiTi SMAs are subjected to thermal cycling treatment [8]. Essentially, such room temperature aging effect is ascribed to dislocation-assisted atomic diffusion of QIDs. In addition, the temperature effect on the thermal cycling–induced instability of MT was found in solution-treated Ni-rich NiTi SMAs [9]. As the maximum temperature of thermal cycling (T_{max}) is increased from 150 to 250 °C, the instability of MT tends to be more obvious due to the enhanced diffusion of QIDs, which leads to nucleation of nano-precipitates with the increasing T_{max}. Although the multiplication of transformation-induced dislocations under thermal cycling condition can not be negligible [5, 6], the relationship between QIDs and transformation-induced dislocations was not considered in the referenced works. Meanwhile, despite the increasing attention to this issue, the complicated influence of QIDs on the instability of MT in NiTi SMAs during thermal cycling is far beyond current understanding.

The present work focuses on clarifying the thermal cycling–induced instability of martensitic transformation in a solution-treated NiTi alloy with a slightly high Ni-content of 51 at.% Ni (nominal) and containing a vast number of QIDs, with special attention to the effect of T_{max} on the instability of MT in the alloy during thermal cycling and TEM

characterization of the microstructure evolution in the alloy during thermal cycling.

Experimental Procedure

Binary Ni-rich NiTi alloy button ingots with a nominal composition of 51 at.% Ni were fabricated through vacuum arc melting. Each ingot was repeatedly melted 6 times to ensure the composition homogeneity. Disc-like samples with a diameter of 3 mm and a thickness of 1 mm were prepared by spark-cut from the center part of the button ingots, and then subjected to solution treatment at 950 °C for 8 h followed by quenching in water. A differential scanning calorimeter (DSC, NETZSCH 214 POLYMA) equipped with a liquid nitrogen refrigeration system was used to analyze the phase transformation behavior of the alloy during thermal cycling. Before the DSC test, all samples were mechanically ground on all faces to remove the oxide layer. The characteristic temperature of martensitic transformation can be determined through the tangential method, e.g., the start temperature of martensitic transformation (M_s) and finish temperature of austenitic transformation (A_f) are shown in Fig. 1a. To study the effect of T_{max} on the thermal cycling–induced instability of martensitic transformation in the solution-treated $Ni_{51}Ti_{49}$ alloy, a series of DSC thermal cycling tests with different T_{max} of 20, 60 and 100 °C, and a constant minimum temperature of thermal cycling

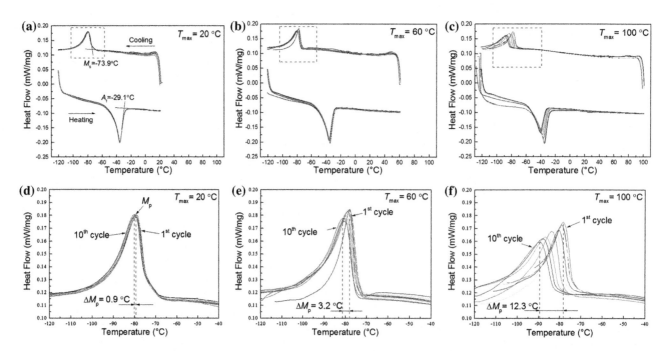

Fig. 1 DSC curves of the solution-treated $Ni_{51}Ti_{49}$ alloy under thermal cycling with different T_{max}: **a** 20 °C, **b** 60 °C and **c** 100 °C; and **d, e, f** the enlarged views of the parts marked in **a, b, c** respectively

$(T_{min}, -120\,°C)$ were performed. It should be indicated that the selected temperature range of all thermal cycling tests can ensure complete phase transformations during heating and cooling. During DSC thermal cycling tests, a heating/cooling rate of 10 °C/min was used, and the dwelling times at T_{min} and T_{max} were both 3 min for achieving thermal equilibrium. Each sample was subjected to the same thermal cycling 10 times. Post-mortem microstructural investigations of the samples subjected to thermal cycling were performed by a transmission electron microscopy (TEM, JEOL 1400 PLUS, 120 kV). Samples for TEM observation were prepared by twin-jet electro-polishing at 5 °C using 20 vol.% H_2SO_4 and 80 vol.% methanol solutions.

Results and Discussion

DSC curves of the solution-treated $Ni_{51}Ti_{49}$ alloy under thermal cycling with different T_{max} (20, 60 and 100 °C) are shown in Fig. 1, in which the characteristic temperatures of martensitic transformation are determined through the tangential method. Clearly, there are one-step martensitic transformation on cooling and one-step austenitic transformation on heating, as shown in Fig. 1a–c, which correspond to the transformations of B2 $\rightleftharpoons$ B19′. The phase transformation path remains unchanged during thermal cycling with different T_{max}, indicating that interphase R-phase is not introduced. The peak temperature of the forward martensitic transformation is represented by M_p, and the variation of M_p (denoted as ΔM_p) is used to estimate the variation of martensitic transformation temperature during thermal cycling. After undergoing 10 thermal cycles, the values of ΔM_p with different T_{max} are shown in Fig. 1d–f, which are 0.9, 3.2 and 12.3 °C. It is clear that the instability of MT in the solution-treated $Ni_{51}Ti_{49}$ alloy during thermal cycling is promoted slightly by the increase of T_{max} from 20 to 60 °C, while being intensified significantly by a higher T_{max} of 100 °C. It is noteworthy that the decrease of martensitic transformation temperature of solution-treated NiTi alloys is

usually about 10 °C after the first 10 thermal cycles [5, 6]. The changing tendency of ΔM_p shown in Fig. 1d–f suggests that the martensitic transformation in the solution-treated $Ni_{51}Ti_{49}$ alloy under thermal cycling with low T_{max} (e.g., 20 °C) is relatively stable, while tending to be obviously unstable with high T_{max} (e.g., 100 °C).

In addition to clarification of the influence of T_{max} on the instability of martensitic transformation in the $Ni_{51}Ti_{49}$ alloy, an important characteristic parameter in MT, transformation hysteresis (H), was measured according to the difference between A_f and M_s (namely $H = A_f - M_s$). Changes of both M_p and H with thermal cycle number at different T_{max} are shown in Fig. 2. Obviously, M_p decreases linearly with thermal cycle number, and the decrease rate of M_p can be obtained from the absolute value of the slope of the linear fitting curves, as shown in Fig. 2a. The decrease rate of M_p is relatively smaller at T_{max} = 20 °C than at T_{max} = 60 °C, and much smaller than at T_{max} = 100 °C; in particular, as T_{max} is increased from 20 to 100 °C, the variation rate of M_p increases from 0.1 to 1.4 °C/cycle. In contrast, H shows a linear increase with cycle number, as shown in Fig. 2b. Clearly, after 10 thermal cycles the variation of H (i.e., ΔH) is very small with increasing T_{max} from 20 °C to 60°C, which is 0.4°C and 0.5°C, respectively, while exhibiting a relatively large increase to 4.3 °C at T_{max} = 100 °C. The variation rate of H increases from 0.05 to 0.48 °C/cycle with increasing T_{max} from 20 to 100 °C. The above DSC results indicate that there exists a temperature dependence of MT instability in the solution-treated $Ni_{51}Ti_{49}$ alloy during thermal cycling.

TEM images of the solution-treated $Ni_{51}Ti_{49}$ alloy before and after 10 thermal cycles with different T_{max} are shown in Fig. 3. Apparently, no dislocations can be seen in the matrix before thermal cycling, except for several Ti-rich particles and grain boundaries, as shown in Fig. 3a, which is consistent with the fact that the density of dislocations is extremely low in the as-cast sample after solution treatment. However, a considerable number of dislocations can be observed in all thermally cycled samples, as shown in Fig. 3b–d. The observed dislocations in the thermally cycled

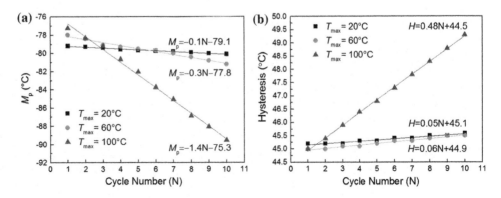

Fig. 2 Changes of **a** M_p and **b** H ($H = A_f - M_s$) with thermal cycling at different T_{max}

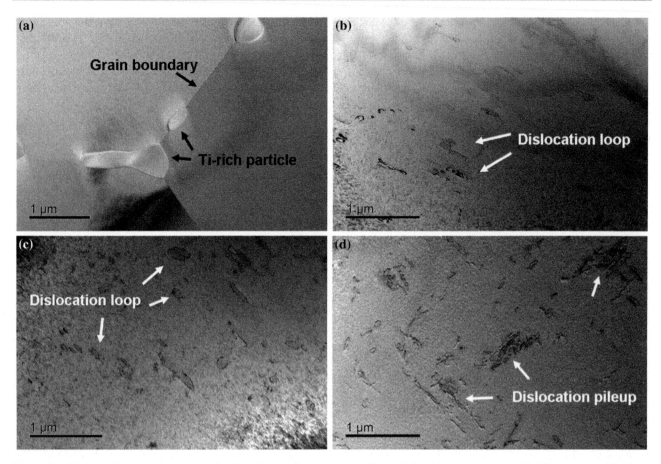

Fig. 3 TEM images of the solution-treated $Ni_{51}Ti_{49}$ alloy **a** before thermal cycling and after 10 thermal cycles with different T_{max} of **b** 20 °C, **c** 60 °C and **d** 100 °C

samples were introduced by martensitic transformation during thermal cycling. The evolution of transformation-induced dislocations under thermal cycling with different T_{max} mainly exhibits following two features: (1) The density of transformation-induced dislocations increases with increasing T_{max}; for example, at $T_{max} = 20$ °C only several individual dislocation loops can be seen, as shown in Fig. 3b, and some more dislocation loops can be observed at $T_{max} = 60$ °C, as shown in Fig. 3c. Remarkably, at $T_{max} = 100$ °C, there are significantly more dislocations formed, as shown in Fig. 3d. (2) The structure of transformation-induced dislocations tends to be more tangled with increasing T_{max}. As shown in Fig. 3d, at $T_{max} = 100$ °C, several dislocation pileups (indicated by white arrows) consisting of numerous tangled dislocation loops can be seen, in addition to some individual dislocation loops.

From the viewpoints of energy equilibrium, irreversible lattice defects, such as dislocations, need to be introduced at the front of the interfaces between austenite and martensite to accommodate additional elastic strain energy induced by MT [10]. As the transformation-induced dislocations can act as obstacles to martensitic transformation, thus lower

temperature is required to provide enough driving force to thermally activate the subsequent martensitic transformation, resulting in a decrease of martensitic transformation temperature during thermal cycling [11, 12]. In addition, the transformation hysteresis in the NiTi alloy is decided by the dissipated energy during MT. The multiplication of transformation-induced dislocations results in the increased dissipation of elastic strain energy, leading to the increase of the hysteresis during thermal cycling [13]. Based on the analysis above, it can be understood that the multiplication of transformation-induced dislocations is mainly responsible for the decreased M_p and increased H during thermal cycling, as demonstrated by DSC results shown in Fig. 1. Moreover, considering that a vast number of quenched-in point defects (QIDs) exist in the matrix of the solution-treated Ni-rich $Ni_{51}Ti_{49}$ alloy, the mechanism of the effect of T_{max} on the instability of MT observed in the present work can be better understood on the basis of the interaction between the high-concentration QIDs and transformation-induced dislocations. Firstly, during the forward martensitic transformation on cooling, the transformation-induced dislocations were introduced at the front of phase interfaces moving from martensite to austenite. Meanwhile, QIDs can be swept up by the movement of interfaces between the phases, resulting in a segregation of quenched-in point defects in the front of interfaces [14, 15]. The concentration of QIDs near the front of the moving interfaces increases associated with the segregation process. When the concentration increases up to a certain level, there is a remarkable pinning effect of QIDs on the transformation-induced dislocations, which is similar to the mechanism of Cottrell atmosphere. Afterwards, during the reverse martensitic transformation on heating, the phase interfaces move from austenite to martensite associated with the shrinking of martensite, and the transformation-induced dislocations pinned by QIDs are left behind. In the case of $T_{max} = 20\ °C$, the diffusion of QIDs at atomic level can hardly occur due to lack of driving force at relatively low temperature. Then, during the subsequent forward martensitic transformation on cooling, phase interfaces move from martensite to austenite again. In situ TEM analysis results indicated that the follow-up transformation-induced dislocations are derived from the initial dislocations through dislocation slips [7, 8]. So, the QIDs pinning effect on the initial transformation-induced dislocations can impede the dislocation movement during the following thermal cycling and further suppress the multiplication of transformation-induced dislocations. The suppression of multiplication of transformation-induced dislocations leads to the relatively stable MT during thermal cycling at $T_{max} = 20\ °C$. On the contrary, when T_{max} is increased to $100\ °C$, the relatively high temperature combined with the assistance of dislocations acting as a "fast channel" can make the activation of atomic diffusion much easier. Associated with the diffusion of QIDs, the pinning effect of QIDs on dislocations is reduced, which can promote dislocation multiplication during the subsequent thermal cycling and consequently lead to a relatively unstable martensitic transformation.

Conclusion

In the present work, the thermal cycling–induced instability of martensitic transformation (MT) in the solution-treated $Ni_{51}Ti_{49}$ alloy was studied comprehensively. It is found that the peak temperature of martensitic transformation (M_p) decreases linearly and the transformation hysteresis (H) increases linearly with increasing cycle number. The martensitic transformation is relatively stable when the maximum temperature of thermal cycling (T_{max}) is $20\ °C$ and becomes instable as T_{max} is increased to $100\ °C$, meaning that there exists a temperature dependence of MT instability during thermal cycling. The multiplication of transformation-induced dislocations is mainly responsible for the decreased M_p and increased H during thermal cycling.

The mechanism of the effect of T_{max} on the thermal cycling–induced instability of MT can be understood by the interaction between quenched-in point defects (QIDs) and transformation-induced dislocations. With a relatively low T_{max}, the pinning effect of QIDs on transformation-induced dislocations results in suppression of dislocation multiplication, which consequently leads to an enhanced stability of MT. With a relatively high T_{max}, the pinning effect is reduced by dislocation-assisted diffusion of QIDs, which enhances the dislocation multiplication, resulting in an obviously unstable martensitic transformation.

Acknowledgements This research was supported by the National Natural Science Foundation of China under Grant Nos. 51401081 and 51571092, and Key Project Program of Guangdong Provincial Natural Science Foundation under Grant No. S2013020012805.

References

1. Jaronie MJ, Leary M, Subic A, Gibson MA (2014) A review of shape memory alloy research, applications and opportunities. Mater Des 56:1078–1113
2. Pelton AR (2011) Nitinol fatigue: a review of microstructures and mechanisms. J Mater Eng Perform 20:613–617
3. Eggeler G, Hornbogen E, Yawny A, Heckmann A, Wagner M (2004) Structural and functional fatigue of NiTi shape memory alloys. Mat Sci Eng A378:24–33
4. Urbina C, De la Flor S, Ferrando F (2009) Effect of thermal cycling on the thermomechanical behaviour of NiTi shape memory alloys. Mat Sci Eng A501:197–206
5. Tong YX, Guo B, Chen F, Tian B, Li L, Zheng YF, Prokofiev EA, Gunderov DV, Valievc RZ (2012) Thermal cycling stability of ultrafine-grained TiNi shape memory alloys processed by equal channel angular pressing. Scripta Mater 67:1–4
6. Pelton AR, Huang GH, Moine P, Sinclair R (2012) Effects of thermal cycling on microstructure and properties in Nitinol. Mat Sci Eng A532:130–138
7. Kustov S, Mas B, Salas D, Cesari E, Raufov S, Nikolaev V, Van Humbeeck J (2015) On the effect of room temperature ageing of Ni-rich Ni–Ti alloys. Scripta Mater 103:10–13
8. Wang XB, Van Humbeeck J, Verlinden B, Kustov S (2016) Thermal cycling induced room temperature aging effect in Ni-rich NiTi shape memory alloy. Scripta Mater 113:206–208
9. Wagner MFX, Dey SR, Gugel H, Frenzel J, Somsen Ch, Eggeler G (2010) Effect of low-temperature precipitation on the transformation characteristics of Ni-rich NiTi shape memory alloys during thermal cycling. Intermetallics 18:1172–1179
10. Cui J, Chu YS, Famodu OO, Furuya Y, Hattrick SJ, James RD, Ludwig A, Thienhaus S, Wuttig M, Zhang ZY, Takeuchi I (2006) Combinatorial search of thermoelastic shape-memory alloys with extremely small hysteresis width. Nature Mater 5:286–290
11. Simon T, Kroger A, Somsen C, Dlouhy A, Eggeler G (2010) On the multiplication of dislocations during martensitic transformations in NiTi shape memory alloys. Acta Mater 58:1850–1860
12. Zhang J, Somsen C, Simon T, Ding XD, Hou S, Ren S, Ren XB, Eggeler G, Otsuka K, Sun J (2012) Leaf-like dislocation substructures and the decrease of martensitic start temperatures: a new explanation for functional fatigue during thermally induced martensitic transformations in coarse-grained Ni-rich Ti-Ni shape memory alloys. Acta Mater 60:1999–2006

13. Hamilton RF, Sehitoglu H, Chumlyakov Y, Maier HJ (2004) Stress dependence of the hysteresis in single crystal NiTi alloys. Acta Mater 52:3383–3402

14. Ibarra A, Juan JS, Bocanegra EH, Nó ML (2007) Evolution of microstructure and thermomechanical properties during superelastic compression cycling in Cu-Al-Ni single crystals. Acta Mater 55:4789–4798

15. Kustov S, Pons J, Cesari E, Van Humbeeck J (2004) Pinning-induced stabilization of martensite—part I: stabilization due to static pinning of interfaces. Acta Mater 52:3075–3081

Influence of Grain Size on Work-Hardening Behavior of Fe-24Ni-0.3C Metastable Austenitic Steel

W. Q. Mao, S. Gao, W. Gong, M. H. Park, Y. Bai, A. Shibata, and N. Tsuji

Abstract

In this study, the effect of grain size on the work-hardening behavior of Fe-24Ni-0.3C metastable austenitic steel was investigated by the use of in situ neutron diffraction during tensile tests in Japan Proton Accelerator Research Complex (J-PARC). The effect of grain size on the work-hardening behavior was considered from viewpoints of martensite formation and stress partitioning between different phases. The result revealed that when the grain size changed within the coarse grained region the influence of the grain size on the stress partitioning was relatively small, thus the work-hardening behavior was mainly determined by the increasing rate of martensite volume fraction. On the other hand, when the grain size decreased down to ultrafine grained scale, the internal stress (phase stress) in martensite significantly increased, which contributed to the increasing work-hardening rate.

Keywords

Deformation induced martensite • Metastable austenitic steel • Neutron diffraction • Work-hardening behavior Grain size

Introduction

In metastable austenitic steels, martensitic transformation can be initiated by applying stress and plastic strain even above M_s temperature, which is termed as deformation induced martensitic transformation (DIMT). The formation of deformation induced martensite can enhance the work-hardening rate and lead to high strength and high ductility in the material. The enhancement of ductility is termed as transformation induced plasticity (TRIP) [1]. In 3rd generation advanced high strength steels (AHSS), more and more attention has been paid recently to the TRIP effect as an effective mechanism to enhance both strength and ductility. Several ways for realizing the TRIP effect in the 3rd generation AHSS have been proposed [2–6]. One of the ways is to optimize the TRIP effect in ultrafine grained structures [2]. It has been widely known that grain refinement increases strength of materials due to the Hall-Petch effect. In case of metastable austenitic steels, the grain size may greatly influence their mechanical properties through affecting the formation of deformation induced martensite [7–9]. In order to maximize the TRIP effect, it is necessary to clarify the effect of grain size on the deformation-induced martensitic transformation and then the work-hardening behavior.

Recent development of in situ neutron diffraction experiments enables us to investigate the deformation and transformation behaviors of the specimen during deformation. There have been several reports studying deformation behaviors of multi-phased materials studied in situ neutron diffraction [10–12], in which it was found that during the plastic deformation a stress partitioning would occur between different phases. Thus, in this study, the effect of the stress partitioning as well as the formation of martensite on work-hardening behavior was investigated using the in situ neutron diffraction. A 24Ni-0.3C metastable austenitic steel was selected as the material which could exhibit favorable TRIP phenomenon during deformation, and 24Ni-0.3C specimens having various mean grain sizes were prepared.

W. Q. Mao (✉) · S. Gao · M. H. Park · Y. Bai · A. Shibata · N. Tsuji
Department of Materials Science and Engineering, Kyoto University, Kyoto, Japan
e-mail: mao.wenqi.26m@st.kyoto-u.ac.jp

W. Gong · M. H. Park · Y. Bai · A. Shibata · N. Tsuji
Elements Strategy Initiative for Structural Materials (ESISM), Kyoto University, Kyoto, Japan

W. Gong
J-PARC Center, Japan Atomic Energy Agency, Tokai, Japan

© The Minerals, Metals & Materials Society 2018
A. P. Stebner and G. B. Olson (eds.), *Proceedings of the International Conference on Martensitic Transformations: Chicago*,
The Minerals, Metals & Materials Series, https://doi.org/10.1007/978-3-319-76968-4_15

Experimental Procedures

A 24Ni-0.3C (mass%) metastable austenitic steel was used in the present study. Sheet specimens with mean grain sizes of 1, 4 and 35 μm were fabricated by cold rolling and subsequent annealing of the 24Ni-0.3C steel. Phase plus grain boundary maps of the specimens with various grain sizes are represented in Fig. 1, in which fully recrystallized austenitic structures were observed.

Tensile tests with in situ neutron diffraction were carried out by the use of an engineering materials diffractometer at beam line 19 (TAKUMI) in J-PARC. The time-of-fight (TOF) source was used in the in situ neutron diffraction experiment, which has a continuous range of velocities and wavelengths. By measuring the flight times of the detected neutrons, the wavelengths can be calculated and diffraction profiles are obtained. More complete descriptions of this technique can be found in the references [13, 14]. The illustration of in situ neutron diffraction is shown in Fig. 2a. The angle between the tensile direction and the neutron incident direction was 45° and two neutron detectors were arranged 90° and 90° from the neutron incident direction. Thus, the reflection of the (hkl) planes perpendicular to the tensile direction and the transverse direction could be recorded. Enlarged evolutions of (111) reflection profiles of austenite in tensile direction are shown in Fig. 2b as an example to represent the effect of stress on diffraction profiles. In Fig. 2b, it is clearly seen that the (111) diffraction peak shifts with increasing the applied stress, which corresponds to the elastic deformation of the crystalline lattice.

The specimens having a gauge part 50 mm in length, 6 mm in width and 1 mm in thickness were tensile deformed in a stepwise manner up to their fracture at room temperature at an initial strain rate of 8.3×10^{-4} s^{-1}. The crosshead was temporarily stopped for 10 min at each step to measure the diffraction profiles. The volume fraction of martensite was measured from the ratio of the integrated intensity of diffraction profiles [15]. The lattice strain was determined by the deviation of lattice plane spacing (d_{hkl}) from the stress free lattice plane spacing d_{hkl}^0 by the use of the Eq. (1) [10].

Then, the evolution of lattice elastic strain can be used to roughly represent the evolution of the phase stress (elastic internal stress) in the corresponding phase [11]. In the present study, the lattice strains of (111) lattice plane in austenite and (110) lattice plane in martensite were used to represent their phase stress, respectively.

$$\varepsilon_{hkl} = \left(d_{hkl} - d_{hkl}^0\right)/d_{hkl}^0 \tag{1}$$

Results and Discussion

Normal stress-strain curves of the fully recrystallized specimens with various grain sizes are shown in Fig. 3. It is seen that with decreasing the mean grain size from 35 to 1 μm, the yield strength increased from 180 to 420 MPa and the total elongation also increased from 83 to 104%, while the ultimate strength slightly decreased from 1100 to 950 MPa. The specimen with the mean grain size of 1 μm exhibited a better combination of strength and ductility than that with the mean grain size of 35 μm. Figure 4a shows the work-hardening rate of the specimens with various grain sizes as a function of the true strain, which was obtained from corresponding true stress-true strain curves. For all the three specimens, the work-hardening rate firstly decreased, then started to increase at different true strains depending on the grain size, and finally decreased again. The increase of the work-hardening rate during deformation is a typical feature for the materials exhibiting the TRIP effect [7–9]. From the change of martensite volume fraction during deformation shown in Fig. 4b, it was found that the increase of the strain hardening rate in the specimens was attributed to the deformation induced martensitic transformation, since the true strain at which the martensitic transformation started corresponded well with the true strain where the increase of the strain hardening occurred. On the other hand, it was found that the enhancement of the strain hardening rate decreased with decreasing the mean grain size of austenite meanwhile the suppression of deformation induced martensitic transformation was confirmed in Fig. 3b. The suppression of DIMT was

Fig. 1 Phase plus grain boundary maps obtained by EBSD analysis of the specimens with different grain sizes: **a** 1 μm; **b** 4 μm; **c** 35 μm

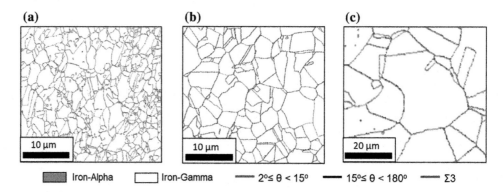

(a)   (b)   (c)

10 μm   10 μm   20 μm

▨ Iron-Alpha   ☐ Iron-Gamma   — 2° ≤ θ < 15°   — 15° ≤ θ < 180°   — Σ3

Fig. 2 **a** Illustration of in situ neutron diffraction. **b** Enlarged change in the reflection of (111) plane of austenite obtained by the detector 2

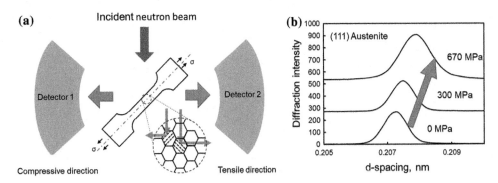

possibly due to the enhanced stability of austenite by the grain refinement [16]. The results confirmed that the enhancement of the work-hardening rate was due to the deformation induced martensitic transformation. More importantly, the increase of the work hardening rate caused by the formation of martensite was proportional to the increasing rate of the martensite volume fraction.

Besides the formation of martensite, the effect of stress partitioning between martensite and austenite phases on the global work-hardening behavior was investigated as well. Figure 5 shows the changes of lattice strains of (111) plane in austenite and (110) plane in martensite of the specimens having different grain sizes, plotted as a function of the true strain. The lattice strain of (111) plane in austenite gradually increased during the plastic deformation, implying a gradual increase of the phase stress and a relatively low work-hardening rate in austenite. On the other hand, the lattice strain of (110) plane in martensite dramatically increased immediately after the DIMT started (as can be compared with Fig. 5b), implying that the phase stress in martensite greatly increased. Thus, it was considered that the enhanced work-hardening rate shown in Fig. 4a was caused by not only

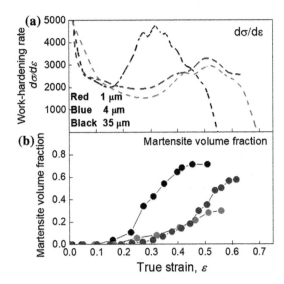

Fig. 4 Work-hardening rate curves (a) and changes in martensite volume fractions (b) in the specimens with different grain sizes, plotted as a function of true stress in tensile test

the increase of the martensite volume fraction but also by the rapid increase of the phase stress in martensite.

With decreasing the mean grain size, the lattice strain, corresponding to the phase stress, of austenite at a given true strain increased due to the grain refinement strengthening [17], while the work-hardening behavior of austenite did not show obvious changes. For martensite, when the grain size of austenite decreased from 35 to 4 μm, the evolution of lattice strain of martensite with the true strain was almost unchanged. However, when the grain size was further reduced to 1 μm, the lattice strain of martensite significantly increased with increasing of the true strain. That is, when the grain size decreased within the relatively coarse grained region, the influence of grain refinement on the evolution of the phase stress in martensite was limited and the work hardening behavior of the 24Ni-0.3C metastable austenitic steel was mainly determined by the increasing rate of the martensite volume fraction. On the other hand, when the grain size decreased down to ultrafine grained scale the phase stress in martensite significantly increased which could contribute to the increase of the global work-hardening

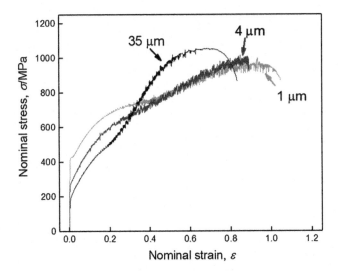

Fig. 3 Nominal stress-strain curves of the specimen with different grain sizes

Fig. 5 Changes in lattice strains in the specimens with different grain size as a function of true strain: **a** lattice strain of (111) plane in austenite; **b** lattice strain of (110) plane in martensite

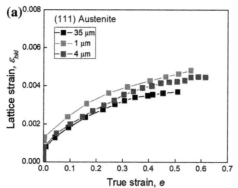
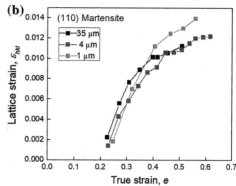

rate. Therefore, although the formation rate of martensite in the 1 μm specimen was much lower than that in the 4 μm specimen, the work-hardening rate of the 1 μm specimen was almost the same as that of the 4 μm specimen due to the higher phase stress in martensite of the 1 μm specimen.

Summary

In this study, the tensile test with in situ neutron diffraction was utilized to study the effect of the grain size on the work-hardening behavior of a 24Ni-0.3C metastable austenitic steel. The effect of the grain size on the work-hardening behavior was considered from viewpoints of martensite formation and stress partitioning. It was found that the enhanced work-hardening caused by DIMT is due to the increasing of martensite volume fraction as well as the increase of the phase stress in martensite. When the grain size changed within the coarse grained region the influence of the grain size on the evolution of phase stress in martensite was relatively small, thus the work hardening behavior of the 24Ni-0.3C metastable austenitic steel was mainly determined by the increasing rate of the martensite volume fraction. On the other hand, when the grain size was decreased down to the ultrafine grained region, the phase stress in martensite significantly increased, which contributed to the increase of the work-hardening rate.

Acknowledgements The neutron experiment at the Materials and Life Science Experimental Facility of the J-PARC was performed under a user program (Proposal No. 2016E0003 and 2017A0136). This work was financially supported by the Elements Strategy Initiative for Structural Materials (ESISM) and the Grant-in-Aid for Scientific Research (S) (No. JP15H05767) both through the Ministry of Education, Culture, Sports, Science and Technology (MEXT), Japan. One of the authors (W.Q.Mao) was financially supported by China Scholarship Council (CSC), China. The supports are gratefully appreciated.

References

1. Zackay VF, Parker ER, Fahr D et al (1967) ASM Trans Quart 60 (2):252–259
2. Wakita M, Adachi Y, Tomota Y (2007) Trans Tech Publ 539:4351–4356
3. Edmonds DV, He K, Rizzo FC et al (2006) Mater Sci Eng A 438:25–34
4. Speer J, Matlock DK, De Cooman BC et al (2003) Acta Mater 51 (9):2611–2622
5. Grässel O, Krüger L, Frommeyer G et al (2000) Int J Plast 16 (10):1391–1409
6. Berrenberg F, Haase C, Barrales-Mora LA et al (2017) Mater Sci Eng A 681:56–64
7. Iwamoto T, Tsuta T (2000) Int J Plast 16(7):791–804
8. Jacques P, Furnémont Q, Mertens A et al (2001) Philos Mag A 81 (7):1789–1812
9. De AK, Speer JG, Matlock DK et al (2006) Metall Mater Trans A 37(6):1875–1886
10. Tomota Y, Tokuda H, Adachi Y et al (2004) Acta Mater 52 (20):5737–5745
11. Tomota Y, Lukáš P, Neov D et al (2003) Acta Mater 51(3):805–817
12. Harjo S, Tomota Y, Lukáš P et al (2001) Acta Mater 49(13):2471–2479
13. Daymond MR, Bourke MAM, Von Dreele RB et al (1997) J Appl Phys 82(4):1554–1562
14. Hutchings MT, Krawitz AD (eds) (2012) Measurement of residual and applied stress using neutron diffraction, vol 216. Springer Science & Business Media
15. De AK, Murdock DC, Mataya MC et al (2004) Scr Mater 50 (12):1445–1449
16. Takaki S, Fukunaga K, Syarif J et al (2004) Mater Trans 45 (7):2245–2251
17. Chokshi AH, Rosen A, Karch J et al (1989) Scr Metall 23 (10):1679–1683

The Influence of α' (bcc) Martensite on the Dynamic and Magnetic Response of Powder Metallurgy FeMnSiCrNi Shape Memory Alloys

M. Mocanu, E. Mihalache, B. Pricop, F. Borza, M. Grigoraş,
R. I. Comăneci, B. Özkal, and L. G. Bujoreanu

Abstract

In FeMnSi-based SMAs the "executive" phase is ε (hexagonal close packed, hcp) stress induced martensite which retransforms to γ (face center cubic, fcc) austenite during heating, causing free-recovery shape memory effect (SME). At low Mn content or high deformation degrees, α' (body center cubic, bcc) martensite can be additionally induced by cooling or deformation, being considered as detrimental for the magnitude of SME. In the case of powder metallurgy (PM) Fe-14Mn-6Si-9Cr-5Ni (wt%) SMAs containing 5 fractions of mechanically alloyed (MA'ed) powders (0–40%$_{vol.}$) solution treated to 5 temperatures (700–1100 °C), large amounts, (20–90%) of α'-bcc martensite were detected by XRD and observed by SEM. Nevertheless, free-recovery SME was obtained and enhanced by training, up to bending strokes of 24 mm, developed with a rate of 1.71 mm/°C. The paper corroborates the qualitative and quantitative evolutions of α', ε and γ phases with DMA and thermomagnetic measurements performed on the 25 sets of specimens, during heating up to 500 °C.

Keywords

FeMnSiCrNi SMA • Martensite • Structural analysis
Dynamic mechanical analysis
Thermomagnetic analysis

Introduction

After the first reports on Fe-Mn-Si Shape Memory Alloys (SMAs) single crystals, [1–3], Fe-(28-34) Mn-(4-6.5) Si (mass%, as all chemical compositions will be listed hereinafter) polycrystalline alloys with almost perfect Shape Memory Effect (SME) were obtained [4–6]. Corrosion resistance was improved by Cr and Ni additions, thus contributing to the development of Fe-28Mn-6Si-5Cr [7] and Fe-14Mn-6Si-9Cr-5Ni SMAs [8], which became of commercial use [9]. A recent overview enumerated: (i) lock rings for bicycle frame pipes, (ii) powder blowing nozzle protection pipes, (iii) waterproof resin coated rings for bulk superconductor reinforcement, (iv) connection segments of rib-shaped rows of steel pipes for underground tunnel digging, (v) plates for crane rail junctions, (vi) concrete pre-straining rods, (vii) plates for controlling the curvature of concrete beams and (viii) 2t-anti-seismic dampers [10] as main applications of Fe-Mn-Si based SMAs. For this alloy system SME is caused by thermally induced reversion to γ-face centered cubic (fcc) austenite of ε-hexagonal close packed (hcp) stress-induced martensite (SIM) [11]. At low Mn content or at high deformation degrees, besides ε-hcp, α'-body centered cubic (bcc) martensite can be additionally induced by cooling or deformation [12]. It was argued that, for shape recovery enhancement, ε-hcp SIM plates should be as narrow as possible, with a single variant orientation and should interact neither with each other nor with pre-existing thermally induced ε martensite, in order not to cause the formation of α'-bcc martensite, which is detrimental for SME [13]. One of the factors that can suppress thermally induced formation of ε-hcp martensite is the paramagnetic-

M. Mocanu · E. Mihalache · B. Pricop · R. I. Comăneci
L. G. Bujoreanu (✉)
Faculty of Materials Science and Engineering, The "Gheorghe Asachi" Technical University of Iaşi, Blvd. Dimitrie Mangeron 67, 700050 Iaşi, Romania
e-mail: lgbujor@tuiasi.ro

F. Borza · M. Grigoraş
National Institute of Research and Development for Technical Physics, Blvd. Dimitrie Mangeron 47, 700050 Iaşi, Romania

B. Özkal
Particulate Materials Laboratory, Metallurgical and Materials Engineering Department, Istanbul Technical University, 34469 Maslak, Turkey

antiferromagnetic transition, occurring at Néel temperature (T_N) [14]. This transition is characterized by an abrupt change in elastic modulus [15] and heat capacity [16] during heating and causes a marked stabilization effect of the antiferromagnetic ordering on the γ-fcc austenite [17].

In contrast to other alloy systems, such as NiTi-based [18] or Cu-based [19] SMAs, very scarce reports are found in literature concerning Fe-Mn-Si based SMAs obtained via powder metallurgy (PM) processing with mechanical alloying (MA) routine. Thus, in spite of porosity-originating brittleness [20], PM enabled accurate chemical composition controlling [21] and MA provided solid state dissolution of alloying elements [22], if oxide formation was prevented [23], which allowed to obtain Fe-Mn-Si alloys with better mechanical properties and performing SME, as compared to conventional casting [24]. These few results refer to "high manganese Fe-Mn-Si SMAs" which are generally ternary. In a series of works on quintenary Fe-Mn-Si-Cr-Ni SMAs, a part of present authors reported thermally induced reversion to γ-fcc austenite of α'-bcc SIM [25], the amount of which experienced an increasing tendency with the increase of heat treatment temperature [26] and the number of mechanical cycles up to 4% maximum strains [27]. The substitution of a fraction of as blended powders with MA'd particles contributed to the reduction of surface oxidation [28] being associated with the presence of amorphous regions in PM-MA powder mixtures [29] and the enhancement of thermally induced formation of martensite [30] at Fe-14Mn-6Si-9Cr-5Ni SMAs. In solution treated state, mechanical alloying enabled the increase of α'-bcc martensite amount [31] but with a suitable selection of MA'd powder fraction, hot rolling temperature [32] and solution treatment atmosphere [33] an increase of shape recovery degree was observed in Fe-14Mn-6Si-9Cr-5Ni SMAs [34] which developed, in the case of an expanded-diameter ring, a free end displacement of 11.3 mm, during heating up to 573 K [35]. Some of present authors recently reported a general trend of internal friction to augment with increasing MA'ed fraction, associated with the decrease in the amount of γ-fcc austenite which transformed to α'-bcc and ε-hcp SIM [36].

Based on above mentioned results, the present work aims to further study the influence of heat treatment temperature and MA'd powder fraction on the structure of a PM Fe-Mn-Si-Cr-Ni SMA and to correlate structural changes with antiferromagnetic-paramagnetic transition emphasized by storage modulus and magnetization variations, observed by dynamic mechanical analysis (DMA) and thermomagnetic analysis, respectively.

Experimental Procedure

Paralellipipedic specimens $(4 \times 10 \times 40 \text{ mm}^3)$ were obtained, with the chemical composition 66 Fe, 14 Mn, 6 Si, 9 Cr and 5 Ni (mass%), by elemental powders blending, compacting and sintering, as previously detailed [37]. Five different groups of specimens were produced, from: (i) as-blended powders (0_MA) and with (ii) $10\%_{vol}$, (iii) $20\%_{vol}$, (iv) $30\%_{vol}$ and (v) $40\%_{vol}$ fractions of MA'ed powders, designated as 10_MA, 20_MA, 30_MA and 40_MA, respectively [38]. Sintered specimens were further hot rolled at 1370 K, until thickness decreased down to 1×10^{-3} m [39]. Rectangular $(4 \times 25 \text{ mm}^2)$ and lamellar $(10 \times 150 \text{ mm}^2)$ specimens were cut by spark erosion and were solution treated to 973, 1073, 1173, 1273 or 1373 K/0.3 ks/water. In the following, the specimens will be designated by their MA'ed fraction and solution treatment temperature in °C (e.g. the specimen with $30\%_{vol}$ MA'ed powder, heat treated at 1173 K will be referred as 30_MA_900).

Rectangular specimens, corresponding to each of the above twenty five solution treated states, were embedded into cold mounting resin and metallographically prepared by grinding, polishing and etching with a solution of 1.2% $K_2S_2O_5$ + 1% NH_4HF_2 in 100 ml distilled water, before being analyzed by X-ray diffraction (XRD) and scanning electron microscopy (SEM). The former were performed on Expert PRO MPD diffractometer with Cu Kα radiation and the latter on a Scanning Electron Microscope SEM JEOL JSM 6390 with EDX and electron beam resolution of 1.1 ÷ 2.5 nm at a voltage U = 20 ÷ 1 kV. One rectangular specimen of each of the twenty five solution treated types was subjected to temperature scans, up to 673 K, performed by dynamic mechanical analysis (DMA), at a rate of 5 K/min and an amplitude of 20 μm, on a DMA 242 Artemis NETZSCH device equipped with "three-point bending" specimen holder. Storage modulus variations versus temperature were smoothened with PROTEUS software, which controls the functioning of DMA device. Fragments of each of the twenty five specimens were cut and subjected to thermomagnetic analysis, at a rate of 2 K/min, on a Lake Shore VSM 7410 Vibrating Sample Magnetometer, with magnetic induction H = 0 ÷ 3.2 T and temperature range T = 4 ÷ 1300 K.

Lamellar specimens were trained to develop free-recovery SME. For this purpose, the specimens were deformed by bending in martensitic condition, with a special cylindrical caliber, were rigidly fixed at one end and the motion of the free end, under growing temperature, was detected by video control [40]. Free end displacement with temperature was recorded in each training cycle [41].

Experimental Results and Discussion

Structural Analysis

The XRD patterns of the twenty five specimens are shown in Fig. 1.

According to crystallographic databases, the main diffraction maxima of the two developing martensitic phases are located at $2\theta = 44.485°$ for α' (110) and $2\theta = 47.041°$ for ε (101). The presence of α' (110), which theoretically overlaps γ (111), at $2\theta = 43.608°$ and ε (002) at $2\theta = 44.383°$, can be noticed due to the shift experienced by this maximum as an effect of internal constraints accompanying martensite formation [42]. Thus α' (110) peak is noticeable at most of the specimens and tends to become prominent with increasing MA'ed fraction.

The amounts of the three phases were semi-quantitatively determined by the ratios of the respective maxima

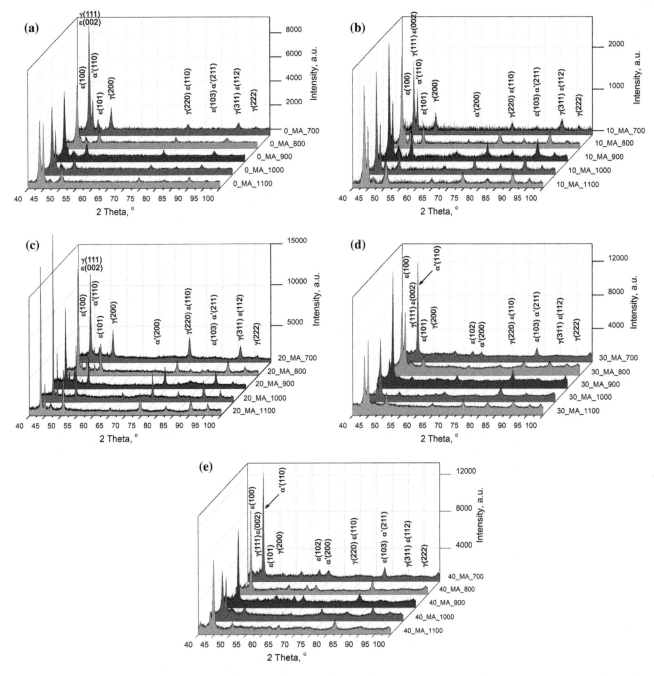

Fig. 1 XRD patterns of the specimens with different MA'ed fractions, subjected to five solution treatment temperatures: **a** 0_MA; **b** 10_MA; **c** 20_MA; **d** 30_MA; **e** 40_MA

intensities, as previously detailed [43]. Based on these amounts Fig. 2 was built.

The amount of α′-bcc ranges between 20–90% (47% of values being observed at specimens at 30_MA and 40_MA). From Fig. 2a, an overall increasing tendency of α′-bcc amount with MA'ed fraction is noticeable. Figure 2b shows a majority decreasing tendency of ε-hcp amount with increasing MA'ed fraction. The amount of ε-hcp martensite ranges between 5 and 38%, 84% of the specimens having ε-hcp martensite amounts below 25%. Finally, the Fig. 2c depicts a general decreasing tendency of γ-fcc austenite, with increasing MA'ed fraction. The amount of austenite ranges between 8 and 69%, 84% of the specimens containing above 10% austenite.

In order to observe the structure of analyzed specimens, SEM micrographs were recorded, the most representative aspects being illustrated in Fig. 3.

At specimen 0_MA_700, in spite of the large amount of martensite (47% α′ and 16% ε) no martensite plates are noticeable. The failure surface from Fig. 3a illustrates the brittle character of the material, emphasized by long crack ribs. Zn stearate binder is shown in the inset. Figure 3b details deformed powder grains, with an average size below 10 μm, joined by connecting bridges formed during MA. The effects of increasing solution treatment temperature to 1373 K are observed at specimen 0_MA_1100. Figure 3c displays a general aspect of ε-hcp martensite with an inset with a grain completely occupied by α′-bcc martensite plates, with marked relief. The magnified detail from Fig. 3d shows 100 nm thin parallel martensite plates of ε-hcp. Increasing MA'ed fraction to 40%vol. caused the refinement of ε-hcp martensite plates, while α′-bcc martensite plates become shorter as shown in the general aspect from Fig. 3e which includes an inset of fine parallel ε-hcp plates. Intergranular cracks were formed along grain boundaries, as observed in Fig. 3f. As an effect of increasing solution treatment temperature, the density of ε-hcp martensite plates increased and large arrays of fine parallel plates are observed in the general aspect from Fig. 3g. The plates completely

cross the grains, stopping at their border, as observed in the detail from Fig. 3h. So, increasing both solid solution temperature and MA'ed fraction contributed to the refinement of ε-hcp martensite plates.

Storage Modulus and Magnetization Variation with Temperature

While heated, common metallic materials experience typical variations of storage modulus (E′) versus temperature displaying continuous decrease of E′, due to the augmentation of thermal agitation that weakens interatomic bounds. "Modulus softening" represents an additional decrease of E′ that can be associated with the reversion of martensite (harder) to austenite (softer), in the case of non-thermoelastic transformations. On the other hand, a "modulus hardening" can be noticed at Fe-Mn-Si based SMAs, due to the antiferromagnetic-paramagnetic phase transition occurring at the Néel temperature. Such behaviors were illustrated in the upper parts of Fig. 4.

The lower parts of the figure show the variation of magnetization vs. temperature from room temperature to 773 K. Néel temperatures were determined as T_N^c for cooling and T_N^h for heating and correspond to magnetization maxima observed during the two respective processes. It should be noticed that the determination of magnetization maximum during heating is sometimes difficult. For this reason, in some cases T_N^h was ascribed to the first inflexion point in magnetization increase, such as the example from Fig. 4a. The starting and the ending temperatures of modulus increasing were designated as T_s^{DMA} and T_f^{DMA} and were marked, on magnetization versus temperature diagrams, with dark and bright ascending arrows. It is noticeable that, in most of the cases, at the temperature where E′ starts increasing, a saturation in magnetization growth on heating can be observed. For instance, at specimen 0_MA_700, Fig. 4a shows two local increases of storage modulus marked with the arrows S_1 and S_2, on magnetization curve.

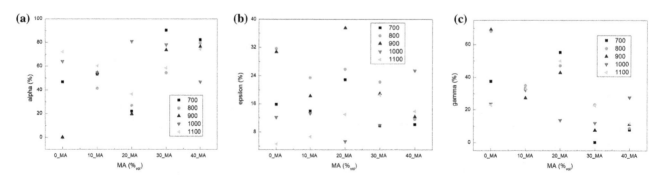

Fig. 2 Summary of the amounts of phases determined on XRD patterns, as a function of MA'ed fraction and solution treatment temperature: **a** α′-bcc martensite amount; **b** ε-hcp martensite amount and **c** γ-fcc austenite amount

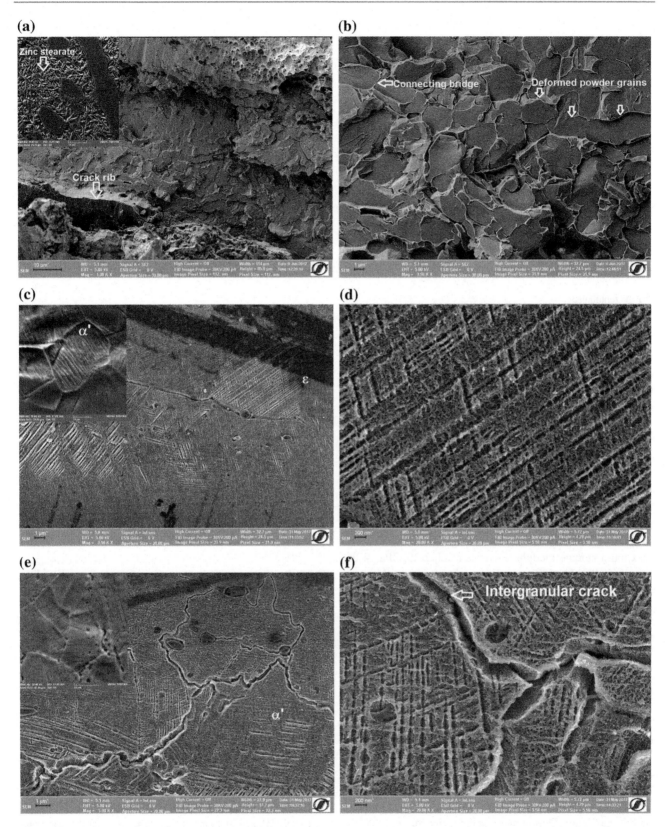

Fig. 3 Representative SEM micrographs illustrating the effects of MA'ed fraction and solution treatment temperature: **a** general aspect of brittle failure surface 0_MA_700 specimen; **b** detail of 0_MA_700 failure surface with connection bridges between deformed powder grains; **c** general aspect of 0_MA_1100 specimen; **d** detail of 0_MA_1100 specimen in the area of parallel ε-hcp martensites plates; **e** general aspect of 40_MA_700 specimen; **f** detail of 40_MA_700 specimen with intergranular crack originating from triple junction between grains; **g** general aspect of 40_MA_1100 specimen, with fine martensite plates; **h** detail of a grain boundary in 40_MA_1100 specimen

(g)   **(h)**

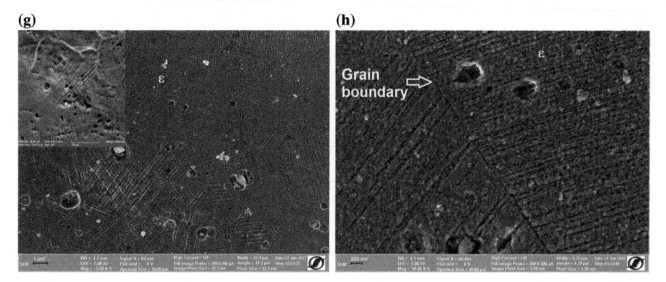

Fig. 3 (continued)

In all of the cases $T_s^{DMA} < T_N^h$ and $T_f^{DMA} < T_N^h$ (with a single exception) suggesting that storage modulus destabilization that causes modulus hardening and the abrupt change in E' starts and finishes before the paramagnetic state is reached.

Table 1 summarizes all the data obtained from the diagrams of storage modulus *vs.* temperature and magnetization *vs.* temperature, for the specimens solution treated at 973 and 1373 K with all the five MA'ed fractions.

With increasing solution treatment temperature and MA'ed fraction T_N generally tends to decrease. In addition, the thermal hysteresis between T_N^h and T_N^c, which is always lower, tends to decrease with increasing MA'ed fraction. The maximum storage modulus increase reaches 19 GPa, at the specimen 40_MA_700, according to the variation illustrated in Fig. 4c.

Thermomechanical Training

The lamellar specimen 40_MA_1000, with a phase structure comprising approx. 46.8 α'-*bcc*, 25.5 ε-*hcp* and 27.7 γ-*fcc*, was selected for free-recovery SME training, owing to its lowest amount of α'-*bcc* among 40_MA specimens. The evolution is summarized in Fig. 5. The specimen was bent on a cylindrical caliber and fastened at one end. The starting point is shown in Fig. 5a, where the position of specimen's free end at 28 °C (301 K) is designated at h_{28}. Heating was applied, along the entire length of the specimen, with a mobile gas lamp. During each heating, the free end moved upwards, the successive positions and instant temperatures being recorded by a camera. Heating was applied until the free end stopped lifting, typically before temperature reached

573 K. After this, the gas was turned off and the specimen was sprayed with cold water. Then it was released and bent again against the caliber. This procedure was repeated five times. In the fifth cycle, during heating up to 236 °C (509 K), a total displacement $\Delta h_5 = 24 \times 10^{-3}$ m was recorded within 14 s, which gives an average displacement rate of 1.71 10^{-3} m s^{-1}. Free end's position at 236 °C (509 K) together with total displacement Δh_5 are shown in Fig. 5b.

Conclusions

The following conclusions can be drawn from the above discussion:

- Low Mn% and the substitution of as blended with MA'ed powders enabled thermally induced formation of α'-*bcc* and ε-*hcp* martensites (up to 90% and 38%, respectively) during solution treatment (ST).
- With increasing MA%$_{vol}$, α'-*bcc* martensite experienced an overall increasing tendency while ε-*hcp* martensite and γ-*fcc* austenite had majority and general trends to decrease, respectively.
- Increasing MA%$_{vol}$ and solution treatment temperature enhanced martensite plate refinement.
- Antiferromagnetic $\rightarrow$ paramagnetic transition, at Néel temperature (T_N), caused storage modulus increases (up to 19 GPa) at critical temperatures that tend to decrease with increasing ST temperature and MA'ed fraction.
- During heating, storage modulus increase takes place before paramagnetic state is fully reached.

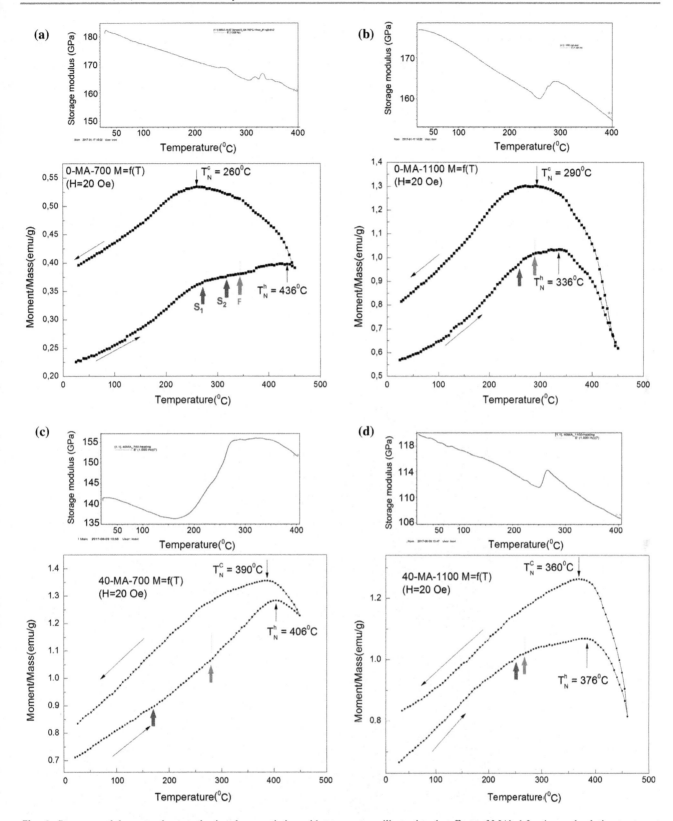

Fig. 4 Storage modulus up and magnetization down variation with temperature, illustrating the effects of MA'ed fraction and solution treatment temperature: **a** 0_MA_700; **b** 0_MA_1100; **c** 40_MA_700; **d** 40_MA_1100

Table 1 Critical variation temperatures associated with the start $\left(T_s^{DMA}\right)$ and finish $\left(T_f^{DMA}\right)$ of storage modulus increase ($\Delta E'$), on DMA thermograms, and with magnetization maxima/saturations during heating $\left(T_N^h\right)$ and cooling $\left(T_N^c\right)$, on thermomagnetic diagrams

MA	ST temperature	Heating				Cooling
		T_N^h	T_s^{DMA}	T_f^{DMA}	$\Delta E'$	T_N^c
$\%_{vol}$	K	K	K	K	GPa	K
0_MA	973	709	578	605	3	533
	1373	609	533	564	4	563
10_MA	973	709	601	614	1	597
	1373	623	565	600	3	592
20_MA	973	669	502	542	12	588
	1373	609	513	545	12	588
30_MA	973	624	518	556	2	573
	1373	579	566	592	4	578
40_MA	973	679	436	548	19	663
	1373	649	523	539	3	633

Fig. 5 Training effect by free-recovery SME at 40_MA_1000 lamellar specimen: **a** at the beginning of the first cycle; **b** at the end of the fifth cycle

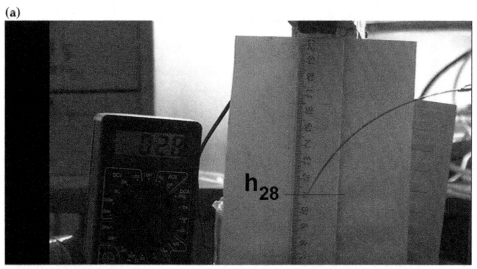

- Free-recovery shape memory effect was obtained at specimen 40_MA_1000 (46.8% α′-bcc, 25.5% ε-hcp and 27.7% γ-fcc), which developed a stroke of 24 mm with a rate of 1.7 mm/s, after 5 training cycles.
- It has been shown that, in spite of the large amount of α′-bcc martensite, considered detrimental for shape memory effect, a fair free-recovery SME can be obtained by corresponding thermomechanical training.

Acknowledgements This research was supported by UEFISCDI through project code PN II-PT - PCE-2012-4-0033, Contract 13/2013. Ministry of Research and Innovation, NUCLEU programme, PN 16370201 project is highly acknowledged.

References

1. Sato A et al (1982) Shape memory effect in γ ↔ ε transformation in Fe-30Mn-1Si alloy single crystals. Acta Metall 30(6):1177–1183
2. Sato A, Soma K, Mori T (1982) Hardening due to pre-existing ε-martensite in an Fe-30Mn-1Si alloy single crystal. Acta Metall 30(10):1901–1907
3. Sato A et al (1984) Orientation and composition dependencies of shape memory effect in Fe-Mn-Si alloys. Acta Metall 32(4):539–547
4. Murakami M, Suzuki H, Nakamura Y (1987) Effect of Si on the shape memory effect of polycrystalline Fe-Mn-Si alloys. Trans ISIJ 27:B-87
5. Murakami M, et al (1987) Effect of alloying content, phase and magnetic transformation on the shape memory effect of Fe-Mn-Si alloys. Trans ISIJ, 27:B-88
6. Murakami M, Otsuka H, Matsuda S (1987) Improvement of shape memory effect of Fe-Mn-Si alloys. Trans. ISIJ, 27:B-89
7. Otsuka H et al (1990) Effects of alloying additions on Fe-Mn-Si shape memory alloys. ISIJ Int 30:674–679
8. Moriya Y et al (1991) Properties of Fe-Cr-Ni-Mn-Si (Co) shape memory alloys. J Phys IV France 01:433–437
9. Maki T (1998) Ferrous shape memory alloys. Shape memory materials. In: Otsuka K, Wayman CM (eds). University Press, Cambridge, pp 117–132
10. Sawaguchi T et al (2016) Design concept and applications of FeMnSi-based alloys from shape-memory to seismic response control. Mater Trans 57(3):283–293
11. Dunne D (2012) Phase transformations in steels, diffusionless transformations, high strength steels, modelling and advanced analytical techniques. In: Pereloma E, Edmonds DV (eds), vol 2. Woodhead Publishing, pp 83–125
12. Arruda GJ, Buono VTL, Andrade MS (1999) The influence of deformation on the microstructure and transformation temperatures of Fe-Mn-Si-Cr-Ni shape memory alloys. Mat Sci Eng A 273–275:528–532
13. Wen YH, Li N, Xiong LR (2005) Composition design principles for Fe-Mn-Si-Cr-Ni based alloys with better shape memory effect and higher recovery stress. Mat Sci Eng A 407:31–35
14. Qin Z, Yu M, Zhang Y (1996) Néel transition and γ ↔ ε transformation in polycrystalline Fe-Mn-Si shape memory alloys. J Mater Sci 31:2311–2315
15. Chen S et al (1999) Effect of f.c.c. antiferromagnetism on martensitic transformation in Fe–Mn–Si based alloys. Mat Sci Eng A 264:262–268
16. La Roca P et al (2016) Composition dependence of the néel temperature and the entropy of the magnetic transition in the fcc phase of Fe-Mn and Fe-Mn-Co alloys. J Alloy Compd 688:594–598
17. Guerrero LM et al (2017) Composition effects on the fcc-hcp martensitic transformation and on the magnetic ordering of the fcc structure in Fe-Mn-Cr alloys. Mater Des 116:127–135
18. Bahador A et al (2017) Mechanical and superelastic properties of laser welded Ti–Ni shape-memory alloys produced by powder metallurgy. J Mater Proces Tech 248:198–206
19. Mazzer EM et al (2017) Effect of dislocations and residual stresses on the martensitic transformation of Cu-Al-Ni-Mn shape memory alloy powders. J Alloy Compd. https://doi.org/10.1016/j.jallcom.2017.06.312
20. Xu Z, Hodgson MA, Cao P (2015) A comparative study of powder metallurgical (PM) and wrought Fe-Mn-Si alloys. Mat Sci Eng A 630:116–124
21. Zhang Z et al (2003) Characterization of intermetallic Fe-Mn-Si powders produced by casting and mechanical ball milling. Powder Technol 137:139–147
22. Liu T et al (1999) Mechanical alloying of Fe-Mn and Fe-Mn-Si. Mat Sci Eng A 271:8–13
23. Oro R et al (2014) Effect of processing conditions on microstructural features in Mn-Si sintered steels. Mater Char 95:105–117
24. Saito T, Kapusta C, Takasaki A (2014) Synthesis and characterization of Fe-Mn-Si shape memory alloy by mechanical alloying and subsequent sintering. Mat Sci Eng A 592:88–94
25. Bujoreanu LG, Stanciu S, Özkal B, Comăneci RI, Meyer M (2009) Comparative study of the structures of Fe-Mn-Si-Cr-Ni shape memory alloys obtained by classical and by powder metallurgy respectively. ESOMAT, p 05003
26. Pricop B et al (2015) A study of martensite formation in powder metallurgy Fe-Mn-Si-Cr-Ni shape memory alloys. Mater Today Proc 2S:S789–S792
27. Pricop B et al (2010) Mechanical cycling effects at Fe-Mn-Si-Cr-Ni SMAs obtained by powder metallurgy. Phys Procedia 10:125–131
28. Pricop B et al (2011) Mechanical alloying effects on the thermal behaviour of a Fe-Mn-Si-Cr-Ni shape memory alloy under powder form. Optoelectron Adv Mater 5(5):555–561
29. Pricop B et al (2012) Thermal behavior of mechanically alloyed powders used for producing an Fe-Mn-Si-Cr-Ni shape memory alloy. J Mater Eng Perform 21(11):2407–2416
30. Pricop B, et al (2015) Powder metallurgy and mechanical alloying effects on the formation of thermally induced martensite in an FeMnSiCrNi SMA. In: Matec web of conferences, vol 33, p 4004
31. Pricop B et al (2013) Influence of mechanical alloying on the behaviour of Fe-Mn-Si-Cr-Ni shape memory alloys made by powder metallurgy. Mater Sci Forum 738–739:237–241
32. Spiridon IP et al (2013) The influence of heat treatment atmosphere and maintaining period on the homogeneity degree of a Fe-Mn-Si-Cr-Ni shape memory alloy obtained through powder metallurgy. J Optoelectron Adv M 15(7–8):730–733
33. Pricop B et al (2016) Structural changes caused by high-temperature holding of powder shape memory alloy 66% Fe—14% Mn—6% Si—9% Cr—5% Ni. Met Sci Heat Treat 57(9–10):553–558
34. Söyler AU, Özkal B, Bujoreanu LG (2014) Improved shape memory characteristics of Fe-14Mn-6Si-9Cr-5Ni alloy via mechanical alloying. J Mater Eng Perform 23:2357–2361

35. Bujoreanu LG (2015) Development of shape memory and superelastic applications of some experimental alloys. J Optoelectron Adv M 17(9–10):1437–1443

36. Mihalache E et al (2017) Structural effects of thermomechanical processing on the static and dynamic responses of powder metallurgy Fe-Mn-Si based shape memory alloys. Adv Sci Tech 97:153–158

37. Söyler AU, Özkal B, Bujoreanu LG (2010) Sintering densification and microstructural characterization of mechanical alloyed Fe-Mn-Si based powder metal system. TMS Suppl Proc 3:785–792

38. Söyler AU, Özkal B, Bujoreanu LG (2011) Investigation of mechanical alloying process parameters on Fe-Mn-Si based system. TMS Suppl Proc 1:577–583

39. Pricop B et al (2014) Influence of mechanically alloyed fraction and hot rolling temperature in the last pass on the structure of Fe-14Mn-6Si-9Cr-5Ni (mass%) shape memory alloys processed by powder metallurgy. Optoelectron Adv Mat 8(3–4):247–250

40. Spiridon I-P et al (2016) A study of free recovery in a Fe-Mn-Si-Cr shape memory alloy. Met Sci Heat Treat 57(9–10):548–5528

41. Bujoreanu LG et al (2008) Influence of some extrinsic factors on the two way shape memory effect of electric actuators. J Optoelectron Adv M 10(3):602–606

42. Xing L et al (2000) Study of the paramagnetic-antiferromagnetic transition and the $\gamma \rightarrow \varepsilon$ martensitic transformation in Fe-Mn alloys. J Mater Sci 35:5597–5603

43. Sawaguchi T et al (2008) Effects of Nb and C in solution and in NbC form on the transformation-related internal friction of Fe–17Mn (mass%) alloys. ISIJ Int 48(1):99–106

Critical Analyses on the Instrumented Ultramicrohardness Results on Aging NiTi Alloy in Distinct Phase Fields

R. S. Teixeira, A. S. Paula, F. S. Santos, P. F. Rodrigues, and F. M. Braz Fernandes

Abstract

A study was carried out on a Ni-rich NiTi alloy solubilized at 850 °C and subjected to two different treatments: annealing for 0.5 h at 500 °C and rolling at room temperature followed by annealing for 0.5 h at 500 °C. In both cases, measurements of instrumented ultramicrohardness showed that the treatment promoted a superelastic effect.

Keywords

NiTi alloy • Instrumented ultramicrohardness
Phase transformation • Mechanical properties

Introduction

The shape memory effect (SME) is defined as recovery of the original shape when a material deformed in the martensitic phase is heated above the temperature of transformation to the austenitic phase. On the other hand, the superelastic effect (SE) is defined as recovery of the original shape when a material deformed in the austenitic phase is converted to martensite at constant temperature and the stress responsible for the deformation is removed [1, 2].

NiTi alloys may exhibit different phases at room temperature, depending on chemical composition and processing history: austenite (B2), martensite (B19' and R) and metastable (Ni_4Ti_3, Ni_3Ti_2) and stable (Ni_3Ti) precipitates in the case of Ni-rich and equiatomic alloys [3].

Thermal and thermomechanical treatments affect the characteristics of SME and SE in NiTi alloys by changing the temperatures of transformation. Superelastic behavior is possible only when the critical stress for conversion to martensite is below the critical stress for dislocation slip in austenite. This can be achieved by thermal or thermomechanical treatments [4–8].

Instrumented ultramicrohardness tests have been widely used to evaluate the mechanical behavior of metallic, ceramic, polymeric and biological materials [9] and seem to be a more accurate method than conventional hardness tests to study NiTi alloys exhibiting SME and SE [10].

R. S. Teixeira · A. S. Paula (✉)
Seção de Engenharia Mecânica e de Materiais, Instituto Militar de Engenharia (IME), Praça General Tibúrcio 80, Urca, Rio de Janeiro, RJ, Brazil
e-mail: andersan@ime.eb.br

R. S. Teixeira
e-mail: rodolfoteixeira@ime.eb.br

F. S. Santos
Departamento de Engenharia Química e de Materiais, Pontifícia Universidade Católica (PUC), Rua Marquês de São Vicente 225, Gávea, Rio de Janeiro, RJ, Brazil
e-mail: santosfabiana.dasilva@gmail.com

P. F. Rodrigues · F. M. Braz Fernandes
CENIMAT/i3N, Faculdade de Ciências e Tecnologia, Departamento de Ciência dos Materiais, Universidade Nova de Lisboa, 2829-516 Caparica, Portugal
e-mail: pf.rodrigues@campus.fct.unl.pt

F. M. Braz Fernandes
e-mail: fbf@fct.unl.pt

Methodology

The material used in this work consists of Ni-rich NiTi alloys, produced by Vacuum Induction Melting (VIM) [11]. The samples were heated to 850 °C for 1 h and immediately transferred from the furnace to a mill to be hot rolled by multiple passes (without reheating between passes) in order to reduce the thickness from 18 to 1.6 mm. After that, the samples were subjected to a solubilization heat treatment at 850 °C for 1 h and quenched in water at 20 °C. Subsequently, a group of samples was aged at 500 °C for 0.5 h (SA) and another group was rolled at room temperature up to a thickness reduction 18% and then aged at 500 °C for 0.5 h (SRA).

Differential scanning calorimetry (DSC) measurements were performed in a DSC 204 F1 Phoenix calorimeter. The

© The Minerals, Metals & Materials Society 2018
A. P. Stebner and G. B. Olson (eds.), *Proceedings of the International Conference on Martensitic Transformations: Chicago*,
The Minerals, Metals & Materials Series, https://doi.org/10.1007/978-3-319-76968-4_17

thermal cycling was performed from 150 °C to −150°, with a heating/cooling rate 10.0 °C/min.

X-ray diffraction patterns at 20 °C were recorded in a PANalytical X'PERT PRO MPD diffractometer with copper anode, operated at 45 kV and 40 mA.

The Instrumented Ultramicrohardness analysis was made at 20 °C with a Vickers indenter (Shimadzu DUH-211S) using two different levels of maximum load (1 gf/10 mN and 20 gf/200 mN).

For the XRD and Instrumented Ultramicrohardness analyses, the samples were measured after being subjected to two different treatments: heated at 100 °C in boiling water for 5 min and cooled to 20 °C (labeled as preheated); and cooled in liquid nitrogen for 5 min and then warmed to 20 °C (labeled as precooled).

Results and Discussion

The DSC results were used to determine the phase compositions of the samples which would be submitted to XRD and instrumented ultramicrohardness analysis. For the samples SA (Fig. 1) and SRA (Fig. 2), after cooling to 20 °C, the R-phase should be present and possibly some minor traces of B2. Heating from −150 to 20 °C, the B19' phase is still present.

It is observed (Figs. 1 and 2) that the temperature at, which the XRD and ultramicrohardness measurements were performed is near the end of the first peak on cooling and the beginning of the first peak on heating. This is an indication that on cooling the transformation B2 → R has not been completed. In heating, the reverse transformation from B19' starts at a lower temperature.

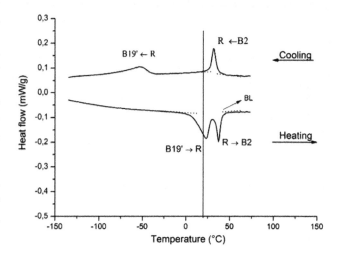

Fig. 2 DSC results showing the phase transformations of sample SRA (BL: Baseline)

Based on the works of Fernandes [12] and Paula [13], that revealed that the DSC technique has limitations to detect transformation events which involve low values of heat flow, it is expected that there are also some limitations to detect overlapped phase transformations.

According to the XRD results (Fig. 3), B2, R and B19' phases and Ni_4Ti_3 precipitates are present in SRA sample at 20 °C, in both precooled and preheated conditions. In the case of the SA sample (Fig. 4), B2 and R phases and Ni_4Ti_3 precipitates are present in both precooled and preheated conditions.

In order to evaluate the superelastic behavior of the alloy under study, analyses were carried out by instrumented ultramicrohardness. In Figs, 5 and 6 the curves obtained in these tests are presented.

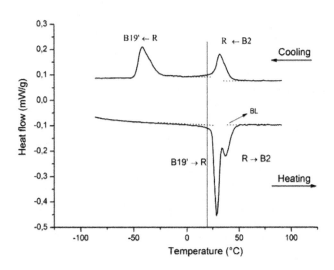

Fig. 1 DSC results showing the phase transformations of sample SA (BL: Baseline)

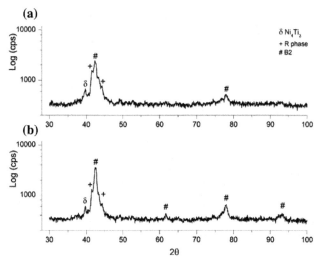

Fig. 3 X-Ray analysis conducted at controlled room temperature (20 °C)—**a** precooled; **b** preheated—Sample SA

(a)

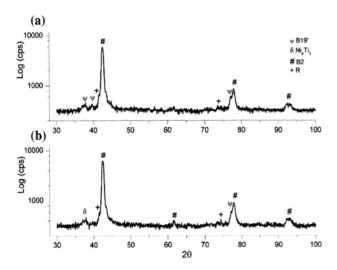

(b)

Fig. 4 X-Ray analysis conducted at controlled room temperature (20 °C)—**a** precooled; **b** preheated—Sample SRA

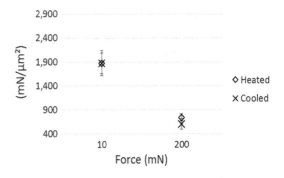

Fig. 5 Plastic dynamic hardness (DHV-2)—Sample SA

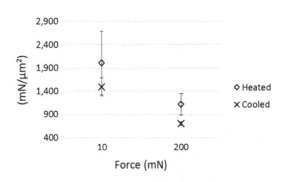

Fig. 6 Plastic dynamic hardness (DHV-2)—Sample SRA

The formula to obtain the plastic dynamic hardness DHV-2 values is DHV-2 $= a \cdot F/(h_r)^2$, where a is a constant which depends on the indenter shape, F is the test force and h_r is the intersection of the tangent line with the unloading curve [14].

The values of DHV-2 for SA (Fig. 5) and SRA (Fig. 6) samples were lower for a precooled condition than for the preheated condition except for the sample SA for a load of 1 gf = 10 mN, where the values for precooled and preheated conditions are the same. Considering that the slopes of the curves in this test were higher for the preheated condition, this influences the value of h_r, making it smaller. The value of DHV-2 is an indication of the superelastic behavior in these alloys, considering that lower values of intersection of the tangent line with the unloading curve (h_r) represents higher values of DHV-2, as a consequence of higher elastic return. When comparing the DHV-2 values for maximum loads associated with 1 gf = 10 mN and 20 gf = 200 mN, the values are approximately 2–4 times higher for the lower load. This is reasonable, since hardness is the result of the ratio of the maximum load to the square of the depth h_r. The mean values of dynamic plastic hardness (DHV-2) for the SRA sample showed values with significant heterogeneity, but with a mean close to that of the SA samples.

The formula to obtain the indentation modulus (E_{it}) is $1/E_r = (1 - v^2)/E_{it} + (1 - v_i^2)/E_i$, where $Ap = 24.50 \times hc^2$ (Vickers indenter), $hc = hmax - 0.75(hmax - hr)$ (depth of contact of the indenter), Er: Young's modulus of system, Ei: Young's modulus of indenter (1.14×106 N/mm^2), v_i: Poisson's ratio of indenter (0.07), Eit: Young's modulus on sample, v: Poisson's ratio of sample, $S = dP/dh = 2 \cdot Er \cdot$

$Ap^{0.5}/\pi^{0.5}$, S: stiffness of the upper portion of unloading data and Ap: Projected area of contact [14].

The values of the indentation modulus are associated with the elastic return capacity of the material, which is closely related to the Young modulus. Considering this characteristic, the values of the mean indentation modulus (E_{it}) for SA (Fig. 7) and SRA (Fig. 8) samples were lower for the precooled condition than for the preheated condition. This is justified by the lower slope of the unloading curves, that is, lower stiffness S for the precooled condition for all loads and for the load condition 20 gf = 200 mN when compared to the load of 1 gf = 10 mN. The higher this value, the greater the elastic recovery capacity of the material. This is reinforced by the behavior of h_c, which was generally larger for precooled samples, because h_c is directly related to elastic stiffness and has an inverse relationship with the indentation modulus (E_{it}). When the SRA and SA samples were compared, the mean values of Eit were lower for the SA sample for both conditions.

The formula to obtain the indentation hardness (H_{it}) values is $H_{it} = F_{max}/A_p$ where F_{max} is the maximum force, A_p is the projected area of contact between the indenter and the test piece. $A_p = 24.50 \times h_c^2$, where h_c is the depth of contact of the indenter with the test piece, calculated as follows: $h_c = h_{max} - \varepsilon(h_{max} - h_r)$ where ε depends on the indenter

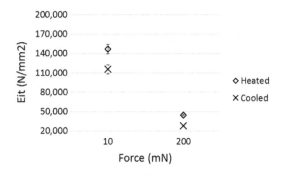

Fig. 7 Indentation modulus (E_{it})—Sample SA

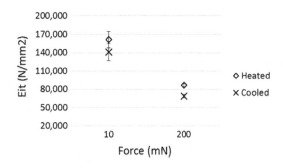

Fig. 8 Indentation modulus (E_{it})—Sample SRA

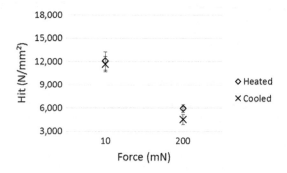

Fig. 9 Indentation hardness (H_{it})—Sample SA

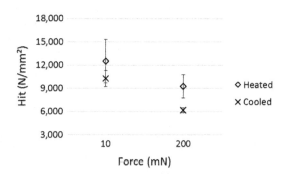

Fig. 10 Indentation hardness (H_{it})—Sample SRA

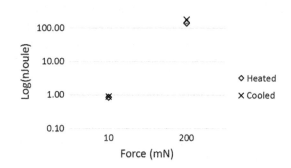

Fig. 11 Total work—Sample SA

geometry Vickers indenter and Triangular indenter $\varepsilon = \frac{3}{4}$, h_r is derived from the force-displacement curve and is the intercept of the tangent to the unloading cycle at F_{max} with the displacement axis [14].

The mean values of H_{it} for SA (Fig. 9) and SRA (Fig. 10) samples were lower for the precooled condition. This hardness is related to the beginning of plastic deformation, that, in qualitative terms, can be related to the yield strength for advanced deformation when using highest maximum load for the material under study [14]. From this point of view, one must first analyse the values of H_{it} associated with the maximum load. This would explain the fact that precooled samples in both load conditions have lower values of H_{it}, since the stress associated with reorientation of the R-phase variants is lower than for the B19' phase and much larger than that associated to the stress-induced transformation from B2 [1]. With respect to lower H_{it} values for loads of 20 gf = 200 mN when compared to 1 gf = 10 mN, it would be justified by the higher values of depth of contact of the indenter (h_c), which is inversely proportional to H_{it}, because at this loading level an advanced deformation level is possibly associated with hardening of the region deformed and transformed by penetration of the indenter, which results in a

lower elastic recovery capacity. When the SRA and SA samples were compared, the H_{it} values were lower for the SA sample, except for 1 gf = 10 mN load in the precooled condition.

Figures 11, 12, 13, 14 show the calculated values of the areas below the loading (*total work* − W_{total}) and unloading (*elastic working* − $W_{elastic}$) curve [14]. In all cases, lower values are observed for the work associated with a maximum load of 1 gf = 10 mN than with a maximum load of 20 gf = 200 mN, due to smaller areas in the initial stages of

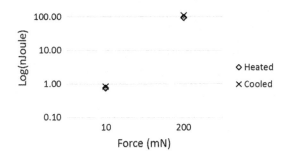

Fig. 12 Total work—Sample SRA

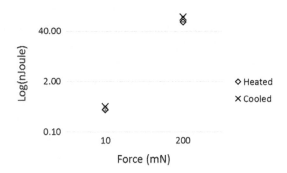

Fig. 13 Elastic work—Sample SA

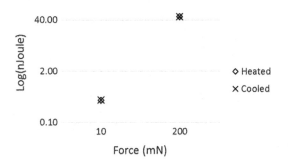

Fig. 14 Elastic work—Sample SRA

material deformation. When the SRA and SA samples were compared, the values of total elastic work were lower for the SRA sample.

Conclusions

When evaluating superelastic behavior through instrumented ultramicrohardness, the following conclusions could be drawn. Both samples exhibited a significant elastic work (lattice relaxation and reverse transformation) that could be related to superelastic behavior, more evident for tests performed at 1 gf due to the small amount of plastic work (plastic deformation) relative to total work.

Based on the 1 gf maximum load and the data collected in the curves, the properties that were portrayed in this work, together with the microstructural characteristics of the samples, suggest that the ASE3 (preheated condition) exhibited better superelastic behavior, besides the absence of the B19' phase and high indentation modulus (E_{it}) and indentation hardness (H_{it}) values and lower plastic dynamic hardness (DHV-2) that were associated with high elastic recovery.

The instrumented ultramicrohardness test in conjunction with microstructural data may be a tool for a prior analysis of superelastic behavior in these alloys.

Acknowledgements P. F. Rodrigues and F. M. Braz Fernandes acknowledge the funding of CENIMAT/I3N by COMPETE 2020, through FCT, under the project UID/CTM/50025/2013.P. F. Rodrigues acknowledges the funding of CAPES (CsF/Brazil – BEX 11943-13-0). R. S. Teixeira, F. S. Santos and A. S. Paula acknowledge the funding of CNPq (master scholarship and research productivity scholarship PQ-2 – Process 307798/2015-1).The authors thank Professor Jorge Otubo for donating the starting materials that were produced in the Instituto Tecnológico da Aeronáutica (São José dos Campos, SP – Brazil).The authors thank Professors Ladário da Silva and José Augusto Oliveira Huguenin which allowed the use of the Laboratório Multiusuário de Caracterização de Materiais do Instituto de Ciências Exatas da Universidade Federal Fluminense (ICEx/UFF), in Volta Redonda/RJ - Brazil, for the use of the instrumented ultramicrohardness equipment for analysis of the samples.The authors acknowledge the funding of European Program IRSES - Marie Curie PIRSES GA-2013-612585 (under the MIDAS project – "Micro and Nanoscale Design of Thermally Actuating Systems").

References

1. Otsuka K, Wayman CM (1998) Shape memory materials. Cambridge University Press
2. Laplanche G, Kazuch A, Eggeler G (2015) Processing of NiTi shape memory sheets—microstructural heterogeneity and evolution of texture. J Alloy Compd 651:333–339
3. Jiang S-Y, Zhang Y, Zhao Y, Liu S, Hu L, Zhao C (2015) Influence of Ni_4Ti_3 precipitates on phase transformation of NiTi shape memory alloy. Trans Nonferrous Metals Soc China (English Edition) 25:4063–4071
4. Aboutalebi MR, Karimzadeh M, Salehi MT, Abbasi SM, Morakabati M (2015) Influences of aging and thermomechanical treatments on the martensitic transformation and superelasticity of highly Ni-rich Ti-51.5 at.% Ni shape memory alloy. Thermochim Acta 616:14–19
5. Saedi S, Turabi AS, Andani MT, Haberland C, Karaca H, Elahinia M (2016) The influence of heat treatment on the thermomechanical response of Ni-rich NiTi alloys manufactured by selective laser melting. J Alloys Compd 677:204–210
6. Karbakhsh Ravari B, Farjami S, Nishida M (2014) Effects of Ni concentration and aging conditions on multistage martensitic transformation in aged Ni-rich Ti-Ni alloys. Acta Materialia 69:17–29
7. Khaleghi F, Khalil-Allfi J, Abbasi-Chianeh V, Noori S (2013) Effect of short-time annealing treatment on the superelastic behavior of cold drawn Ni-rich NiTi shape memory wires. J Alloys Compd 554:32–38
8. Mirzadeh H, Parsa MH (2014) Hot deformation and dynamic recrystallization of NiTi intermetallic compound. J Alloy Compd 614:56–59

9. Vanlandingham MR (2003) Rev Instrum Indentation 108:249–265
10. Gall K, Juntunen K, Maier HJ, Sehitoglu H, Chumlyakovy Y (2001) Instrumented micro-indentation of NiTi shape-memory alloys. Acta Materialia 49:3205–3217
11. Oliveira GCM (2010) Caracterização Microestrutural e Mecânica da Liga NiTi com EMF Produzida em Forno de Indução a Vácuo. 2010. Dissertação (Mestrado em Engenharia Aeronáutica e Mecânica - Área de Física e Química dos Materiais Aeroespaciais) - Instituto Tecnológico de Aeronáutica, São José dos Campos
12. Fernandes FMB, Paula AS, Canejo J, Mahesh KK, Silva RJC (2004) Kinetics characterization of martensitic transformation on Ti-Rich NiTi SMA. In: Proceedings of the international conference on shape memory and superelastic technologies, 3–7 Oct, Kurhaus Baden-Baden, Baden-Baden, Germany
13. Paula AS, Mahesh KK, Santos CML, Canejo JPHG, Fernandes FB (2006) One- and two-step phase transformation in Ti-rich NiTi shape memory alloy. Int J Appl Electromagnet Mech 23:25–32
14. Dynamic ultra-micro hardness tester instruction manual (2009)

Effect of Thermomechanical Treatment on Mechanical Properties of Ferromagnetic Fe-Ni-Co-Ti Alloy

Anatoliy Titenko, Lesya Demchenko, Larisa Kozlova, and Mustafa Babanli

Abstract

The work studies the effect of different thermomechanical treatment regimes of Fe-Ni-Co-Ti alloy on its microstructure and mechanical properties. The characteristic temperatures of martensitic transformation, the mechanical characteristics, such as microhardness, ultimate tensile strength and ductility of martensite and austenite phases have been determined. All these parameters non-monotonously vary with the increase of plastic deformation degree. Plastic deformation under drawing results in stabilization of austenite.

Keywords

Iron alloy • Thermomechanical treatment
Martensite transformation • Deformation
Aging • Tensile properties • Thermoelasticity

Introduction

The functional ferromagnetic shape memory materials attract an increasing attention of researchers due to their unusual deformation behavior resulting in a reversible size change under the influence of temperature, applied external stresses,

magnetic fields and combination of all the above, demonstrating the shape memory effect (SME), pseudoelasticity or superelasticity (SE), plasticity of transformation, magneto-elastic deformation, etc. The aforementioned properties have various practical applications as actuators in robotics, automotive, aerospace and biomedical industries.

The phenomenon of pseudoelasticity consists in an ability of the material to restore its original shape after a plastic deformation that significantly exceeds elastic one. Mainly, copper (Cu-Al-Ni, Cu-Zn-Al, etc.) and nickel (NiTi) based pseudoelastic functional materials were studied and found practical applications.

The investigation of deformations induced by external stresses was also carried out for ferromagnetic iron-based alloys, among which Fe-Pd, Fe-Pt and Fe-Ni-Co-Ti are of the greatest concern.

This research is devoted to the study of mechanical behavior of ferromagnetic Fe-Ni-Co-Ti alloy in various structural states.

The composition of a ferromagnetic Fe-Ni-Co-Ti alloy determines the character of martensitic transformation (MT) and the alloy properties. An increase of nickel and titanium content reduces the martensite transformation start temperature (M_s) and decreases the $\gamma \leftrightarrow \alpha$ transformation hysteresis width. Adding the titanium causes the alloy solid solution decomposition during dispersion hardening, while an increase of cobalt content results in growing the Curie temperature (T_c) of austenite as well as reducing the volume effect and elasticity modulus [1, 2].

The general criteria for MT occurrence in metals, as follows from [3–5], include: (1) the presence of a low driving force of transformation and an insignificant value of shear deformation; (2) a crystallographic reversibility of MT; (3) a reversibility of the lattice defect motion; (4) the formation of self-accommodation complexes in martensite crystals.

To get thermoelastic MT in iron-nickel alloys the following conditions [1, 5–8] should be met: (1) a high degree of martensite crystal lattice tetragonality (c/a); (2) a high

A. Titenko · L. Kozlova
Institute of Magnetism, National Academy of Sciences and Ministry of Education and Science of Ukraine, 36-B, Vernadskoho Blvd, Kiev, 03142, Ukraine
e-mail: titenko@imag.kiev.ua

L. Kozlova
e-mail: kozlova@imag.kiev.ua

L. Demchenko (✉)
National Technical University of Ukraine "Kyiv Polytechnic Institute", 37, Peremohy Ave, Kiev, 03056, Ukraine
e-mail: lesyademch@gmail.com

M. Babanli
Azerbaijan State University of Oil and Industry, Azadliq Ave., 20, Baku, AZ1010, Azerbaijan
e-mail: mustafababanli@yahoo.com

© The Minerals, Metals & Materials Society 2018
A. P. Stebner and G. B. Olson (eds.), *Proceedings of the International Conference on Martensitic Transformations: Chicago*,
The Minerals, Metals & Materials Series, https://doi.org/10.1007/978-3-319-76968-4_18

yield strength of γ and α—phases to preserve an interface coherence during the martensite crystals nucleation and their growth; (3) a magnetic ordering which helps reduce the volume effect ($\Delta V/V^{\gamma \leftrightarrow \alpha}$) of MT.

The 1st, 2nd conditions are achieved by precipitation of γ'-phase nanoparticles of the $(CoNi)_3Ti$ composition under aging, and the 3rd condition are realized as a result of a large volume magnetostriction of austenite below T_c. The above-mentioned factors contribute to the reduction of temperature hysteresis of MT in alloys with particles as compared to single-phase alloys and cause the change in the kinetics of MT from nonthermoelastic to thermoelastic.

It should be noted that austenite in Fe-Ni-Co-Ti alloy is disordered, however, during its aging the ordered thin coherent $L1_2$—type nanoparticles precipitate, which favor MT and are responsible for its crystallographic reversibility as well as undergo no spontaneous MT at cooling [9].

The phenomena of martensitic inelasticity [10, 11] could be observed in SME, SE and effect of transformation plasticity. The mechanisms of martensitic inelasticity can vary depending on loading temperature and deformation degree. The main structural mechanisms for reversible shape change in the alloys with thermoelastic MT are: (1) martensite formation with preferable orientation under the external load; (2) martensite reorientation; (3) martensite crystals detwinning and monodomainization.

The reduction of MT thermal hysteresis width (ΔT), which is equivalent to the increase of thermoelasticity, in polycrystalline alloys of Fe-Ni-Co-Ti type, is achieved by lowering the austenite phase shear modulus, which creates conditions for the reduction of the elastic energy of martensite crystals coherently conjugated to austenitic matrix. High thermoelasticity occurs due to austenite strengthening, usually as a result of deformation or dispersion hardening, where the role of precipitated particles is to significantly increase the resistance to dislocation movement by sliding, what is accompanied by strength improvement of the high-temperature phase [8–10].

The most common way to increase the strength and plastic properties of ferrous alloys (steels) is a thermomechanical treatment (TMT), which combines plastic deformation in the austenitic state followed by quenching and annealing. The main result of preliminary tensile strain is the decrease of M_s, as considerable as high the deformation degree is, with MT being activated by stresses arising in the material under deformation, and austenite phase stabilization being associated with crystal structure distortion resulting in strengthening. The influence of hydrostatic pressure (p) on temperature of phase transition (T_t) is described as the derivative: $\frac{\partial T_t}{\partial p}$ which, in the case of these alloys, is negative and results in the decrease of M_s and T_c [12, 13].

Another concept of nanostructured alloys hardening which allows gaining a high ductility of iron alloys was developed in [14] based on three steps: (1) a selection of the alloy chemical composition for the lattice softening, (2) a severe plastic deformation using high torsion stresses for structure treatments, and (3) annealing for nanotwinned structure smoothing. As a result, these alloys have high strength (up to 2.3–3 GPa) and elongation (about 10–20%) before a rupture [14].

This work studies the effect of preliminary plastic deformation followed by quenching and then aging of austenite on the characteristics of thermoelastic martensite transformation as well as on the elastic deformation behavior of martensite and austenite phases to develop regimes of thermomechanical treatment (TMT) and to improve the mechanical properties of Fe-Ni-Co-Ti shape memory alloys.

Materials and Methods

The polycrystalline ferromagnetic Fe-27.2%Ni-17.4% Co-5.2%Ti (wt%) alloy was chosen as a subject of this investigation. The alloy was produced by way of induction melting in an inert atmosphere and pouring into a copper mould. The TMT of samples was conducted by multiple drawing through dies with the diameter ranging from 20 mm to 5 mm at room temperature, followed by cold water quenching from 1373 K (1100 °C) and then aging at 923 K (650 °C) for 5, 10, 20 min. As a result of multiple drawing operations, the samples in the form of rods having various compression degrees from 3.8 to 75.3% were obtained. The degree of plastic deformation (ψ) at drawing was defined by the below formula:

$$\psi = \frac{F_0 - F_1}{F_0} = \frac{d_0^2 - d_1^2}{d_0^2} = 1 - \frac{d_1^2}{d_0^2}, \qquad (1)$$

where d_0 and d_1 are rods diameters before and after drawing; F_0 and F_1 are respective cross-section areas of rods before and after drawing.

The characteristic temperatures (M_s, M_f, A_s, A_f) and thermal hysteresis width (ΔT) of MT, the Curie temperature (T_c) of austenite and relative values of volume fraction of martensite were determined from temperature dependences of low-field magnetic susceptibility $\chi(T)$ and electrical resistance measured by four-point method. The ΔT of transformation was determined from the difference of temperatures corresponding to the middle of direct and reverse transformation. The Curie point was defined as the temperature at the intersection point of tangent to the sloping part of the $\chi(T)$ curve and straight line extrapolated from ferromagnetic low-temperature phase curve section.

The microhardness (H_V) of investigated samples was measured by Vickers method with loading on indenter of 0.49 N. The austenite grain size (d) was determined by optical metallography as an average of ten measurements. The fractographic investigations were performed using scanning electron microscopy (SEM). The alloy structural-phase state was investigated directly by metallography, and indirectly by simultaneous measurements of uniaxial tension and low-field magnetic susceptibility at fixed temperatures.

Mechanical characteristics such as yield strength (σ_y), ultimate tensile strength (σ_{ut}), strain at rupture (ε_{max}) of the alloy austenitic and martensitic phases were studied under the uniaxial tension with the deformation rate of $\approx 4 \times 10^{-4} c^{-1}$ at room temperature (300 K) and liquid nitrogen temperature (77 K). Their values were obtained by averaging the results of three measurements using a tensile testing machine. The testing specimens were of cylindrical shape with the diameter of 0.5 mm and working part length of 10 mm. Prior to test the samples after TMT were electropolished to remove surface strain.

Results and Discussion

Microhardness and Austenite Grain Size

The dependences of H_V and d values on preliminary plastic deformation degree ψ are shown in Fig. 1. The microhardness of quenched samples was about 2.5 GPa and its initial grain size was 40 μm. The plastic deformation by drawing results in the decrease of grain size from 25 μm at $\psi = 3.8\%$ to 8 μm at $\psi = 59.5\%$, while the microhardness is 3 GPa for 3.8% deformation and remains practically constant over the strain range up to 40%, reaching its maximum value of 3.8 GPa at 59.5%.

Aging at 923 K for 10 min results in the austenite grain size growth up to their initial value of 40 μm due to the recrystallization, that is accompanied by dispersion hardening with the γ'–$(Ni,Co)_3Ti$ phase particles. At the same time, after such annealing of deformed (with $\psi = 59.5\%$) samples, the austenite grain size sharply decreases down to 15 μm, but the microhardness remains practically the same. It is approximately about 4 GPa. The ageing lasting 20 min leads to the appearance of a maximum on the $H_V = f(\psi)$ dependence and it should be noted that such aging eliminates the impact of a high degree deformation on H_V.

The photographs showing a grain structure of samples in the drawing plane and in the plane perpendicular to the drawing direction at $\psi = 22.5\%$ are presented in Fig. 2a, b. According to optical microscopy data, the appearance of a large number of twin boundaries related to annealing twins [15] at low deformation degrees is observed. This can affect

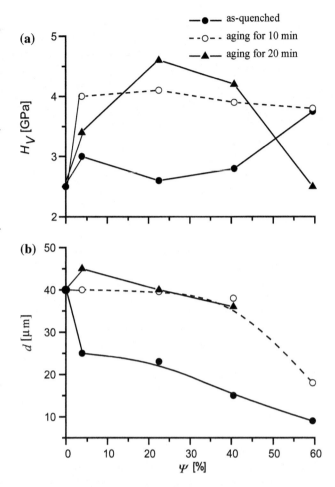

Fig. 1 The dependences of microhardness, H_V (**a**) and austenite average grain size, d (**b**) on preliminary plastic deformation degree (ψ) of quenched samples after the aging for different time

the following simultaneously occurring processes of solid solution decomposition and recrystallization. With the increase of deformation degree, the process of dislocation density growth becomes dominant as compared to twinning that can substantially modify the processes of solid solution decomposition and recrystallization, resulting in smoother change of mechanical properties of the alloy (Figs. 1a, 4). Whereas the MT temperature decrease caused by the deformation increase can be explained by the decrease of austenite grain size, the increase of T_c is the most probably a result of composition homogenization over the sample volume.

The Characteristics of Phase Transformations

According to the results of the present work, MT is not observed in quenched and plastically deformed samples of studied Fe-Ni-Co-Ti alloy within the whole temperature range. There is only one anomaly caused by the

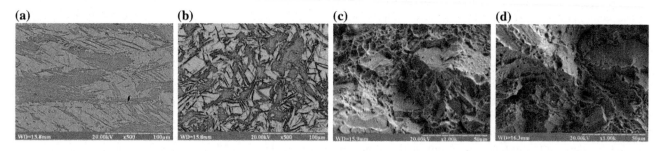

Fig. 2 The microstructures of samples (deformed with $\psi = 22.5\%$) in the drawing plane (**a**) and in the plane perpendicular (**b**) to drawing direction as well as fractographs of fractured surface (**c, d**) after a rupture at room temperature

ferromagnetic ordering of austenite on the temperature dependence of low-field magnetic susceptibility. As a result of aging, austenite becomes unstable in respect to MT, and a temperature hysteresis loop caused by the transformation appears on the temperature dependences of magnetic susceptibility. It should be noted that M_s temperature increases and T_c decreases as the aging time extends. Such tendencies are observed for cold-deformed and then aged samples (see Fig. 3a, c). It is a usual behavior when the temperature

hysteresis of the $\gamma \leftrightarrow \alpha$ transformation decreases with the increase of the aging time (Fig. 3b) at $\psi = 0\%$. Depending on ψ value (Fig. 3b), the hysteresis achieves a minimum at low (3.8–7.4%) ψ values.

According to the susceptibility anomalies, the relative volume fraction of formed martensite grows with the increase of the aging time by over a quarter for each ψ value. It is worth noting a quite significant increase in the ferromagnetic ordering temperature with the ψ growth (Fig. 3c). The compression with the degree of $\psi = 75.3\%$ completely suppresses MT, and transformation does not occur down to the temperature -197 °C.

Tensile Properties

The mechanical behavior of samples before their fracture was studied for ultimate strength at uniaxial tension. The tests were carried out at room temperature (in austenitic state) and liquid nitrogen temperature (tempered martensite state with residual austenite for the aged samples). The graphs of changes in the ultimate tensile strength (σ_{ut}) and plasticity (ε_{max}) of the alloy as a function of the preliminary deformation degree for relevant time of subsequent annealing are presented in Fig. 4. It should be noted that only as-quenched samples had quite evident plasticity (Fig. 4e), so the yield strength $\sigma_y = f(\psi)$ dependences (Fig. 4a) are given only for these samples. The ultimate tensile strength and yield stress of quenched austenite increase almost twice with the increase of the preliminary deformation up to 55%; if the temperature decreases down to 77 K (martensite state), the similar increase of σ_{ut} and σ_y makes up 25% more as compared to the austenite state (Fig. 4a). At the same time, plasticity sharply decreases (Fig. 4e). Aging of the quenched alloy causes a multiple increase in σ_{ut} and decrease in ε_{max}, respectively. Moreover, the mechanical properties of austenite and martensite become very similar (see Fig. 4b–d, f–h). Significant changes take place when austenite is coldly deformed prior to being quenched and aged. In this case, the curves of $\sigma_{ut}(\psi)$ and $\varepsilon_{max}(\psi)$ dependences for austenite and martensite first sharply diverge, and then converge with the

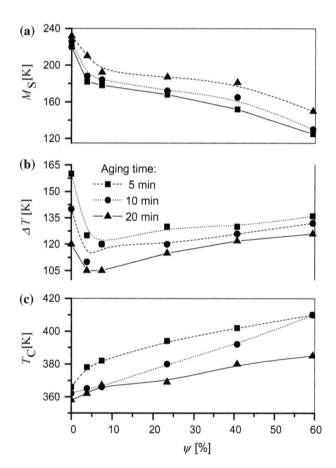

Fig. 3 The characteristics of martensitic and ferromagnetic transformations, such as M_s (**a**), ΔT (**b**), T_c (**c**), depending on plastic deformation degree (ψ) of quenched samples after the aging for different time

ψ growth. Figure 4 shows that the maximum plasticity of austenite with a sufficient strength corresponds to insignificant (7.4%) preliminary deformation and the shortest (5 min) aging time (Fig.4b, f).Martensite is always more durable than austenite, however more brittle, with a maximum of σ_{ut} at small deformation, considerably reducing with the aging time increase. The fractographs of the sample fractured surface (Fig. 2d), after a rupture at room temperature ($\psi = 22.5\%$), indicate the viscous character of destruction of the samples tested at room temperature.

In the case of quenched austenite, there is a possibility of quantitative comparison with the existing concepts. A monotonic increase in the yield strength (σ_y) depending on the deformation degree of quenched austenite can be attributed to the decrease of average grain size (d) according to the Hall-Petch relationship [16, 17]:

$$\sigma_y = \sigma_0 + k_y d^{-1/2} \qquad (2)$$

where σ_0 is a material constant for the starting stress for dislocation movement, the lattice friction stress, k_y is the strengthening coefficient (a constant specific to each material).

Figure 4i shows that the relation (2) can be well satisfied for the following parameters $\sigma_0 = 27$ MPa and $k_y = 2.36$ MN/m$^{3/2}$. The dependences of MT characteristics and Curie point obtained in the present work for quenched (unstrained, with $\psi = 0\%$) austenite on the aging time well correspond to the previously observed ones for other Fe-Ni-Co-Ti alloys [1, 8, 18].

Moreover, the increase in the number of precipitated particles during the aging contributes to the decrease of elastic energy of coherent martensite crystal in the austenite matrix. This in turn causes the rise of M_s of the direct MT (Fig. 3a) and the decrease of ΔT (Fig. 3b) and is in line with the results of [19–22]:

$$\Delta T / T_0 = E_m (1 - \alpha) / L, \qquad (3)$$

where T_0 is the temperature of the equilibrium between austenitic and martensitic phases, E_m is the elastic energy of coherent martensite crystal in the austenite matrix, L is the MT heat, α is a coefficient which depends on strength characteristics of the material and determines the degree of coherent stress relaxation.

Depending on the chosen TMT regime, the following tendencies are observed: the increase in strength characteristics with the increase of the aging time at the same deformation degree (ψ) and the increase of strength characteristics with the increase of ψ at the same aging time, which is explained by relevant change in the austenite grain size after the chosen treatment.

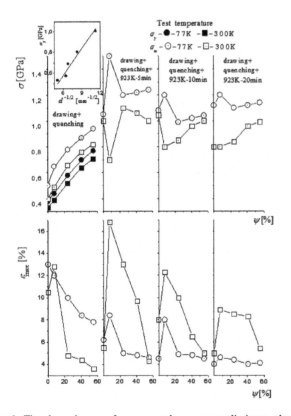

Fig. 4 The dependences of σ_y, σ_{ut} and ε_{max} on preliminary plastic deformation degree (ψ) of the samples: as-quenched (**a, e**) and then aged for 5 (**b, f**), 10 (**c, g**) and 20 min (**d, h**); the Hall-Petch dependence (**i**)

Conclusion

Introduction of a new treatment, namely, cold deformation before quenching of austenite followed by ageing non-monotonically changes mechanical properties of Fe-Ni-Co-Ti alloy austenite and martensite, as well as the martensite transformation characteristics. The investigations of MT characteristics and elastic deformation behavior of samples at different temperatures lead to the following conclusions:

1. The increase of the preliminary plastic deformation degree (ψ) up to 59.5% results in the gradual decrease of the austenite grain size and MT characteristic temperatures. At the same time, the transformation hysteresis temperature increases by 40–70 K. The transformed volume fraction slightly varies up to $\psi = 40.7\%$, and then sharply decreases with further ψ growth.

2. The experimentally obtained ultimate tensile strength $\sigma_{ut}(\psi)$ dependences show that the strong austenite strain

hardening occurs with the increase of preliminary deformation ψ, while the strain before fracture ε_{max} decreases 1.5–2 times. At the same time, the strain hardening reduction with 2–4 time decrease of ε_{max} is observed for the martensite phase.

3. The high average ultimate tensile strength (more than 1 GPa) of austenite and martensite is caused by strain and precipitation hardening. According to the optical microscopy investigations, an appearance of a large number of twin boundaries and, consequently, elastic stresses in grain volume is observed at insignificant deformations, which can affect the subsequent simultaneously occurring processes of solid solution decomposition and recrystallization.

4. An optimal complex of the mechanical properties of Fe-27.2%Ni-17.4%Co-5.2%Ti alloy was obtained as a result of TMT, consisting of drawing ψ = 3.8–22.5% and subsequent aging at 923 K (650°C) for 5–10 min, which contributes to the achievement of high strength values at high level of material ductility.

Acknowledgements This research was supported by the laboratories of the Institute of Magnetism and the National Technical University of Ukraine "KPI" of the National Academy of Sciences of Ukraine and the Ministry of Education and Science of Ukraine.

References

1. Kokorin VV (1987) Martensitic transformations in heterogeneous solid solutions. Naukova Dumka, Kiev (in Russian)
2. Maki T, Furutani S, Tamura I (1989) Shape memory effect related to thin plate martensite with large thermal hysteresis in ausaged Fe-Ni-Co-Ti alloy. ISIJ Int 29:438–445
3. Christian JW (1975) Theory of phase transformations in metals and alloys. Pergamon Press, New York
4. Tong HC, Wayman CM (1974) Characteristic temperatures and other properties of thermoelastic martensites. Acta Met 22:887–896
5. Otsuka K, Wayman, CM (1998) Shape memory materials. In: Otsuka K, Wayman CM (ed), Cambridge Univ Press, Cambridge
6. Hornbogen E, Jost N (1991) Alloys of iron and reversibility of martensitic transformations. J Phys IV France, European symposium on martensitic transformation and shape memory properties, 01(C4):199–210
7. Jost N, Escher K, Donner P, Sade M, Halter K, Hornbogen E (1990) Steels with shape memory. Wire 40:639–640
8. Jost N (1990) Reversible transformation and shape memory effects due to thermomechanical treatments of Fe-Ni-Co-based austenites. Mater Sci Forum 56–58:667–672
9. Maki T, Kobayashi K, Minato M, Tamura I (1984) Thermoelastic martensite in an ausaged Fe-Ni-Ti-Co alloy. Scripta Met 18:1105–1109
10. Maki T, Namura I (1986) Shape memory effect in ferrous alloys. In: Proceedings of the international conference on martensitic transformation, Japan, pp 963–970
11. Kokorin, VV, Chernenko VA (1997) Reversion stress evaluation in Fe-Ni-Co-Ti alloys. In: Proceedings of the second international conference on SMST pp. 119–124
12. Kakeshita T, Shimizu K (1997) Effects of hydrostatic pressure on martensitic transformations. Mater T JIM 38:668–681
13. Zakrevskiy IG, Kokorin VV, Chernenko VA (1989) Baroelastic martensitic transformation in Fe-Ni-Co-Ti alloy. Sov Phys 34:73–74
14. Edalati K, Toh S, Furuta T, Kuramoto S, Watanabe M, Horita Z (2012) Development of ultrahigh strength and high ductility in nanostructured iron alloys with lattice softening and nanotwins. Scripta Mater 67:511–514
15. Titenko AN, Kozlova LE, Chernenko VA (2001) The influence of degree of preliminary plastic deformation and annealing time on mechanical properties of Fe-Ni-Co-Ti shape memory alloy. Metallofiz Nov Tekh 23:1513–1524 (in Russian)
16. Petch NJ (1953) The cleavage strength of polycrystals. J Iron Steel Inst 174:25–28
17. Hall EO (1951) The deformation and ageing of mild steel: III. Discussion of result, P Phys Soc B 64:747–753
18. Kokorin VV, Kozlova LE, Titenko AN, Perekos AE, Levchuk YuS (2008) Characteristics of thermoelastic martensitic transformation in ferromagnetic Fe-Ni-Co-Ti alloys alloyed with Cu. Phys Met Metallogr 105:598–601
19. Kozlova LE, Titenko AN (2006) Stress-induced martensite transformation in polycrystalline aged Cu-Al-Mn alloys. Mat Sci Eng A 438–440:738–742
20. Titenko AN, Demchenko LD (2012) Superelastic deformation in polycrystalline Fe-Ni-Co-Ti-Cu alloys. J Mater Eng Perform 21:2011–2206
21. Titenko A, Demchenko L (2016) Effect of annealing in magnetic field on ferromagnetic nanoparticle formation in Cu-Al-Mn alloy with induced martensite transformation. Nanoscale Res Lett 11:237
22. Titenko A, Demchenko L, Perekos A, Gerasimov O (2017) Effect of thermomagnetic treatment on structure and properties of Cu-Al-Mn alloy. Nanoscale Res Lett 12:285

Reversed Austenite for Enhancing Ductility of Martensitic Stainless Steel

Sebastian Dieck, Martin Ecke, Paul Rosemann, and Thorsten Halle

Abstract

Quenching and partitioning (Q&P) heat treatment increases the deformability of high-strength martensitic steels. Therefore, it is necessary to have some metastable austenite in the microstructure, which transforms in martensite during plastic deformation (TRIP effect). The austenitic-martensitic microstructure is gained by an increased austenitization temperature, water quenching and additional partitioning. The partitioning enables local carbon diffusion, which stabilizes retained austenite and leads to partial reversion of martensite to austenite. The influence of partitioning time was studied for the martensitic stainless steel AISI 420 (X46Cr13, 1.4034). In line with these efforts, metallographic, XRD and EBSD measurements were performed to characterize the microstructural evolution. The mechanical properties were tested using tension and compression loading. Additional corrosion investigations showed the benefits of Q&P heat treatment compared to conventional tempering. The reversion of austenite by the partitioning treatment was verified with EBSD and XRD. Furthermore, the results of the mechanical and corrosion testing showed improved properties due to the Q&P heat treatment.

Keywords

Reversed austenite • Quenching and partitioning Stainless steel • Corrosion resistance

Introduction

The usability of products is defined by the material properties. Whenever a product is used at its application limit, a further development, including material development and material fabrication, is necessary to exceed these limitations [1]. An optimised heat treatment offers a wide range of property enhancements regarding advanced steels.

The novel concept of "Quenching and Partitioning" (Q&P) heat treatment is known to enhance the formability of high-strength steels [2, 3]. It is a two-step heat treatment, which is illustrated in Fig. 1a. At first, the steel is austenitized to get a homogeneous distribution of alloying elements in an austenitic matrix, and quenched to a temperature between martensite start (M_s) and martensite finish (M_f) to retain a certain amount of austenite. During the directly following second step, the Partitioning, the material is tempered on low temperature, so thermal activated carbon (C) diffusion takes place. In this process, the retained austenite enriches with C due to its higher C solubility, which results in stabilising the retained austenite during the cooling down to room temperature (RT) [4–6]. A peculiarity of this stabilised austenite is quality of strain-induced transformation into martensite (TRIP effect), which results in enhanced strength and formability [7, 8]. Whereas previous investigations on Q&P were focussed on low alloyed steels, there are recent studies of applying Q&P on martensitic stainless steel. The steel X46Cr13 is an especially promising candidate because M_f is below RT [9]. For applying Q&P, there is no special equipment needed, such as a special furnace.

Furthermore, a combination of very high strength ($\sim 1,800$ MPa) and good strain-to-rupture ($\sim 20\%$) are

S. Dieck (✉) · M. Ecke · P. Rosemann · T. Halle
Otto-von-Guericke-University Magdeburg, Institute of Materials and Joining Technology, Magdeburg, Germany
e-mail: sebastian.dieck@ovgu.de

M. Ecke
e-mail: martin.ecke@ovgu.de

P. Rosemann
e-mail: paul.rosemann@bam.de

T. Halle
e-mail: thorsten.halle@ovgu.de

P. Rosemann
Department 7.6 Corrosion and Corrosion Protection, BAM Federal Institute for Materials Research and Testing, Berlin, Germany

© The Minerals, Metals & Materials Society 2018
A. P. Stebner and G. B. Olson (eds.), *Proceedings of the International Conference on Martensitic Transformations: Chicago*,
The Minerals, Metals & Materials Series, https://doi.org/10.1007/978-3-319-76968-4_19

Fig. 1 Q&P process—
a schematic heat treatment
diagram [11], **b** diffusion paths of
C during partitioning of X46Cr13
(red arrows), **c** resulting
microstructure
(martensite = white, stabilised
austenite = red) [9]

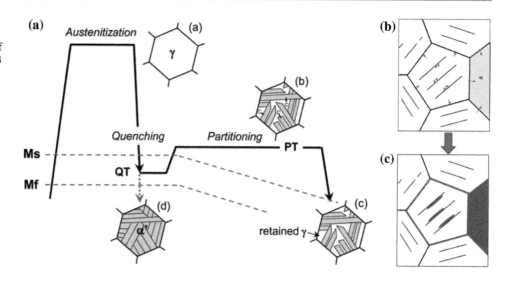

possible [10]. The mechanical behaviour is effected by the combination of TRIP and an increased fraction of stabilised austenite. The latter is formed during partitioning on small and large angle grain boundaries (see Fig. 1b). Consequently, there is a local transformation of martensite to austenite, the so-called reversed austenite (revA), in these C-enriched zones (see Fig. 1c) [10]. Additionally, small carbides precipitate during the partitioning, which increases the strength due to precipitation hardening [9, 11]. This work aims to gain a better understanding of the properties of X46Cr13 after an appropriate Q&P heat treatment, including microstructure, mechanical properties and corrosion behaviour of this martensitic stainless steel.

Materials and Methods

The martensitic stainless steel X46Cr13 (AISI 420, 1.4043) was used for the investigations of the Q&P heat treatment. The chemical composition of the cold rolled and soft annealed initial state is shown in Table 1.

The selection of heat treatment parameters was based on the investigations of Yuan and Raabe [10]. The material was austenitized at 1150 °C (T_A). The time of austenitizing (t_A) had to be enlarged to 15 min because of the large carbides (~ 10 μm) within the raw material. Quenching was performed in water and the partitioning temperature (T_P) was 400 °C. The partitioning time (t_P) was varied (0 min—as-quenched, 5 and 30 min) to investigate the time-dependent forming of revA and its influence on the properties. The mechanical behaviour was examined by tension and compression testing (referred to DIN 50125 and

DIN 50106). The microstructure was studied using optical microscopy (OM), X-ray diffraction (XRD) and electron backscatter diffraction (EBSD). Therefore, sample preparation included grinding, polishing and etching. Corrosion behaviour was investigated with the KorroPad test, which detects stainless steel surfaces with a high susceptibility to pitting corrosion by a colour change reaction [12, 13]. This test allows furthermore to visualize the effect of various heat treatment parameters on the pitting corrosion resistance of martensitic stainless steels [14–16]. All samples were grinded (P180) and passivated for 24 h at 95% relative humidity before the KorroPad-test. Afterwards, the KorroPad's were applied on sample surfaces for 15 min. Pitting corrosion results in blue colouring, which can be compared for different heat treatment states.

Results and Discussion

Mechanical Properties

The influence of partitioning time on the mechanical behaviour of X46Cr13 can be seen in Fig. 2. The as-quenched state shows brittle fracture under tensile loading before reaching the yield strength. Higher strength and formability can be seen after 5 min of partitioning at 400 °C, and when t_P is raised up to 30 min (Fig. 2b), a tensile strength of 1,700 MPa and a total elongation of 13% are reached. The different elastic behaviour of the tensile specimen is in conflict to literature [17] and can be ascribed to experimental setup.

Table 1 Chemical composition of X46Cr13 in ma.−%

Element	C	Cr	Mn	Si	Ni	Mo	N	P	S
Content	0.45	13.95	0.65	0.34	0.12	0.02	0.02	0.012	<0.001

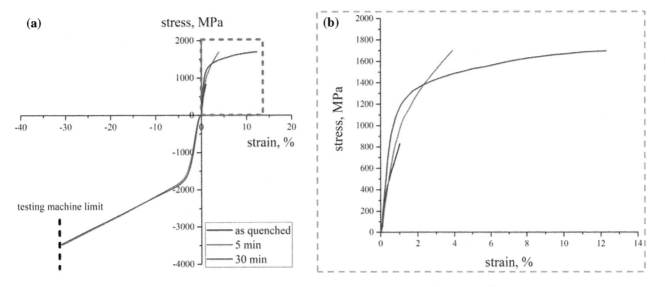

Fig. 2 Results of mechanical testing: **a** combined graph for tension and compression testing, **b** tension testing

In contrast to tensile loading, a similar behaviour of 5 and 30 min partitioned specimen is achieved under compression loading. It was not possible to determine maximum stresses or strains due to testing machine limits. Yet, before reaching the break-off criteria, it is obvious that the characteristic values (yield strength, maximum strength and formability) of compression loading exceed the properties of tension loading. However, expecting that plastic deformation is caused only by shear stresses, the materials behaviour should be the same for compression and tension [18]. But this is not the case for high-strength martensitic steels. There are experimental results which have shown that high-strength steels are able to tolerate higher stresses under compression than under tension, which is known as the strength differential effect (SD effect) [19, 20]. The reason for these loading direction–dependent behaviours is not clarified so far. An interaction of diverse effects is expected, although there is no consensus about the exact emphasis. Firstly, it is possible that micro cracking limits the possible deformation under tensile loading. Further possibilities are the Bauschinger

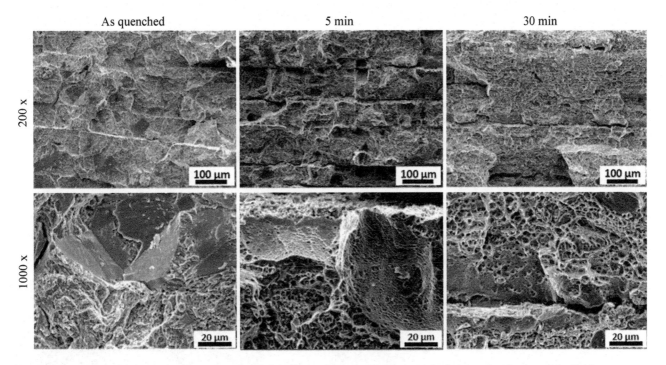

Fig. 3 SEM pictures (SE contrast) of fracture surfaces of tensile specimen; increasing t_P results in increasing fraction of ductile fracture

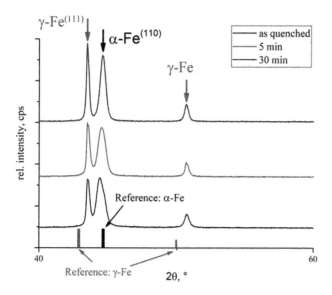

Fig. 4 X-ray diffraction diagram (2θ angle 40°–60°), higher partitioning times increase the austenite fraction

effect, interactions of interstitial atoms and dislocation moving as well as varying macro residual stresses [19–21]. Additionally, Ellermann and Scholtes determined for 100Cr6 that the TRIP effect is supressed under compression [21].

The latter could explain the high formability in the compression tests. Particularly, the magnitude of the SD effect, which is determined for martensitic steel (∼ 10%), is not comparable with the results of the present investigations. Furthermore Singh et al. reported that SD effect decreases with increasing tempering time [19]. All in all, more investigations are needed to explain the different mechanical behaviour under compression and tensile loading.

Figure 3 shows the fracture surfaces of tensile specimen. The as quenched sample shows the expected brittle fracture, whereas the partitioned specimen is characterised by mostly ductile fracture. An increasing of t_P leads to a higher fraction of ductile fracture, which correlates with the results of mechanical testing (Fig. 2). Additionally, all tested samples

have terraced fracture surfaces. Crack propagation seems to be located preferential in lines transverse to the tensile direction. The reason therefore will be explained in section "Microstructural Analysis".

Microstructural Analysis

The development of phase fractions due to Q&P is displayed in Fig. 4. The as-quenched state consists mostly of distorted α-Fe (martensite), expressed by wide α-Fe peaks, which are shifted compared to the reference. Furthermore, retained austenite is detected. Increasing t_P leads to a decrease of α-Fe width, which means the decrease of residual stresses caused by C depletion of martensite. The integral peak intensities of austenitic phase fraction increase with the partitioning time due to local carbon enrichment and the reversion of austenite.

EBSD measurements (Fig. 5) confirm the XRD results. The as-quenched state is characterised by embedded islands of retained austenite in a martensitic matrix. The 5 min partitioned sample shows additional small austenitic needles (revA) inside the martensite. The time-dependent growing of revA is proven by the comparison of 5 and 30 min partitioned samples. Referred to Yuan et al., revA forms on large angle grain boundaries, too [10], but due to the small size of some nanometres, it can't be detected by XRD and EBSD.

The EBSD pictures, especially of the 5 min sample, suggest a line-like distribution of retained austenite. This is proven by OM (Fig. 6a) and corresponds to the morphology of the fracture surfaces (Fig. 3). An investigation with SEM using BSE contrast and EDX analysis, see Fig. 6b and d, demonstrates the location of Cr rich carbide agglomerations inside the austenitic lines. These carbides are already existing within the raw material and the result of inappropriate fabrication, including hot and cold rolling as well as soft annealing. The austenitizing, respectively the chosen parameters in the context of Q&P, is insufficient because the carbide agglomerations could not be dissolved in the austenite. The partial solving of carbides leads to the

Fig. 5 EBSD measurements (bcc: red, fcc: blue), increasing t_P results in increasing austenite fraction

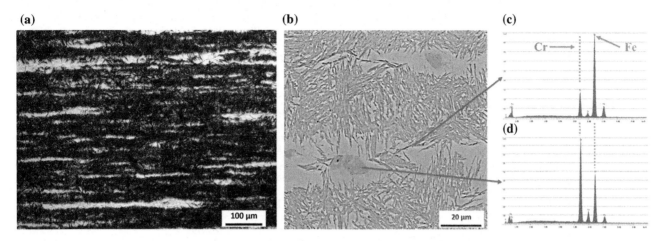

Fig. 6 Analysis of an austenitic line of a 5 min partitioned sample—**a** OM (etchant: Beraha II), **b** SEM picture (BSE contrast) and EDX analysis of **c** austenite as well as **d** dark phase (carbide)

remaining of austenite in the C enriched surrounding areas after the first quenching. For now it is not clear whether C diffusion during partitioning is sufficient to enable the TRIP effect under loading condition for these large austenitic fractions.

Corrosion Resistance

The applicability of stainless steels depends on the corrosion resistance, and the standard heat treatment for the steel X46Cr13, quenching and tempering (Q&T), is disadvantageous for pitting corrosion resistance. The as-quenched state shows no susceptibility to pitting corrosion due to the homogeneous Cr distribution. Thus, a stable passive layer forms on the surface and protects the steel. The conventional tempering (2 h between 400 and 650 °C) leads to high pitting corrosion susceptibility, as can be seen by the blue dots in the KorroPad in Fig. 7. This is caused by carbide forming and local Cr depletion, which destabilizes the passive layer. In contrast, there is no pitting corrosion detectable on Q&P heat treated samples. This is the result of lower T_P

and t_P during the partitioning, which inhibit carbide forming and Cr depletion.

Conclusion

The present work proves the applicability of the Q&P heat treatment for martensitic stainless steel X46Cr13. The investigations focussed on the influence of partitioning time on materials properties. Therefore, the mechanical behaviour under different loading directions was characterised as well as the microstructural evolution and the corrosion behaviour. The main results are as follows:

1. The as-quenched state is characterised by brittle fracture under loading. The partitioning treatment enhances the strength and the deformation behaviour. A partitioning time of 30 min enables tensile strength of 1,700 MPa and a maximum elongation of 13%.
2. The enhancement of mechanical behaviour is caused by carbon diffusion during partitioning, which stabilises

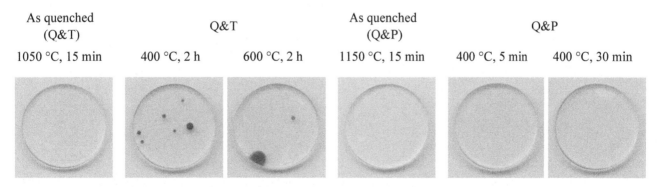

Fig. 7 Results of KorroPad testing—comparison of usual Q&T and Q&P

retained austenite and leads to the formation of reversed austenite.

3. A distinctive SD effect was detected, but the cause could not be clearly identified so far.
4. The corrosion resistance after Q&P is enhanced compared to common Q&T condition, because there is no Cr depletion due to partitioning treatment.

Furthermore, it was determined that the austenitizing is insufficient. Line-like carbide agglomerations, remaining from the raw material, lead to large lines of retained austenite. Under tensile loading, crack propagation initiates preferably along these lines, which is expected to decrease the deformability. An optimized raw material or an adjustment of the austenitizing could further enhance the mechanical behaviour by achieving a more homogeneous distribution of alloying elements.

Acknowledgements The authors would like to acknowledge financial support by the GKMM 1554. Further thanks to the Federal Institute of Materials Science and Testing (BAM) for corrosion testing. The authors are grateful to Mr. Sebastian Fritsch (Technical University Chemnitz) for performing the compression testing.

References

1. Dieck S, Baumann T, Hasemann G, Rannabauer S, Krüger M (2014) Magdeburg
2. Edmonds DV, He K, Rizzo FC, de Cooman BC, Matlock DK, Speer JG (2006) Mater Sci Eng A 438–440, S 25–34
3. Wang Li, Speer JG (2013) Metall Microstruct Anal 2:Nr 4, S 268–281
4. Speer J, Matlock DK, de Cooman BC, Schroth JG (2003) Acta Mater 51:Nr 9, S 2611–2622
5. Clarke AJ, Speer JG, Miller MK, Hackenberg RE, Edmonds DV, Matlock DK, Rizzo FC, Clarke KD, de Moor E (2008) Acta Mater 56:Nr 1, S 16–22
6. Santofimia MJ, Zhao L, Sietsma, J (2011) Metall Mater Trans A 42:Nr 12, S 3620–3626
7. Arlazarov A, Bouaziz O, Masse JP, Kegel F (2015) Mater Sci Eng A 620:S 293–300
8. Arlazarov A, Ollat M, Masse JP, Bouzat M (2016) Mater Sci Eng A 661:S 79–86
9. Dieck S, Rosemann P, Kromm A, Halle T (2017) IOP conference series: materials science and engineering, vol 181, pp S 12034
10. Yuan L, Ponge D, Wittig J, Choi P, Jiménez JA, Raabe D (2012) Acta Mater 60:6–7, S 2790–2804
11. Tsuchiyama T, Tobata J, Tao T, Nakada N, Takaki S (2012) Mater Sci Eng A 532:S 585–592
12. Burkert A, Klapper HS, Lehmann J (2013) Mater Corros 64:Nr 8, S 675–682
13. Lehmann J, Burkert A, Mietz J (2016) Mater Corros 67:Nr 1, S 84–91
14. Rosemann P (2017) 1. Auflage. Herzogenrath: Shaker (Berichte aus der Werkstofftechnik)
15. Burkert A, Lehmann J, Müller T, Bohlmann T (2014) Schlussbericht AiF Forschungsvorhaben 17136 N/1
16. Lehmann J, Burkert A, Steinhoff U-M (2012) Auflage für den Nachweis von korrosionsempfindlichen Metalloberflächen und Verfahren zum Nachweis von korrosionsempfindlichen Metalloberflächen. Bundesanstalt für Materialforschung und -prüfung (BAM). Anmeldenr. 102010037775, Deutschland. 29.03.2012. Deutschland. Veröffentlichungsnr. 102010037775. IPC G01 N 17/00
17. Lei Y (2012) Aachen: Shaker (Berichte aus der Materialwissenschaft)
18. Bridgman PW (1964) s.l.: Harvard University Press
19. Singh AP, Padmanabhan KA, Pandey GN, Murty GMD, Jha S (2000) J Mater Sci 35:Nr 6, S 1379–1388
20. Rauch GC, Leslie WC (1972) Metall Mater Trans B 3:Nr 2, S 377–389
21. Ellermann A, Scholtes B (2015) Mater Sci Eng A 620:S 262–272

Tough Ductile Ultra High Strength Steels Through Direct Quenching and Partitioning—An Update

Mahesh C. Somani, David A. Porter, Jukka I. Kömi, L. P. Karjalainen, and Devesh K. Misra

Abstract

The TMR-DQP* processing route comprising thermomechanical rolling followed by direct quenching and partitioning, has shown huge potential for the development of tough, ductile ultra-high-strength steels, both for structural and wear-resistant applications. The approach comprised designing suitable chemical compositions, establishing appropriate DQP processing conditions with the aid of physical simulation, and finally testing laboratory rolled DQP material with the emphasis on cost-effective process development, amenable for industrial hot strip production. Evaluation of DQP processed samples cooled slowly following DQP processing, thus simulating coiling, confirmed achieving the desired martensite-austenite microstructures and targeted mechanical properties. Ausforming in no-recrystallization regime (T_{nr}) resulted in extensive refining and randomization of the martensite packets/laths besides fine division of interlath austenite, thus resulting in an all-round improvement of mechanical properties. Preliminary investigations on alloys designed with 0.2 C have shown promising properties not only for structural applications, but also wear-resistance purposes.

Keywords

Direct quenching and partitioning • Thermomechanical processing • Martensite • Bainite • Austenite

Introduction

In general, quenching and tempering treatment may impart reasonable toughness and acceptable ductility in high-strength structural steels, but their uniform elongation (i.e., strain hardening capacity) is relatively low. In recent years, a novel concept of quenching and partitioning (Q&P) has been proposed as a potential processing route for improving the balance of elongation to fracture and tensile strength for advanced high-strength steels [1–3]. In the Q&P processing, the steel is austenitized, quenched to a temperature between the martensite start (M_s) and finish (M_f) temperatures, and held at a suitable temperature for a suitable time to allow the partitioning of carbon from martensite to austenite, which can thereby be partly or fully stabilized down to room temperature. The formation of iron carbides and the decomposition of austenite are intentionally suppressed by the use of Si, Al, or P alloying [1–3].

This study presents a recently developed novel processing route comprising thermomechanical rolling followed by direct quenching and partitioning (TMR-DQP) based on the physical and laboratory rolling simulation studies for improving the work hardening capacity and uniform elongation of high-strength hot-rolled structural steel [4, 5]. The specific aim of this study was to develop a steel with yield strength on the order of ~1100 MPa combined with good ductility and impact toughness. The approach involved designing suitable chemical compositions based on high silicon and/or aluminum contents, selecting appropriate DQP processing conditions with the aid of physical simulation on a Gleeble simulator, and, finally, testing DQP material processed on a laboratory rolling mill. This paper presents a brief account of the salient results encompassing the

M. C. Somani (✉) · D. A. Porter · J. I. Kömi · L. P. Karjalainen
University of Oulu, Centre for Advanced Steels Research,
P.O. Box 4200 90014 Oulu, Finland
e-mail: mahesh.somani@oulu.fi

D. A. Porter
e-mail: david.porter@oulu.fi

J. I. Kömi
e-mail: jukka.komi@oulu.fi

L. P. Karjalainen
e-mail: pentti.karjalainen@oulu.fi

D. K. Misra
Department of Metallurgical, Materials and Biomedical Engineering, University of Texas at El Paso, 500 W. University Avenue, El Paso, TX 79968-0521, USA
e-mail: dmisra2@utep.edu

© The Minerals, Metals & Materials Society 2018
A. P. Stebner and G. B. Olson (eds.), *Proceedings of the International Conference on Martensitic Transformations: Chicago*,
The Minerals, Metals & Materials Series, https://doi.org/10.1007/978-3-319-76968-4_20

dilatation behavior, hot rolling simulations, various microstructural mechanisms and mechanical properties.

Experimental Procedures

Table 1 shows the chemical compositions of the vacuum melted steels that have been used in this study together with their codes as used in this paper. Preliminary experiments were made using steels High-Si and High-Al, and steels A to D were designed based on the results of those experiments. Cylindrical specimens of dimensions Ø6 × 9 mm or Ø5 7.5 mm were cut for physical simulation on a Gleeble 1500 simulator. Two types of dilatation tests were made starting with either undeformed or deformed austenite prior to quenching to roughly simulate industrial rolling with high and low finish-rolling temperatures (FRT). For instance, in the case of High-Si steel, to conduct dilatation tests with deformed austenite, samples were compressed at 850 °C with three hits of 0.2 strain each at a strain rate of 1 s^{-1} with 25 s time interval after each hit, prior to cooling at 30 °C/s to a quenching temperature (T_Q) below M_s, giving initial martensite fractions of 75–85%, followed by partitioning at a temperature T_P ($\geq T_Q$) for a given time (Pt).

Laboratory rolling trials were made in a two-high, reversing laboratory rolling mill with blocks of 110 × 80 × 60 mm cut from cast ingots. Some examples of recorded rolling and cooling cycles during TMR-DQP processing are shown in Fig. 1. The samples were heated at 1200 °C for 2 h in a furnace prior to two-stage rolling. In the first stage, hot rolling was carried out in the recrystallization regime in 4 passes to a thickness of 26 mm with about 0.2 strain/pass with the temperature of the fourth pass at about 1030–1050 °C. The second stage comprised waiting for the temperature to drop to ≈900 °C and then rolling in the no-recrystallization regime to a thickness of 11.2 mm with 4 passes of about 0.21 strain/pass with FRT 800–820 °C. Immediately after rolling, the samples were quenched in a tank of water close to the desired T_Q and then subjected to partitioning treatment in a furnace at T_Q, which was switched off to give very slow cooling over 30 h to simulate industrial coiled strip cooling.

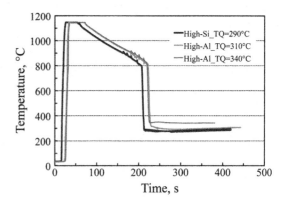

Fig. 1 Some examples of recorded rolling and cooling cycles during TMR-DQP processing

The rolled samples were characterized with respect to microstructures and mechanical properties. In addition, some samples from steels A–D were laboratory rolled and direct quenched to room temperature RT in order to be able to compare the mechanical properties with those of TMR-DQP processed samples.

Results and Discussion

Retained Austenite

A linear analysis of the Gleeble dilatation curves for High-Si steel indicates that quenching at 340, 320, and 290 °C should produce initial martensite fractions of 60, 75, and 84%, respectively [6, 7]. Likewise, for High-Al steel, quenching to 340, 310, and 290 °C gave initial martensite fractions of 70, 86, and 91%, respectively. Subsequent partitioning of Q&P specimens at T_P ($\geq T_Q$) for different durations, however, did not show any systematic variation in measured final fractions of retained austenite (RA) [4, 6, 7]. Figure 2 presents an example showing the variation of average RA (partitioned for 10–1000 s) with partitioning temperature for both strained and unstrained matrices in the case of High-Si steel quenched at different T_Q. Accordingly,

Table 1 Chemical compositions of experimental steels used for Q&P/TMR-DQP experiments

Steel code	C	Si	Mn	Al	Cr	Mo	Ni	Nb
High-Si	0.21	1.48	2.04		0.60			
High-Al	0.19	0.51	1.98	1.06	0.52	0.21		
Steel A	0.21	0.56	1.90	0.92	1.10			
Steel B	0.21	0.53	1.80	0.82	1.10			0.035
Steel C	0.22	0.55	1.50	0.85	1.20		0.79	
Steel D	0.22	0.53	1.50	0.84	1.10	0.16		
(O: 10–17 ppm, N: 5–13 ppm, S: 7–16 ppm)								

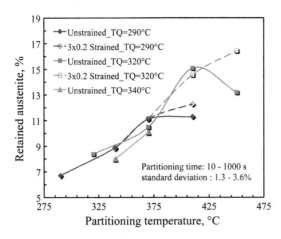

Fig. 2 Variation of average retained austenite (partitioned 10–1000 s) with partitioning temperature in the case of High-Si steel quenched at different T_Q temperatures

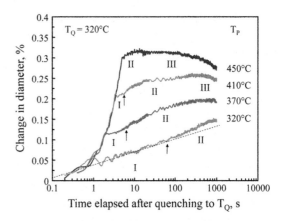

Fig. 3 Change in specimen diameter with time at different partitioning temperatures (T_P) for High-Si steel following quenching at $T_Q = 320\ °C$

the average RA contents for Q&P samples partitioned for 10–1000 s, as estimated by X-ray diffraction (XRD), varied in the range 7–15% and 9–16% for unstrained and strained conditions, respectively, generally increasing with increase in T_P in accord with the literature [3]. In the case of High-Al steel, the austenite contents varied in a narrow range of 7–9%, regardless of the T_P (= T_Q) or austenite state. Irrespective of the steel type and austenite state, the average carbon content of the Gleeble Q&P specimens (0.7–1.2%), is lower than would be expected, suggesting that not all carbon is partitioning to austenite.

It is expected that the amounts of retained austenite will generally be low in TMR-DQP samples, as partitioning temperatures would normally be ≤T_Q and for long durations in the case of coiled strips on a hot strip mill. Limited Q&P tests done on Steels A–D corresponding to $T_Q = T_P$ of 280 and 260 °C (≈80 and 90% martensite, respectively) did not show any significant influence of quench-stop temperature with respect to retained austenite at RT, which varied in a narrow range of 6–7% for unstrained condition. Further straining in austenite, however, resulted only in a marginal increase in the austenite content (7–8.5%). Except for the main peak, it was difficult to discern the other peaks for austenite and their positions; hence, the carbon content for these specimens could not be determined.

Microstructural Mechanisms During Partitioning

An example of gradual expansion during partitioning is depicted in Fig. 3 for High-Si steel, which shows how the specimen diameter varies with time up to 1000 s after reaching $T_Q = 320\ °C$. Specimens subjected to shorter holding times also showed similar trends. After 1 s at T_Q, heating to T_P occurs at 30 °C/s, which explains most of the expansion

immediately after 1 s. It has been shown that linear dilatation due to complete carbon partitioning will only be 0.027–0.033% [6, 7]. However, the isothermal growth of martensite together with carbon partitioning constitutes the expansion observed in region I of Fig. 3 [6–8]. Bainite forms in region II, while the contraction in region III is probably due to martensite tempering. No evidence for carbide precipitation in martensite was found in transmission electron microscopy (TEM) studies, indicating that carbon in the martensite laths is perhaps clustering. During the final cooling at about 20 °C/s, some bainite formation is possible in the case of high T_P, followed by the formation of untempered and in some cases twinned high carbon martensite at temperatures close to 200 °C [2]. Unlike in the case of High-Si steel, Steels A–D in the unstrained condition contract over ≈20 s [9]. This behavior has been observed for all the specimens held for 10–1000 s partitioning times, suggesting that the short period of contraction is connected with interface migration from austenite to martensite despite low mobility, as also modeled by Santofimia et al. [10]. The bidirectional movement, however, disappears in the case of prior strained austenite.

Microstructures of Dilatation Specimens

It has been shown earlier that straining in austenite at 850 °C prior to Q&P processing results in finer packets and blocks of fine martensite laths, shortened and randomized in different directions [4, 5]. Whereas, field-emission scanning electron microscopy—electron back scatter diffraction (FESEM-EBSD) examinations showed that the retained austenite might be present as large pools as well as interlath films, TEM examinations clearly depicted that the specimens contain highly dislocated lath martensite separated by thin films of austenite. Lower bainite and twinned and isothermal martensite were also revealed by TEM examinations [6, 7].

Evaluation of Rolled TMR-DQP Samples

Microstructural Features

The microstructural features of laboratory rolled DQP samples were quite similar to those seen in optical microstructures of Gleeble simulated specimens. Typical austenite contents as measured by XRD were in the range 6–9% and 4.5–7% for High-Si and High-Al steels, respectively. These values are somewhat lower than those obtained on Gleeble samples. Steels A–D processed through the TMR-DQP route followed by slow cooling also showed retained austenite in the range 4–10%, irrespective of the steel type and/or DQP parameters, generally akin with or marginally lower than those obtained on limited Gleeble specimens subjected to Q&P tests.

Figure 4 presents typical examples of microstructures recorded on TMR-DQP samples of Steel D subjected to slow cooling from T_Q in a furnace. These micrographs show dark austenite between the martensitic laths (Figs. 4a, c), more clearly seen as white films in dark field images (Figs. 4b, d). It can be seen in some locations that the original austenite films between martensite laths (appearing as dark in bright field images) developed a carbon concentration gradient during partitioning because of the slow diffusion as the temperature decreased during slow cooling. During final cooling, a section of the mid-rib region of original austenite transformed to untempered high C martensite, leaving thin austenite films between the tempered and untempered martensite laths. Unlike in the case of Gleeble samples, carbides could be seen in some martensite laths of TMR-DQP samples, suggesting loss of carbon due to tempering during slow cooling in a furnace. Similar features have been observed for other experimental steels subjected to coiling simulation.

Mechanical Properties

Table 2 presents a summary of mechanical properties for TMR-DQP samples compared to TMR-DQ samples. The 0.2% yield strength ($R_{p0.2}$) is marginally lower than the target level ($\approx$1100 MPa) in some High-Si, High-Al, and Steel B samples, but as the T_Q (=T_P) is lowered, the $R_{P0.2}$ seems to improve significantly and meet the target. Also, it is expected that temper rolling should be able to raise the yield strength to the required level (e.g., 1.0% proof strength ($R_{p1.0}$) values are well above 1300 MPa in most cases). Almost all the samples exhibited reasonable values of R_m ($\approx$1500 MPa and above). As expected, $R_{p0.2}$, $R_{p1.0}$, and R_m strengths generally increase with decrease in T_Q for all steels. While the total elongation (A) and reduction of area to fracture (Z) vary in a narrow range, except for DQ samples, the plastic component of the uniform elongation (A_g) is generally higher for all steels (3.2–4.9%), except for Steel A (2.7%), when compared to that of DQ samples (1.4–2.9%). Referring to Table 2, the RA fractions are only about 4–5% in DQP samples quenched at low temperature ($\approx$200 °C); hence, the corresponding hardness values are quite high (470–490 HV10), comparable to their DQ versions (495–505 HV10). Referring to the impact properties in Table 2, the temperatures corresponding to 28 J impact energy (T28J) are quite low and distinctly improved over that of the DQ samples. Both the alloying and T_Q seem to influence the T28J as well as the upper shelf energy (KV (US)), but there is no systematic trend. Notwithstanding the variation, it seems that both the T28J and KV(US) are quite high in the Ni variant (Steel C), particularly at low T_Q temperature ($\approx$200 °C). Nb and Mo variants (Steels B and D, respectively) also seem to hold promise at high T_Q (275–285 °C), with comparable tensile properties and T28J temperatures (about −126 °C) but with marginally lower KV (US) (64–76 J).

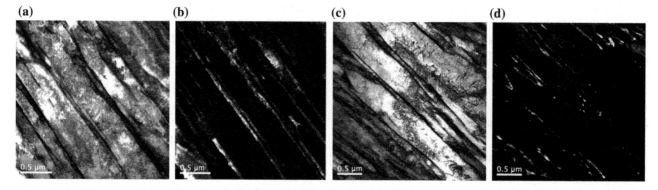

(a) **(b)** **(c)** **(d)**

Fig. 4 Typical examples of microstructures recorded on Steel D following TMR-DQP processing and slow cooling in furnace: (**a**, **b**) DQP 300 °C, (**c**, **d**) DQP 285 °C

Table 2 Tensile[a] and Charpy V impact properties of laboratory rolled and TMR-DQP/DQ plates. Hardness data and retained austenite (RA) fractions are also included

Steel code	T_Q/T_P (°C)	$R_{p0.2}$ (MPa)	$R_{p1.0}$ (MPa)	R_m (MPa)	A (%)	A_g (%)	Z (%)	T28J (°C)	KV(US) (J)	Hardness HV10	RA (%)
High-Si	290 CS	1029	1355	1505	12.8	3.9	54	−108	100	442	8.4
High-Al	320 CS	1020	1271	1393	11.1	3.7	48	−90	66	410	5.1
	260 CS	1180	1413	1495	9.9	3.2	46	−87	63	447	3.9
Steel A	270 CS	1206	1380	1541	13.0	2.7	56	−80	84	466	5.1
	205 CS	1107	1373	1540	12.1	4.2	50	−128	96	471	4.8
	DQ	1365	1580	1609	10.9	1.9	42	−43[b]	70[b]	495	ND
Steel B	275 CS	1040	1374	1577	12.1	4.4	44	−126	64	465	9.8
	205 CS	1188	1430	1548	11.6	4.2	45	−79	95	491	4.1
	DQ	1367	1526	1534	7.6	1.4	34	−17[b]	66[b]	501	ND
Steel C	265 CS	1130	1448	1560	11.8	3.5	56	−108	101	466	6.7
	210 CS	1096	1367	1551	13.4	4.5	57	−137	131	488	4.2
	DQ	1355	1623	1686	11.8	2.9	50	−63[b]	82[b]	503	ND
Steel D	285 CS	1148	1372	1631	14.0	4.9	58	−127	76	468	8.3
	DQ	1331	1608	1672	11.3	2.2	46	−52[b]	71[b]	504	ND

[a]Average of 3 tests according to EN-10002-I standard
[b]Corrected for 7.5 mm thickness
CS Coiling simulation
ND Not detectable

Summary and Conclusions

Based on preliminary Q&P simulations, a novel TMR-DQP processing route has been designed that should be applicable to industrial hot strip production. Metallographic studies clearly revealed presence of extended pools and laths of retained austenite between martensite laths. Retained austenite contents increased with increase in T_Q and T_P temperatures, though prior straining in austenite had only little effect. Dilatometer measurements and electron microscopy have clearly revealed that besides carbon partitioning, isothermal martensite and bainite form at or close to the partitioning temperature. In the case of Steels A–D, short periods of volume contractions seen in unstrained Q&P specimens at the start of partitioning suggest a possibility of bidirectional movement of the γ-α′ interfaces until equilibrium is reached. Laboratory rolling simulations suggest that even at relatively low quenching temperatures (T_Q 200 °C), most of the untransformed austenite could be retained (≈4–5%) at RT. High values of $R_{p0.2}$, $R_{p1.0}$, and R_m and hardness combined with improved A_g and A and remarkable values for T28J and KV(US) suggest that there is good potential for the TMR-DQP route not only for tough ductile structural steels but also for hard abrasion-resistant steels. The results also indicate that it is not the amount of austenite but presumably its size and distribution that decide its influence on uniform and total elongation and low-temperature impact toughness.

Acknowledgements The work was carried out under the auspices of Light and Efficient Solutions Program of FIMECC (Finnish Metals and Engineering Competence Cluster) Ltd. The authors gratefully acknowledge the financial support from Tekes (Finnish Funding Agency for Technology and Innovation) and SSAB Europe Oy (formerly, Rautaruukki Oyj).

References

1. Speer JG, Edmonds DV, Rizzo FC, Matlock DK (2004) Partitioning of carbon from supersaturated plates of ferrite, with application to steel processing and fundamentals of the bainite transformation. Current Opin. Solid State Mater. Sci 8:219–237
2. Li HY, Lu XW, Li WJ, Jin XJ (2010) Microstructure and mechanical properties of an ultrahigh-strength 40SiMnNiCr steel during the one-step quenching and partitioning process. Metall Mater Trans A 41A:1284–1300
3. De Moor E, Lacroix S, Clarke AJ, Penning J, Speer JG (2008) Effect of retained austenite stabilized via quench and partitioning on the strain hardening of martensitic steels. Metall Mater Trans A 39A:2586–2595
4. Somani MC, Porter DA, Karjalainen LP, Misra DK (2013) Evaluation of DQ&P processing route for the development of ultra-high strength tough ductile steels. Int J Metall Eng 2(2): 154–160

5. Somani MC, Porter DA, Karjalainen LP, Suikkanen PP, Misra DK (2014) Innovation and processing of novel tough ductile ultra-high strength steels through TMR-DQP processing route. Mater Sci Forum 783–786:1009–1014

6. Somani MC, Porter DA, Karjalainen LP, Misra DK (2012) On the decomposition of austenite in a high-silicon steel during quenching and partitioning. In: Proceedings of the international symposium recent developments in steel processing, MS&T'12, Pittsburgh, PA, MS&T Partner Societies, pp 1013–1020

7. Somani MC, Porter DA, Karjalainen LP, Misra DK (2014) On various aspects of decomposition of austenite in a high-silicon steel during quenching and partitioning. Metall Mater Trans A 45A:1247–1257

8. Kim D, Speer JG, De Cooman BC (2011) Isothermal transformation of a CMnSi steel below the Ms temperature. Metall Mater Trans A 42A:1575–1585

9. Somani MC, Porter DA, Kömi JI, Karjalainen LP, Misra DK (2017) Recent advances in TMR-DQP processing for tough, ductile high strength strip steels. In: Proceedings of the international symposium advanced high strength sheet steels, Keystone, Co, 30 May-02 June 2017, AIST, Warrendale, PA (in press)

10. Santofimia MJ, Speer JG, Clarke AJ, Zhao L, Sietsma J (2009) Influence of interface mobility on the evolution of austenite-martensite grain assemblies during annealing. Acta Mater 57:4548–4557

Physical Simulation of Press Hardening of TRIP Steel

Hana Jirková, Kateřina Opatová, Martin F.-X. Wagner, and Bohuslav Mašek

Abstract

Deformation-induced martensitic transformation is used for improving mechanical properties of AHS steels which contain metastable retained austenite. TRIP steels are one of the categories that fall into this group. Their microstructures consist of proeutectoid ferrite, bainite, and metastable retained austenite. Cold working causes retained austenite in these steels to transform to deformation-induced martensite. A technical complication to their treatment routes is the isothermal holding stage. At this stage, bainite forms and retained austenite becomes stabilized which is the key aspect of the process. A CMnSi-type low-alloy steel with 0.2% carbon was subjected to various experimental cooling sequences which represented press hardening operations at tool temperatures ranging from 500 °C to room temperature, followed by isothermal holding in the bainitic transformation region. By varying the cooling parameters, one can obtain a broad range of mixed martensitic-bainitic structures containing retained austenite, with strengths in the vicinity of 1300 MPa, and A_{20} elongation levels of 10%.

Keywords

Press hardening · TRIP steel · Retained austenite
Two step etching · EBSD

H. Jirková (✉) · K. Opatová · M. F.-X. Wagner · B. Mašek
University of West Bohemia, RTI - Regional Technological Institute, Univerzitní 22, 306 14 Pilsen, Czech Republic
e-mail: h.jirkova@email.cz

K. Opatová
e-mail: opatovak@rti.zcu.cz

M. F.-X. Wagner
e-mail: martin.wagner@mb.tu-chemnitz.de

B. Mašek
e-mail: masekb@kmm.zcu.cz

Introduction

TRIP steels are belong to the group of multiphase low-alloy high-strength steels. The desired properties of these steels are achieved not only by appropriate alloying but also by applying suitable heat treatment parameters [1]. The key aspects are, first, the rate of cooling from the soaking temperature to the bainitic transformation hold, because pearlite formation must be prevented and, second, the hold and the bainitic transformation temperature, during which retained austenite becomes stabilized thanks to carbon diffusion. Even in steels with 0.2% carbon and no other alloying elements than manganese and silicon, a correct choice of processing parameters can provide ultimate strengths of more than 800 MPa and tensile strains of about 20% [2, 3].

Press hardening of these steels enables high-precision sheet parts to be made, for instance in the automotive industry [4]. This method involves forming and subsequent quenching and offers high effectiveness [4]. Press hardening of high-strength steels finds use predominantly in the automotive industry, for making A-pillars, B-pillars, bumpers, roof rails, rocker rails and tunnels and other parts [5, 6]. The numbers of parts manufactured by this method have been increasing exponentially recently. These parts are also increasingly used outside the automotive sector. For these reasons, it was necessary to explore the possibility of press hardening of hot-drawn TRIP steels which had been cooled in the tool at the controlled temperature of stabilization of retained austenite and then cooled in air to room temperature (RT).

Experimental Programme

This experimental programme involved CMnSi steel with 0.2% carbon (Table 1). The alloying of this steel is typical of TRIP steels. In addition to carbon, the steel is alloyed with manganese and silicon. These elements have an important role in controlling phase transformations and stabilizing

© The Minerals, Metals & Materials Society 2018
A. P. Stebner and G. B. Olson (eds.), *Proceedings of the International Conference on Martensitic Transformations: Chicago*, The Minerals, Metals & Materials Series, https://doi.org/10.1007/978-3-319-76968-4_21

retained austenite. Silicon prevents carbide precipitation during bainite transformation and manganese retards the formation of pearlite [1, 7] Both elements strengthen steel by solid solution strengthening. The as-received experimental material was soft annealed sheet of 1.5 mm thickness. The initial microstructure consisted of ferrite and pearlite with a hardness of 180 HV10.

Determination of Phase Transformation Temperatures

Temperatures of phase transformations in this experimental steel were determined by several methods. One of them involved a Bähr dilatometer with an inert atmosphere. Nine cooling rates from 0.1 to 89 K/s were tested. The effect of deformation on the kinetics of phase transformations was mapped by means of constructing not only a conventional CCT diagram but also the CCCT deformation diagram. For the latter diagram, the specimen was subjected to a deformation of $\varphi = 0.7$ applied at the rate of 10 s^{-1}. It was found that deformation shifts the onset of ferrite transformation toward higher cooling rates. At the same time, deformation speeds up pearlite transformation. Consequently, the time interval for pearlite formation during isothermal cooling expands (Fig. 1). Martensite transformation responds in an opposite manner and starts at somewhat lower temperatures when deformation is applied.

These diagrams were compared with a CCT diagram calculated by means of JMatPro [8], (Fig. 2, Table 2). In addition, the M_s temperature was calculated using the phenomenological model of Andrews [9] (Eq. 1). The results of this model were in very good agreement with the dilatometer data (Table 2).

$$M_s = 539 - 423C - 30,4Mn - 17,7Ni - 12,1Cr - 11Si - 7Mo$$

(1)

Modelling of Press Hardening Process

Physical simulation of press hardening was carried out in a thermomechanical simulator with the aid of material-technological modelling (Fig. 3). The data for the model were measured in a real-world process at specific temperatures of the tool. Since the thermomechanical

simulator offers rapid and precise control of cooling, it also enables cooling in a tool at room temperature to be simulated accurately.

Physical simulation comprised several steps, starting with heating to 937 °C and holding for 100 s. In the next step, the sheet blank was transferred to the tool which took 10 s and involved cooling in air. Once in the tool, the blank cooled rapidly. The kinetics of this cooling was governed by the tool temperature. After short equalization of temperature ending at the tool temperature, the process ended with a cooling step that corresponded to air cooling.

The first stage of the experiment explored the effects of various tool temperatures on microstructural evolution and evolution of mechanical properties. A total of five tool temperatures were chosen: room temperature, 200, 300, 425 and 550 °C. As a reference, a water cooling step was added to the experimental programme as well (Fig. 4). At the second stage of the experiment, a variant with isothermal holding for austenite stabilization was tested. The tool temperature of 425 °C was chosen for this variant, on the basis of dilatometer data and outcomes of earlier experiments [2]; the holding times were 0, 100, 300 and 600 s (Fig. 5).

The microstructures of specimens obtained by the experimental treatment were examined by optical microscopy (OM) and scanning electron microscopy (SEM). Tescan VEGA 3 and Zeiss EVO MA 25 scanning electron microscopes were used. EBSD and EDX analyses were carried out in the Zeiss Crossbeam 340-47-44 microscope. The amount of retained austenite was measured by X-ray diffraction phase analysis. An automatic powder diffractometer AXS Bruker D8 Discover with a HI-STAR position-sensitive area detector and a cobalt X-ray source ($\lambda K\alpha = 0.1790307$ nm) was employed for this measurement. Mechanical properties were measured by HV10 hardness testing and tensile testing.

Results and Discussion

After routes with tool temperatures between RT and 300 °C (Table 3), the specimen microstructures consisted of a mixture of ferrite and martensite. Their hardness was 250 HV10 (Fig. 6a). Keeping the tool at RT led to retained austenite volume fractions of a mere 3%. The ultimate strength was between 861 and 939 MPa and the A_{20} elongation levels were approximately 14%. When water

	C	Mn	Si	Al	Nb	P	S	Ni	Cu	Mo	W
Table 1 Chemical composition of the experimental steel [wt%]	0.21	1.4	1.8	0.006	0.002	0.007	0.005	0.07	0.06	0.02	0.02

Fig. 1 CCT and CCCT diagrams for the experimental steel constructed from dilatometric measurement data

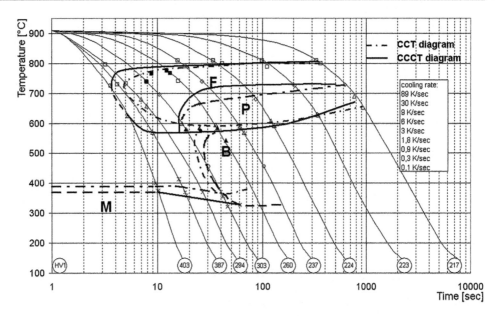

Fig. 2 CCT diagram calculated using JMatPro software

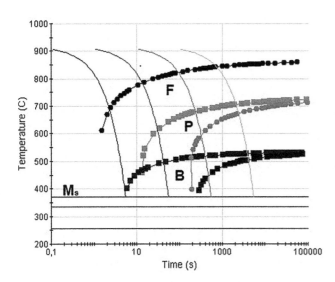

Fig. 3 Material-technological modelling by means of thermomechanical simulator

quenching was applied immediately after the pause in the process, no changes in the microstructure or mechanical properties were detected (Fig. 6b), with the exception of an increase of hardness to 290 HV10.

Bainite was first detected upon a route with the temperature of 425 °C. It is the desired phase in the microstructure of TRIP steels (Fig. 6c). The resulting microstructure comprised ferrite, martensite and bainite. The retained austenite fraction remained very low: 4%. The presence of bainite and

the reduced amounts of martensite raised the tensile strain levels to 21%, whereas the ultimate strength decreased only slightly to 844 MPa. A further rise in the tool temperature to 550 °C led, for the first time, to pearlite formation, with the resulting microstructure of a mixture of ferrite, pearlite and martensite (Fig. 6d). The reduction in the volume fraction of a hardening phase led to a hardness decrease to 229 HV10 and to a lower ultimate strength of 720 MPa (Table 3).

Typical treatment routes for TRIP steels include intercritical annealing as well as holding at the isothermal bainitic transformation temperature. During the hold, not only

Table 2 Phase transformation temperatures found by various calculation and measuring methods

JMatPro Calculated		Andrews 30 °C/s Calculated	Dilatometry −30 °C/s Measurement
M_s [°C]	M_f [°C]	M_s [°C]	M_s [°C]
370	256	387	387

Fig. 4 Thermal profile of press hardening in the material-technological model for several tool temperatures

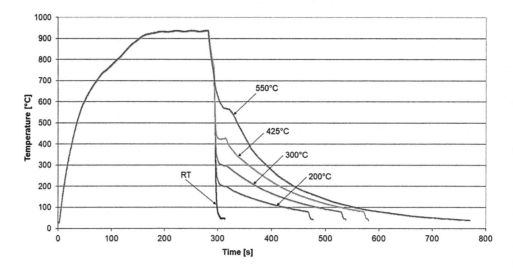

Fig. 5 Thermal profile of press hardening in the material-technological model with several holding times at 425 °C

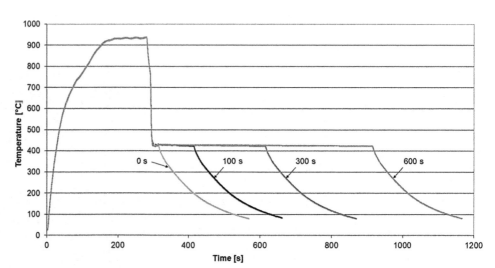

Table 3 Effects of tool temperature on mechanical properties and retained austenite fraction

Heating temperature [°C]	Tool temperature [°C]	Cooling	$R_{p0.2}$ [MPa]	R_m [MPa]	A_{20} [%]	HV10 [–]	RA [%]
937	Water cooling		407	875	13	291	–
	RT		410	876	14	241	3
	200	Air cooling	437	939	13	277	–
	300		416	861	14	257	–
	425		374	844	21	250	4
	550		311	720	20	229	–

bainite forms but also retained austenite becomes stabilized. Therefore, an experimental treatment was tested which comprised holding at 425 °C (Table 4). Holding times of 100 and 300 s led to microstructures that, however, still consisted of a mixture of martensite and ferrite and a small amount of bainite (Fig. 7a). Holding led to tempering and reduced the hardness to 225 HV10. This was also reflected in the decrease in the ultimate strength from 844 MPa to

754 MPa and 735 MPa respectively, whereas the elongation remained around 20% (Table 4).

Only an extension of the holding time to 600 s brought about an appreciable change in the microstructure: Bainite formation became more intensive, with the result of a microstructure of bainite, martensite and a small amount of free ferrite (Fig. 7b). In addition, the volume fraction of retained austenite increased significantly to 13%. Owing to

Fig. 6 Microstructures of physical simulation specimens: **a** with the tool at room temperature, **b** water quenching, **c** tool temperature of 425 °C, **d** tool temperature of 550 °C

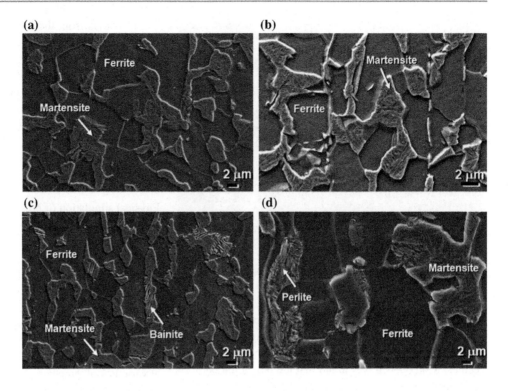

Table 4 Effects of holding time at the bainitic transformation temperature on mechanical properties and retained austenite fraction

Heating temperature [°C]	Tool temperature [°C]	Holding time [s]	$R_{p0.2}$ [MPa]	R_m [MPa]	A_{20} [%]	HV10 [–]	RA [%]
937	425	0	374	844	21	250	4
		100	352	754	25	228	–
		300	328	735	22	225	6
		600	695	1300	9	408	13

Fig. 7 Microstructures upon simulated press hardening with a tool temperature of 425 °C and isothermal bainitic transformation: **a** time at temperature: 300 s, **b** time at temperature: 600 s

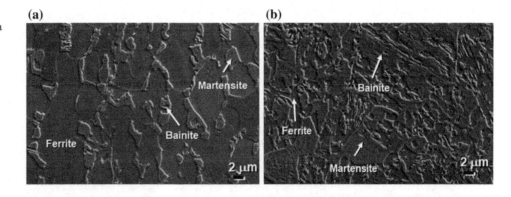

the majority of hardening constituents, the ultimate strength rose to 1300 MPa while elongation decreased to 9%.

The distribution and morphology of retained austenite is one of the parameters that are of key importance for its stability during cold working. It is the stability of retained austenite which affects the progress of the subsequent strain-induced martensitic transformation. Retained austenite

distribution and particle sizes and morphology were examined using two-stage etching (stage 1: −3% nital, stage 2: 10% aqueous solution of $Na_2S_2O_5$). This reagent leaves retained austenite white. The two-stage etching was carried out on the specimen whose route involved the tool temperature of 425 °C and the time at temperature of 600 s (Fig. 8). In this specimen, X-ray diffraction identified

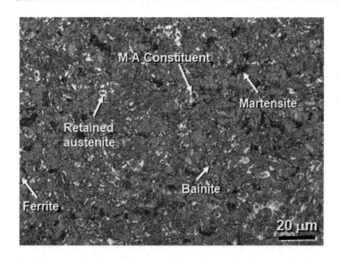

Fig. 8 Tool temperature: 425 °C—600 s—two-stage etching

13% of retained austenite. Both polygonal grains and foils of retained austenite were identified by two-stage etching. The thicker foils were found predominantly between bainite needles. Moreover, it was found that the polygonal particles of retained austenite contained the M-A constituent more often than others. This means that the retained austenite in the centre of such grains had insufficient stability and began to transform to martensite during cooling to room temperature. These islands of the M-A constituent were found mainly in particles larger than 5 μm.

Since the resolving power of optical microscopy proved insufficient, detailed examination was carried out using EBSD. The specimens were prepared by a mechanical route which was adjusted for the purpose of this analysis. The last preparation step involved colloidal silica in water in order to remove the traces of mechanical preparation. EBSD analysis confirmed the presence of retained austenite between bainite needles and along prior austenite grain boundaries (Fig. 9a).

For comparison, the EBSD analysis was also carried out on a specimen whose process route involved cooling in a tool at room temperature, in which X-ray diffraction identified only 3% of retained austenite. In this case, too, retained austenite was detected along prior austenite grain boundaries. It was clear that some austenite grains had not fully transformed to martensite (Fig. 9b).

Conclusions

Physical simulation of press hardening was conducted with a CMnSi TRIP low-alloyed steel using various parameter fully transformed to martensite settings. It included a process which is used to obtain microstructures typical of TRIP steels. The parameters of the physical simulation were identical to those found in a real-world press hardening route involving a heated tool. Two routes were tested. The difference between them was in the isothermal hold at the bainitic transformation temperature following the press hardening step. The first one was press hardening simulation at various tool temperatures without the isothermal hold. The tool temperatures included room temperature, 200, 300, 425 and 550 °C. From the perspective of microstructural evolution suitable for obtaining the TRIP effect, the most appropriate temperature appeared to be 425 °C. At lower temperatures, hardening was too intensive, whereas higher temperatures caused pearlite to form. For these reasons, the temperature of 425 °C was used for the second route, the purpose of which was to stabilize retained austenite during an isothermal hold at the bainitic transformation temperature. The main aspect of interest was the length of the hold. Holding times of 0, 100, 300 and 600 s were tested. It was found that to obtain a higher volume fraction of retained austenite and bainite by means of the specific conditions of cooling of sheet blank in the tool, holding at 425 °C should

(a)

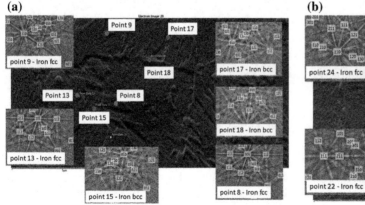

(b)

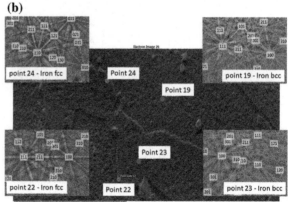

Fig. 9 Distribution of retained austenite observed by EBSD analysis: **a** tool temperature: 425 °C—time at temperature: 600 s, **b** tool at room temperature

be used. Microstructural evolution and evolution of mechanical properties depended substantially on the length of this hold. Intensive bainite formation took place at holds no shorter than 600 s. One can assume that alloying with silicon plays an important role in this process. Holding for 600 s led to stabilization of 13% of retained austenite. Retained austenite was present in the form of both polygonal grains and foil-like particles. The critical size of retained austenite particles was approximately 5 μm. In particles larger than this, the M-A constituent was detected. The ultimate strength was 1300 MPa and the elongation reached 9%. When compared to the route with 300 s holding time, the ultimate strength was almost twice as high. The decrease in the A_{20} elongation, as a consequence of the higher bainite fraction at the expense of ferrite, was from 22 to 9%. This study has shown that in the steel in question, a mere alteration of the length of the hold applied after press-hardening in a heated tool can lead to a broad range of mechanical properties with ultimate strengths ranging from 750 to 1300 MPa and elongations of 25 to 9%.

Acknowledgements This paper includes results created within the project TG02010011 Promoting Commercial Opportunities of UWB, sub-project TRIPIAL Transformation Induced Plasticity Isothermal Annealing Less. The project belongs to the GAMA programme and is subsidised from specific resources of the state budget for research and development through the Technology Agency of the Czech Republic.

References

1. Bleck W (2002) Using the TRIP effect—the down of a promising group of cold formable steels. Paper presented at the international conference on TRIP—aided high strength ferrous alloys, Belgium, 19–21 June 2002, p 13
2. Mašek B et al (2009) The influence of thermomechanical treatment of TRIP steel on its final microstructure. J Mater Eng Perform 18 (4):385–389
3. Kučerová L, Jirková H, Mašek B (2017) The effect of alloying elements on microstructure of 0.2%C TRIP steel. Mater Sci Forum 5(1):209–213
4. Ariza EA (2016) Characterization and methodology for calculating the mechanical properties of a TRIP-steel submitted to hot stamping and quenching and partitioning (Q&P). Mater Sci Eng A 671:54–59
5. Karbasian H, Tekkaya AE (2010) A review on hot stamping. J Mater Process Technol 210:2103–2118
6. Mori K et al. (2017) Hot stamping of ultra-high strength steel parts. CIRP Ann Manuf Technol (in press). http://dx.doi.org/10.1016/j.cirp.2017.05.007
7. Baik SCh et al (2001) Effects of alloying elements on mechanical properties and phase transformation of cold rolled TRIP steel sheets. ISIJ Int 41(3):290–297
8. JMatPro, Release 9.0, Sente Software Ltd. (2016)
9. Andrews KW (1965) Empirical formulae for the calculation of some transformation temperatures. J Iron Steel Inst 203:721–727

Effect of Carbon Content on Bainite Transformation Start Temperature on Fe–9Ni–C Alloys

Hiroyuki Kawata, Toshiyuki Manabe, Kazuki Fujiwara, and Manabu Takahashi

Abstract

Upper bainite in steels has many common features with lath martensite in steels. But there are some studies that indicate that bainite transformation start temperature (B_s) is greater than T_0 on steels containing high carbon content. We measured B_s on Fe–9 mass% Ni alloys containing 0.003–0.89 mass% C. In low carbon alloys, B_s is below T_0, and the increasing of carbon content decreases B_s standing in a line that is parallel to T_0. On the other hand, in high carbon alloys, carbon content does not affect B_s which stands around 753 K. The border between these two tendencies and B_s in high carbon alloys seem to correspond to the intersection point between the line of B_s on low carbon alloys and the calculated $\gamma/(\gamma + \theta)$ phase boundary.

Keywords

Bainite • Transformation • T_0 • Cementite
Driving force

Introduction

Bainite in steel is a useful structure because it has a relatively high strength with good formability and toughness. Many types of high-strength steels contain bainite with various carbon content; therefore, it is important to know the bainite transformation start temperature (B_s) with various carbon content.

The microstructure of bainite is similar to that of martensite in steel [1]. The shape of bainitic ferrite (BF) transfers from lath to plate with decreasing transformation temperature

between B_s and the martensite transformation temperature (M_s). The bainite consisting of lath-shaped ferrite is called upper bainite (UB) [2]. Figure 1 shows the schematic image of the crystallographic and morphological features of UB [2, 3]. The orientation relationship between prior austenite and BF is near the Kurjumov-Sachs relationship [1–3]. The prior austenite grain divides to some packets, which consists of parallel elongated ferrite laths. And these packets are subdivided to some blocks, which consist of BF laths having almost the same crystallographic orientation. These morphological and crystallographic features of UB are similar to those of lath martensite in steel [4]. This conformity suggests that their transformation mechanisms are similar.

According to the bainite transformation model proposed by Bhadeshia [1, 5], the bainite transformation can progress if the driving force of partitionless transformation from fcc to bcc is larger than the free energy for the displacive (martensitic) growth of lath-shaped ferrite. The driving force depends on the degree of supercooling from T_0, at which fcc and bcc of the same composition have the same free energy. Because T_0 decreases with the increasing of carbon content, it is guessed that B_s depends on carbon content.

However, the relationship between B_s and carbon content is not clear. There are many data which show that B_s does not decrease with an increase in carbon [6–8], and sometimes it becomes higher than T_0 [6, 8]. Bainite transformation sometimes occurs during the quench and partitioning process [9, 10]. Therefore, it is important for advanced high-strength steels to clarify the mechanism of bainite transformation start. We evaluated the effect of carbon content on Bs in Fe–9Ni–C alloys [11, 12], in which nickel does not exhibit a solute drag-like effect [7, 13].

Experimental Procedure

Table 1 shows the chemical compositions of the nine Fe–9Ni–C alloys used in this study. Ingots of the alloys melted in vacuum were hot-rolled and cold-rolled to 1 mm

H. Kawata (✉) · T. Manabe · M. Takahashi
Steel Research Laboratories, Nippon Steel & Sumitomo Metal Corporation, 20-1 Shintomi, Futtsu, Chiba, 293-8511, Japan
e-mail: kawata.z84.hiroyuki@jp.nssmc.com

K. Fujiwara
Advanced Technology Research Laboratories, Nippon Steel & Sumitomo Metal Corporation, 1-8 Fuso-Cho, Amagasaki, Hyogo 660-0891, Japan

© The Minerals, Metals & Materials Society 2018
A. P. Stebner and G. B. Olson (eds.), *Proceedings of the International Conference on Martensitic Transformations: Chicago*, The Minerals, Metals & Materials Series, https://doi.org/10.1007/978-3-319-76968-4_22

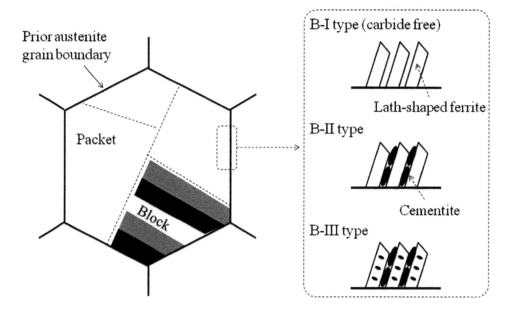

thick sheets. These plates contain nickel segregation. We evaluated the local nickel content by the electron probe micro analyzer, and estimated the effective nickel content on the bainite transformation start [11]. Table 2 shows the effective nickel content in each sheet and T_0 calculated via Thermo-Calc with these effective nickel content values.

The transformation behavior was measured via dilatometer. On the low carbon alloys (A, B, C, D, and E), we made the CCT diagrams and TTT diagrams. On the high carbon alloys (F, G, H, and I), we made TTT diagrams. Figure 2 shows the schematic diagrams. In these diagrams, B_s corresponds to the terrace if the C-curve of the diffusion transformation (ferrite and pearlite) is much slower than that of bainite transformation [1]. To confirm B_s, we observed the microstructure that generated around the terrace.

Results and Discussion

Figure 3 shows B_s and M_s evaluated in this study. The effect of carbon content on B_s was not steady. In low carbon alloys [11], B_s decreased with increasing carbon content from 0.003 to 0.1 mass%. However, in high carbon alloys [12], B_s was constant and seems to be independent of the carbon content, from 0.3 to 0.9 mass%. On the other hand, M_s decreased from 765 to 403 K with increasing carbon content from 0.003 to 0.9 mass%.

In low carbon alloys, the dependence of B_s on the carbon content seems to correspond to the model proposed by Bhadeshia [1, 5]. Figure 4 shows the driving force of the partitionless transformation from fcc to bcc at B_s in each low carbon alloy. The driving force was calculated via Thermo-Calc with the effective nickel content. Regardless of the carbon content, the driving force is almost constant, and its value, about 400 J/mol, corresponds to the driving force needed for the partitionless growth of lath-shaped ferrite [1].

Figure 5 shows B_s with T_0, T_0', and the phase boundaries calculated in Fe–7.6Ni–C ternary system. 7.6 mass%Ni represents the effective nickel content in the alloys used herein. B_s in low carbon alloys exists along the T_0' line, at which the driving force of the partitionless transformation is 400 J/mol. On the other hand, in high carbon alloys, B_s is independent of the carbon content between 0.3 and 0.9 mass %, and it is higher than T_0'. Especially, over 0.3 mass%C, B_s is higher than T_0. Figure 6 shows the microstructure in Alloy G (0.5 mass%C) specimen held at 748 K for 1000 s. The volume fraction that was transformed during isothermal holding in this specimen was assumed to be less than 5% based on the dilatation curve. Although the holding temperature was higher than T_0, 729 K, we can observe UB consisting of lath-shaped ferrite around prior austenite grain boundaries with some pearlite (P) and/or degenerated pearlite (DP) islands.

Table 1 The chemical compositions of the alloys used in this study

Alloy	A	B	C	D	E	F	G	H	I
C/mass%	0.0031	0.052	0.094	0.050	0.099	0.30	0.50	0.69	0.89
Ni/mass%	9.06	9.08	9.09	9.03	9.06	9.03	9.01	8.95	9.00
B/mass-ppm	23	24	24	w/o	w/o	w/o	w/o	w/o	w/o

Table 2 The effective nickel contents and T_0 in the alloys used

Alloy	A	B	C	D	E	F	G	H	I
$W_{Ni,E}$/mass%	7.40	7.49	7.62	7.50	7.42	7.67	7.60	7.75	7.59
T_0/K	923	900	879	900	882	799	729	660	593

Fig. 2 The schematic image of CCT and TTT diagrams in this study

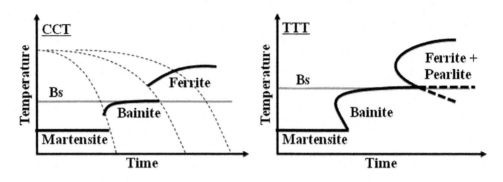

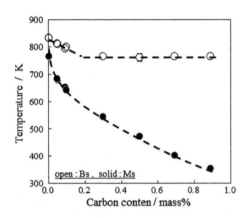

Fig. 3 B_s and M_s in Fe–9Ni–C alloys [11, 12]

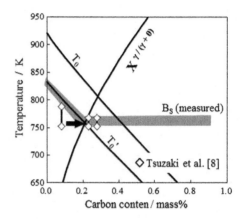

Fig. 5 B_s plotted on T_0, T_0' and $\gamma/(\gamma + \theta)$ phase boundary of the Fe–7.6Ni–C ternary system [12]

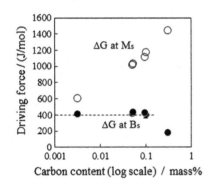

Fig. 4 The driving force of the partitionless transformation from fcc to bcc at B_s [11]

It is impossible to understand the behavior of B_s in high carbon alloys only by Bhadeshia's model. In Fig. 5, B_s in high carbon alloys correspond to the temperature of the intersection point between T_0' and the $\gamma/(\gamma + \theta)$ phase boundary. This intersection point, arrowed in Fig. 5, is

located on 0.205 mass%C, 755 K. Figure 7 shows the B_s in Fe–C alloys [6, 14] with the T_0' line and the $\gamma/(\gamma + \theta)$ phase boundary. The behavior of B_s with carbon content in the Fe–C system resembles that in the Fe–9Ni–C system described in Fig. 5.

This result proposed the simple model for B_s in high carbon alloys [12]. If the fully austenitic specimen containing higher carbon content than that of the intersection point is held between the temperature of the intersection point and T_0', the bainite transformation cannot start because of the shortage of the driving force. However, the cementite can precipitate because the holding temperature is lower than the $\gamma/(\gamma + \theta)$ phase boundary. When the cementite precipitation occurs, the local carbon content around the cementite should decrease toward the $\gamma/(\gamma + \theta)$ phase boundary. If the local carbon content decreases enough, it will become lower than the carbon content of T_0' line at holding temperature. Therefore, the bainite transformation will become able to

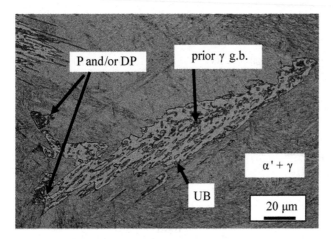

Fig. 6 The optical micrograph of Alloy G (0.5 mass%C) specimen held at 748 K for 1000 s after austenitization [12]

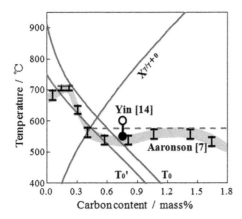

Fig. 7 B_s in previous studies [6, 14] plotted on T_0, T_0' and the $\gamma/(\gamma + \theta)$ phase boundary of the Fe–C system

In this model, the bainite transformation rate is limited by the cementite precipitation. This process looks like the so-called "inverse bainite" [1, 15]. We proposed this model on the assumption that the kinetics of cementite nucleation and growth is fast enough. The alloys used in this study do not contain the elements that delay cementite formation. If the cementite precipitation is depressed, the proposed process cannot proceed smoothly, and B_s in high carbon alloys would become lower than the temperature of the intersection point between T_0' and the $\gamma/(\gamma + \theta)$ phase boundary. The effect of carbon content on B_s will depend on the super-cooling from T_0' in low carbon alloys, and will depend on the cementite precipitation behavior in high carbon alloys.

Summary

1. In low carbon Fe–9Ni alloys, B_s decreased with the increasing of carbon content. Bs existed along with T_0', at which the driving force of the partitionless transformation from fcc to bcc is 400 J/mol. This driving force is as much as that satisfying the martensitic growth of ferrite lath proposed by Bhadeshia.

2. In high carbon Fe–9Ni alloys, B_s was constant regardless of the carbon content, and it could exist over T_0. The inflection point of the behavior of B_s corresponded to the intersection between T_0' and the $\gamma/(\gamma + \theta)$ phase boundary. It suggests that the bainite transformation in high carbon alloys is preceded by the cementite precipitation in austenite, which makes a carbon-poor region around itself.

occur around the cementite because the driving force of the partitionless transformation from fcc to bcc in the local carbon-poor region around the cementite increases to over 400 J/mol. If BF laths nucleate and/or grow, the local carbon content increases toward the T_0' line at holding temperature. By repeating this process, the bainite transformation with the cementite precipitation proceeds continuously over T_0' line. On the other hand, if the holding temperature is higher than the intersection point, the local carbon content around cementite cannot reach the T_0' line; therefore, the bainite transformation cannot occur after the cementite precipitation. This is the reason that B_s in high carbon alloys corresponds to the intersection point of T_0' and the $\gamma/(\gamma + \theta)$ phase boundary.

Acknowledgements The authors would like to express their sincere thanks to Dr. Goro Miyamoto (Tohoku University, Japan) and Dr. Tadashi Maki (Honorary Professor of Kyoto University, Japan) for their valuable comments and stimulating discussions.

References

1. Bhadeshia HKDH (2001) Bainite in steels, 2nd edn. IOM Communications Ltd, London
2. Furuhara T, Kawata H, Morito S, Maki T (2006) Mater Sci Eng A A431:228
3. Ohmori Y, Ohtani H, Kunitake T (1971) Trans ISIJ 11:250
4. Morito S, Tanaka H, Konishi R, Furuhara T, Maki T (2003) Acta Mater 51:1789
5. Bhadeshia HKDH (1981) Acta Mater 29:1117

6. Aaronson HI, Domian HA, Pound GM (1966) Trans Met Soc AIME 236:753
7. Tsuzaki K, Fujiwara K, Maki T (1991) Mater Trans JIM 32:667
8. Ohmori Y, Ohtsubo H, Georgima K, Maruyama N (1993) Mater Trans JIM 34:216
9. Kawata H, Hayashi K, Sugiura N, Yoshinaga N, Takahashi. M (2010) Mater Sci Forum 638–642:3307
10. Santofimia MJ, Zhao L, Sietsma. J (2011) Metall Mater Trans A 42A:3620
11. Kawata H, Fujiwara K, Takahashi M (2017) ISIJ Int 57:1866–1873
12. Kawata H, Manabe T, Fujiwara K, Takahashi M (2018) ISIJ Int 58 (in press)
13. Goldenstein H, Aaronson HI (1990) Metall Trans A 21A:1465
14. Yin J, Hillert M, Borgenstam A (2017) Metall Mater Trans A 48:4006
15. Kinsman KR, Aaronson HI (1970) Metall Trans 1:1485

Influence of Manganese Content and Finish Rolling Temperature on the Martensitic Transformation of Ultrahigh-Strength Strip Steels

Antti Kaijalainen, Mahesh Somani, Mikko Hemmilä, Tommi Liimatainen, David A. Porter, and Jukka Kömi

Abstract

The effects of manganese content and finish rolling temperature (FRT) on the transformed microstructures and properties of two low-alloyed thermomechanically rolled and direct-quenched (TM-DQ) steels were investigated. The materials were characterized in respect of microstructures and tensile properties. In addition, microhardness measurements were made both at the surface and centerline of the hot-rolled strips to help characterize the phase constituents. Detailed microstructural features were further revealed by laser scanning confocal microscopy (LSCM) and field emission scanning electron microscopy combined with electron backscatter diffraction (FESEM-EBSD). It was apparent that a decrease in the temperature of controlled rolling, i.e., the finish rolling temperature (FRT), resulted in reduced martensite fractions at the surface, as a consequence of strain-induced fine ferrite formation. The centerline of the strip, however, comprised essentially martensite and upper bainite. In contrast, high FRT and higher manganese content resulted in essentially a fully martensitic microstructure due to enhanced hardenability. The paper presents a detailed account of the hot rolling and hardenability aspects of TM-DQ ultra-high-strength strip steels and corresponding microstructures and properties.

Keywords

Direct quenching • Hardenability • Microstructure Prior austenite morphology • Ultrahigh-strength

A. Kaijalainen · M. Somani (✉) · D. A. Porter · J. Kömi
Materials and Production Engineering, University of Oulu, Oulu, Finland
e-mail: mahesh.somani@oulu.fi

M. Hemmilä · T. Liimatainen
SSAB Europe Oy, Raahe, Finland

Introduction

In recent years, low-carbon high-strength steels with yield strength in the range 800–1100 MPa produced using the TM-DQ processing route have become interesting materials for structural applications, because these steels can exhibit good combinations of mechanical properties and weldability [1–3]. The microstructures of these steels often comprise bainite and/or auto-tempered martensite [1, 4, 5]. In the case of TM-DQ strip steels, cold bending is the most important method of forming in applications such as containers and crane booms. The bendability improves remarkably when the steel hardness just below the surface is marginally lower than in the bulk owing to the presence of a mixture of ferritic and granular bainitic microstructure near the surface, in contrast to the generally bainitic and/or martensitic microstructure in the bulk of the steel, as the condition for the onset of strain localization and shear band formation is thereby significantly averted [6]. It has also been shown that the near-surface properties, i.e., the properties of the steel down to a depth of approximately 5% of the total sheet thickness, govern the bendability [6].

Hence, it is important to ascertain the factors and underlying mechanisms leading to the formation of a relatively soft microstructure at the surface in order to impart the best possible combinations of yield strength, toughness, ductility and bendability to these high-strength steels. The phase transformation characteristics have been found to be dependent not only on the chemical composition [7–10], but also on the extent of austenite pancaking and FRT [7, 11]. Specific TM processing steps were developed in order to obtain a relatively hard core with a bainite/martensite microstructure and a softer ferrite/granular bainite surface layer resulting from controlled strain-induced transformation.

The main aim of this study is to establish the influence of FRT and manganese content on the hardenability characteristics and subsequent phase transformation characteristics, microstructural features and resultant properties. In particular, the circumstances leading to the desired manifestation of a relatively soft surface microstructure have been established, in accord with some pilot scale processed and direct-quenched high-strength steel strips recently processed at the authors' laboratory [6]. This paper presents a comprehensive summary of the results with special emphasis on the hardenability and phase transformation aspects verified through CCT simulations to understand the microstructure development and related hardness as a function of austenite state and cooling rate.

Materials and Experimental Procedures

The experimental materials were produced as 210-mm-thick continuously cast slabs that were reheated to 1250 °C and then hot rolled at pilot scale under thermomechanical control to final strip thickness (t) of 8 mm, followed by direct quenching to room temperature at a rate of $\sim$50–70 °C/s. The finish rolling temperature (FRT) was varied in the range 800 920 °C. Manganese was varied between two levels and the chemical compositions of the two steels are given in Table 1. Also included in the table are the non-recrystallization temperatures (T_{NR}) of the two compositions calculated using the formula given in Ref. [12], as well as the martensite and bainite start temperatures (M_S and B_S) determined using JMatPro® simulation software (Sente Software Ltd.). The software was also used for plotting the CCT diagrams of the two steels. The material identification codes were so applied in order to describe the chemical composition of the steel (Low Mn or High Mn) and the finish rolling temperature (920–800 °C), e.g., Low Mn-800, etc.

A general characterization of the transformation microstructures was performed with a laser scanning confocal microscope (LSCM VK-X200, Keyence Ltd.) and a FESEM (Ultra plus, Zeiss) microscope on specimens etched with nital or picric acid [13]. The typical prior austenite grain structure was quantified at the quarter-thickness position by measuring the mean linear intercepts along the rolling (RD), normal (ND) and transverse directions (TD). Based on these measurements, the total reduction below the recrystallization temperature (R_{tot}) were determined using the equation given in Ref. [14]. Supplementary microstructure characterizations

were performed using the Oxford-HKL acquisition and analysis software following the microstructural classification described in Ref. [15]. For the EBSD measurements, the FESEM was operated at 10 kV and the step size was 0.2 μm. Tensile tests were conducted in accord with the European standard EN 10002. Microhardness was measured using a Micro-Hardness Tester (CSM) under 1 N load with ten measurements at eight depths below the surface and the centerline.

Results and Discussion

Microstructure

A summary of the austenite and effective grain size (d, EBSD high-angle boundaries > 15°) measurements is presented in Table 2. An example of the influence of FRT on the austenite morphology at the quarter-thickness of the strips in the high-Mn steel is depicted in Fig. 1, showing an increase in the total reduction below T_{NR} (R_{tot}) with a decrease in FRT. The Low Mn-800 sample essentially showed the formation of mainly granular bainite, and hence, the measurement of prior austenite grain shape could not be performed for this sample. The chemical composition of the steels did not affect the total rolling reduction below the recrystallization temperature (R_{tot}) when comparing the same finish rolling temperatures: R_{tot} was approx. 53–55% with high FRT and approx. 66–68% with low FRT for both steels. Similarly, the surface area of the austenite grain boundaries per unit volume (S_V) increased and the effective grain size decreased while the prior austenite grain size decreased, indicating that the sizes of the grains are clearly refined by lowering FRT, as reported in previous studies [3].

The microstructures at the centerline and subsurface of the specimens, as ascertained on the basis of FESEM observations, are listed in Table 3. The transformation microstructures of the specimens consisted of mixtures of quasi- or polygonal ferrite (F), granular bainite (GB), upper bainite (UB) and auto-tempered martensite (ATM). In Table 3, "main phase" means that the phase constituted more than 50% of the microstructure and "minor phase" less than 50%. Microstructures at the centerline consisted of mostly auto-tempered martensite with some bainite. A decrease of FRT increased the fraction of softer

Table 1 Chemical compositions (in wt%) of the investigated steels along with their T_{NR}, M_S and B_S temperatures

Steel	C	Si	Mn	Cr	Ti	B	Nb	V	T_{NR} (°C)	M_S (°C)	B_S (°C)
Low Mn	**0.07**	0.2	**1.4**	1.0	0.02	0.0013	0.04	0.011	987	439	602
High Mn	**0.08**	0.2	**1.8**	1.0	0.02	0.0015	0.04	0.017	997	419	586

Table 2 Mean linear intercept measurements[a] of the prior austenite grain structure (μm) along the three principal directions. Measurements of R_{tot}, S_V, and d based on EBSD data including the 95% confidence limits of the means are also given

Material	$\bar{L}_{RD}$ (μm)	$\bar{L}_{TD}$ (μm)	$\bar{L}_{ND}$ (μm)	R_{tot} (%)	S_V (mm²/mm³)	d (μm)
Low Mn-920	18.2 ± 1.5	12.8 ± 0.9	3.7 ± 0.2	55	335	1.29 ± 0.03
Low Mn-880	23.1 ± 2.1	10.9 ± 0.7	3.3 ± 0.1	62	376	1.24 ± 0.03
Low Mn-840	33.3 ± 3.7	14.5 ± 1.1	3.5 ± 0.1	67	335	1.49 ± 0.05
Low Mn-820	25.5 ± 2.5	10.6 ± 0.7	3.0 ± 0.1	66	405	1.23 ± 0.04
Low Mn-800	Granular bainitic, hence cannot measure					1.56 ± 0.07
High Mn-920	19.4 ± 1.6	11.0 ± 0.7	4.2 ± 0.2	53	310	1.53 ± 0.06
High Mn-880	21.2 ± 1.9	10.3 ± 0.6	2.9 ± 0.1	63	423	1.21 ± 0.03
High Mn-820	22.8 ± 2.1	9.4 ± 0.7	2.5 ± 0.1	67	482	1.09 ± 0.03
High Mn-800	25.5 ± 2.5	10.2 ± 0.6	2.6 ± 0.1	68	461	1.07 ± 0.02

[a]At the quarter-thickness of the strip

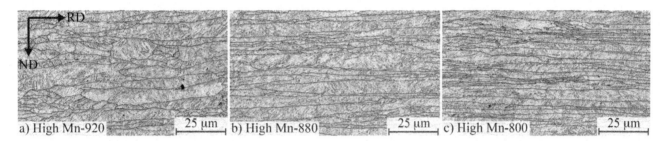

Fig. 1 LSCM images of prior austenite morphologies following etching with picric acid: High Mn steel with **a** 920, **b** 880 and **c** 800 FRT at the quarter-thickness as seen in RD-ND sections

Table 3 Microstructural characterization of investigated materials at the centerline and between 50 and 400 μm below the surface

Material	Centerline		Subsurface	
	Main phase	Secondary phase	Main phase	Secondary phase
Low Mn-920	ATM 80%	UB 20%	UB	GB, ATM
Low Mn-880	ATM 50%	UB 40%, GB 10%	UB	GB, F, ATM
Low Mn-840	UB 50%	GB 40%, ATM 10%	GB	F, UB
Low Mn-820	UB 45%	GB 40%, F 10%, ATM 5%	GB	F
Low Mn-800	GB 50%	UB 30%, F 15%, ATM 5%	GB	F
High Mn-920	ATM 100%	–	ATM	UB, GB
High Mn-880	ATM 100%	–	UB	ATM, GB
High Mn-820	ATM 90%	UB 10%	UB	ATM, GB
High Mn-800	ATM 70%	UB 20%, GB 10%	UB	ATM, GB

microstructures like ferrite and granular bainite. As with the subsurface microstructures, a decrease of FRT increased the incidence of GB and F at the expense of ATM and UB.

Tensile Properties and Microhardness

Tensile testing in the longitudinal direction showed reasonably high strength levels for the two steels, as showed in

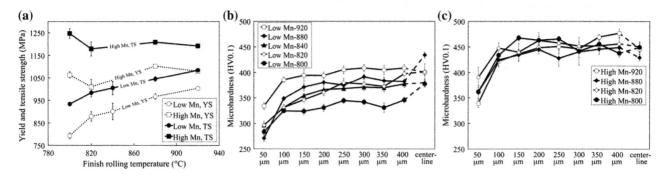

Fig. 2 Effect of FRT on mechanical properties (in the longitudinal direction): **a** yield stress (YS) and tensile strength (TS). Mean values of microhardness vs. depth below the surface for different FRT's in **b** Low Mn and **c** High Mn steel

Fig. 2a. The yield strength and tensile strength of the studied steels vary in the ranges 790–1180 MPa and 930–1250 MPa, respectively, depending on the FRT. In general, the higher Mn version (1.8% Mn) showed higher strength compared to that of the low-Mn steel (1.4% Mn), irrespective of the FRT, obviously as a consequence of the differences in the transformed microstructures.

The subsurface and centerline microhardness profiles presented in Fig. 2b, c show that the hardness throughout the thickness increases with increasing FRT. It can also be seen that the shapes of the hardness profiles are very similar. The polygonal ferrite in the surface layer (depth of 50 μm) is the softest phase, with a microhardness in the range 270–300 HV (Fig. 2b). Furthermore, the difference in the hardness levels between the steels comprising different manganese content can be seen. In the high-Mn steel, the finish rolling temperature has no significant effect on the hardness profiles and the hardness is higher than in the low-Mn steel. Also, the hardness increases more rapidly with depth (at about 100 μm) from approx. 340–390 HV to 430–470 HV (Fig. 2c) corroborating the higher tensile strengths seen in these samples (Fig. 2a). The low-Mn steel samples comprised mainly auto-tempered martensite and upper bainite from 100 μm below the surface to the centerline, thus corroborating lower strengths seen in these samples, as seen in Table 3.

JMatPro® Simulations

Figure 3 shows JMatPro® predicted CCT diagrams for recrystallized austenite with the investigated compositions. The phase transformation start temperatures were quite similar for both steels at the high cooling rates, 20–100 °C/s, where M_S temperatures are 439 °C and 419 °C in the low-Mn and high-Mn steels, respectively. However, with a further decrease in cooling rate, there is an appreciable difference in the phase transformation start temperatures, with

the high-Mn steel showing lower transformation temperatures due to the presence of higher Mn (1.8%) and slightly higher C (0.08%) contents, Table 1. As a consequence, simulated Vickers hardness data showed comparable hardness for high-Mn steel at all cooling rates. Furthermore, the bainite and ferrite curves were shifted to the right. Among those results, high-Mn steel provides mainly martensitic microstructure in the cooling rate range of interest for direct quenching of strips.

Although the CCT diagrams are not valid for deformed austenite, they do agree with the relative effects seen in the microstructures of the hot-rolled strips both at the core and the subsurface layers. The results are particularly interesting for the low-Mn steel, which showed lower yield and tensile strengths.

Relationships Between Chemical Composition, Microstructure and Tensile Properties

Finish rolling temperature is important through its effect on the austenite grain structure, which strongly influences the mechanical properties of the final product. As can be observed from Fig. 1 and Table 2, a higher FRT leads to less pancaking, higher effective grain size and hence to a coarser final microstructure, in contrast to the greater degree of pancaking observed at lower FRTs. Manganese has no effect on the predicted T_{NR} temperature [12]; therefore the T_{NR} temperature should be similar for the two steels, and thus explain the observed independence of the austenite pancaking level (R_{tot}) on manganese content at a given FRT.

Despite the small differences in respect of carbon and manganese contents of the two steels, i.e., 0.4 wt% manganese and 0.01 wt% carbon, the hardenability has increased appreciably, thus influencing the phase transformation kinetics and lowering the B_S and M_S temperatures marginally [8, 10]. The effect of FRT on the phase fractions is illustrated in Fig. 4. The effect of low-temperature finish

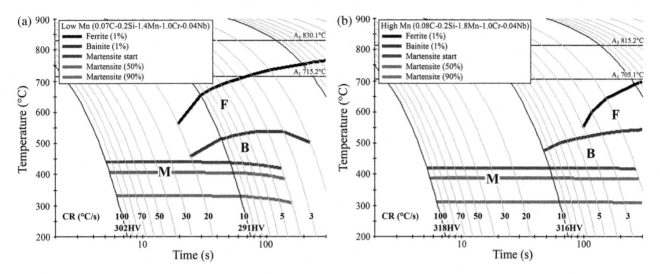

Fig. 3 Simulated CCT diagrams for a 20 μm recrystallized austenite grain size and 950 °C austenitization temperature, plotted using JMatPro® software for **a** low-Mn and **b** high-Mn steel. (Abbreviations: *F* ferrite, *B* bainite and *M* martensite)

rolling, i.e., pancaking, depends on the hardenability of the steel. It is well established that diffusion-controlled transformations are strongly affected by austenite deformation, such that if the steel composition and/or cooling rate result in the formation of ferrite from recrystallized austenite, the phase transformation start temperature increases, i.e., the hardenability decreases, when the austenite is deformed [7].

During hot rolling, the steel temperature near the surface fluctuates strongly as the material flows into and out of the roll gap. Contact with the colder rolls rapidly chills the subsurface regions of the strip [16] as they enter the roll gap, while on leaving the roll gap, heat flow from deeper in the material rapidly reheats the subsurface layers. Thus, it is possible that in the case of low FRT, the surface temperature can drop momentarily to levels where the nucleation and limited growth of ferrite and/or granular bainite can occur even during hot rolling, whereas this would not occur for

higher FRT. Such effects can be responsible for the complex nature of the microstructures nearest to the strip surface.

YS and TS initially increase with a decrease in FRT as a consequence of the increased pancaking and finer packets of martensite, but subsequently at lower FRTs, YS and TS decrease due to the formation of higher temperature transformation products.

Although, the bendability has not been reported in this study since the aim was to investigate hardenability, on the basis of the conclusions reported in Ref. [17], the present results suggest that bendability will be better in the case of the low-Mn steel due to its more beneficial subsurface hardness profile and microstructure. Similarly, a decrease in FRT should improve the bendability of both steels.

Conclusions

The purpose of this research was to investigate the effects of manganese content on the microstructure of hot-rolled and direct-quenched ultra-high-strength steels. The main observations and conclusions of the work can be summarized as follows:

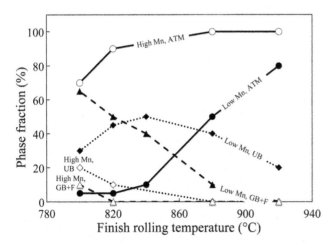

Fig. 4 Effect of FRT on phase fractions at the centerline

- Predicted CCT diagrams for undeformed, recrystallized austenite indicated that the compositional differences between the two studied steels should lead to significant differences in hardenability with the lower manganese and carbon contents promoting bainite and ferrite formation at the cooling rates of interest in direct quenching.
- An increase in the total reduction in the non-recrystallization temperature region (R_{tot}) in conjunction with a lowering of the finishing rolling temperature (FRT) increased austenite pancaking.

- A decrease of FRT increased the formation of softer microstructures such as ferrite (F) and granular bainite (GB) in the subsurface layers. At high FRTs, the microstructures at the centerline consisted mainly of auto-tempered martensite (ATM), especially in the case of higher manganese content. Lowering FRT increased the fractions of GB and F at the expense of ATM and upper bainite (UB) in both the central and subsurface parts of the strip thickness.
- There was a tendency for the yield stress and tensile strength of the steel sheets to decrease on lowering FRT when F and GB formed.

Acknowledgements This work was performed as a part of the Breakthrough Steels and Applications program of DIMECC Oy (the Digital, Internet, Materials & Engineering Co-Creation), Finland. The financial support of the Finnish Funding Agency for Technology and Innovation (Tekes) is gratefully acknowledged.

References

1. Suikkanen PP, Kömi JI (2014) Mater Sci Forum 783–786:246
2. Kaijalainen AJ, Suikkanen P, Karjalainen LP, Jonas JJ (2014) Metall Mater Trans A 45:1273
3. Kaijalainen AJ, Suikkanen PP, Limnell TJ, Karjalainen LP, Kömi JI, Porter DA (2013) J Alloys Compd 577:S642
4. Asahi H, Tsuru E, Hara T, Sugiyama M, Terada Y, Shinada H, Ohkita S, Morimoto H, Doi N, Murata M, Miyazaki H, Yamashita E, Yoshida T, Ayukawa N, Akasaki H, Macia ML, Petersen CW, Koo JY, Bangaru NV, Luton MJ (2004) Int J Offshore Polar Eng 14:36
5. Hemmilä M, Laitinen R, Liimatainen T, Porter DA (2005) Proceedings of 1st international conference on "Super-High Strength Steels". Associazone Italiana di Metallurgica – AIM, Rome
6. Kaijalainen AJ, Suikkanen PP, Karjalainen LP, Porter DA (2016) Mater Sci Eng A 654:151
7. Kozasu I (1997) In: Chandra T, Sakai T (eds) International conference on thermomechanical processing of steel and other materials. The Minerals, Metals & Materials Society, Wollongong, pp 47–55
8. Steven W, Haynes AG (1956) J Iron Steel Inst 183:349
9. Pickering FB (1977) In: Korchysky M (ed) Microalloying '75. Union Carbide Corporation, Washington DC, pp 9–31
10. Stuhlmann W (1954) Härterei Tech Mitteilungen 6:31
11. Wang GD, Wang ZD, Qu JB, Jiang ZY, Liu XH (1997) In: Chandra T, Sakai T (eds) International conference on thermomechanical processing of steel and other materials. The Minerals, Metals & Materials Society, Wollongong, pp 717–723
12. Boratto F, Barbosa R, Yue S, Jonas JJ (1988) In: Tamura I (ed) THERMEC-88. Iron and Steel Institute of Japan, Tokyo, pp 383–390
13. Brownrigg A, Curcio P, Boelen R (1975) Metallography 8:529
14. Higginson RL, Sellars CM (2003) Worked examples in quantitative metallography. Maney, London
15. Krauss G (2015) Steels processing, structure, and performance, 2nd edn. ASM International, Materials Park
16. Pyykkönen J, Suikkanen P, Somani MC, Porter DA (2012) Matériaux Tech 100:S1
17. Kaijalainen AJ, Liimatainen M, Kesti V, Heikkala J, Liimatainen T, Porter DA (2016) Metall Mater Trans A 47:4175

In Situ Neutron Diffraction Study on Microstructure Evolution During Thermo-Mechanical Processing of Medium Manganese Steel

Yoshihiko Nakamura, Akinobu Shibata, Wu Gong, Stefanus Harjo, Takuro Kawasaki, Atsushi Ito, and Nobuhiro Tsuji

Abstract

The microstructure evolution of medium manganese steel (Fe-5Mn-2Si-0.1C (wt%)) during thermo-mechanical processing in ferrite + austenite two-phase region was investigated by in situ neutron diffraction analysis and microstructure observations. When the specimens were isothermally held at a temperature of 700 °C, the fraction of reversely transformed austenite increased gradually with an increase in the isothermal holding time. However, it did not reach the equilibrium fraction of austenite even after isothermal holding for 10 ks. On the other hand, the fraction of reversely transformed austenite increased rapidly after the compressive deformation at a strain rate of 1 s^{-1} at 700 °C and reached the equilibrium state during subsequent isothermal holding for around 3 ks. Moreover, microstructure observations suggested that the austenite, which was reversely transformed during isothermal holding at 700 °C, exhibited film-like shape and existed between pre-existing martensite laths. In contrast, when the compressive deformation was applied during isothermal holding at 700 °C, most of the reversely transformed austenite had globular shapes with grain sizes less than 1 μm.

Keywords

Medium manganese steel • Neutron diffraction
Austenite reverse transformation • Thermo-mechanical processing

Introduction

In recent years, structural metallic materials, especially automotive steel sheets, have been required to manage both high strength and good ductility/toughness, because the demands for improvement of formability, fuel efficiency and passenger safety of automobiles are increasing more and more. As a result, advanced high strength steels (AHSS) that manage both high strength and good ductility/toughness have been investigated and developed actively in the last decade. The first-generation AHSS, such as dual-phase steels, complex-phase steels, etc., show high tensile strength with limited uniform ductility. The second-generation AHSS, so-called high-manganese transformation-induced plasticity (TRIP) and twinning-induced plasticity (TWIP) steels having FCC austenite structure, exhibit both high strength and large uniform elongation. However, disadvantages of these steels are high production cost and low recycling efficiency, because they contain large amounts of manganese, more than about 20 wt%. Based on such a background, medium-manganese steels, which contain 3 ∼ 10 wt% manganese, have received much attention as a new class of AHSS, that is, the third-generation AHSS [1]. Several papers have reported that medium-manganese steels containing a certain amount of retained austenite exhibit good balance of strength and ductility due to the TRIP effect of retained austenite [2–6]. In order to obtain the retained austenite in medium-manganese steels, thermo-mechanical processing, which involves annealing at intercritical temperature (α + γ two phase temperature), has been utilized [7–9].

However, the microstructure evolution in medium-manganese steels, particularly austenite reverse transformation

Y. Nakamura · A. Shibata (✉) · A. Ito · N. Tsuji
Department of Materials Science and Engineering,
Kyoto University, Kyoto, Japan
e-mail: nakamura.yoshihiko.83u@st.kyoto-u.ac.jp

A. Shibata · W. Gong · N. Tsuji
Elements Strategy Initiative for Structural Materials (ESISM),
Kyoto University, Kyoto, Japan

S. Harjo · T. Kawasaki
J-PARC Center, Japan Atomic Energy Agency (JAEA),
Ibaraki, Japan

A. Ito
Department of Materials and Synchrotron Radiation Engineering,
University of Hyogo, Himeji, Japan

© The Minerals, Metals & Materials Society 2018
A. P. Stebner and G. B. Olson (eds.), *Proceedings of the International Conference on Martensitic Transformations: Chicago*,
The Minerals, Metals & Materials Series, https://doi.org/10.1007/978-3-319-76968-4_24

behavior, during thermo-mechanical processing is still unclear. It is very difficult to systematically analyze the microstructure evolution by conventional experimental methods, because austenite usually transforms to ferrite or martensite during cooling to room temperature, and austenite is retained at room temperature in only limited conditions in the medium-manganese steels [9–11]. An in situ neutron diffraction analysis is one of the most promising techniques to directly investigate transformation and deformation behaviors of austenite at elevated temperatures. The present study aimed to investigate microstructure evolution in a medium-manganese steel (Fe-5Mn-2Si-0.1C (wt%)) during thermo-mechanical processing at intercritical temperatures by in situ neutron diffraction analysis.

Experimental Procedure

The material used in this study is a 5Mn-2Si-0.1C (wt%) steel. The chemical composition of this steel is shown in Table 1. In order to eliminate manganese segregation, as-received plates were homogenized at 1100 °C for 36 h and then ice brine quenched. The homogenized specimens with fully martensite structure were used as the starting materials.

The in situ neutron diffraction experiment during thermo-mechanical processing at elevated temperatures was conducted at BL19 "TAKUMI" in Material and Life Science Experimental Facility (MLF)/Japan Proton Accelerator Research Complex (J-PARC), equipped with a physical simulator for thermo-mechanical processing (Thermecmastor-Z). As shown in Fig. 1, the thermo-mechanical process used in this study was a combination of uniaxial compressive deformation and isothermal holding. The cylindrical specimen 11 mm in height and 6.6 mm in diameter was heated up to 700 °C at a heating rate of 10 °C s^{-1} by induction heating and held for 180 s. Then, a uniaxial compression by 60% reduction in height was applied at a strain rate of 1 s^{-1}. The specimen was subsequently held for 10 ks at 700 °C after the compressive deformation. Neutron diffraction was measured throughout the thermo-mechanical process.

Microstructures of the specimens were observed by scanning electron microscopy (SEM, JEOL JSM-7800F) at an accelerating voltage of 12 kV. For the microstructure observation, the specimens were electrolytically polished in a solution of 10% HClO$_4$ and 90% CH$_3$COOH and etched in a 3% nital solution.

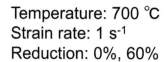
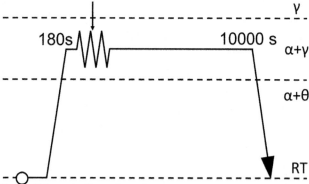

Fig. 1 Schematic illustration of the thermo-mechanical process used in this study

Results

Figure 2 shows neutron diffraction profiles at 700 °C (a) without compressive deformation and (b) with 60% compressive deformation. The starting materials had a fully martensite structure without retained austenite. In both cases with or without compressive deformation, FCC peaks corresponding to reversely transformed austenite could be observed after isothermal holding for 15 s, and the intensities of the FCC peaks increased with an increase in the isothermal holding time. The volume fraction of austenite reversely transformed during thermo-mechanical processing was evaluated quantitatively by Rietveld method using a Z-Rietveld software, and the results are summarized in Fig. 3. Gray and black points show the volume fractions of austenite reversely transformed during the processes with or without compressive deformation, respectively. The horizontal broken line indicates the equilibrium fraction of austenite at 700 °C (50.9%) calculated by Thermo-Calc software and the vertical dotted line represents the time when the compressive deformation was applied (total period of compressive deformation was 0.9 s). From Fig. 3, it can be seen that the fraction of reversely transformed austenite increased gradually when the compressive deformation was not applied. However, the fraction was much less than the equilibrium fraction even after isothermal holding for 10 ks. On the other hand, the fraction of reversely transformed

Table 1 Chemical composition of the steel used in present study (wt%)

C	Si	Mn	P	S	Al	N	O	Fe
0.100	2.04	4.91	<0.005	0.0027	0.008	0.0005	<0.0005	Bal

(a)

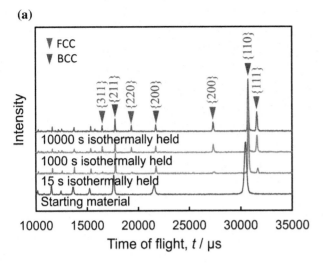

(b)

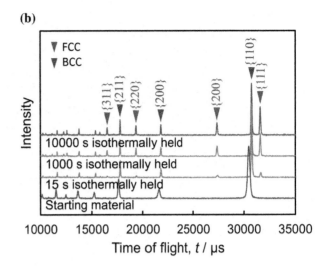

Fig. 2 Neutron diffraction profiles during the thermo-mechanical processing at 700 °C. **a** Without compression and **b** with 60% compression

austenite increased rapidly after the compressive deformation and approached to the equilibrium fraction during subsequent isothermal holding. This result suggests that the

Equilibrium fraction of γ: 50.9%

Fig. 3 Volume fraction of reversely transformation austenite as a function of isothermal holding time at 700 °C. Gray and black points are the data obtained in the processes with or without compressive deformation, respectively. The horizontal broken line indicates the equilibrium fraction of austenite at 700 °C (50.9%) calculated by Thermo-Calc software and the vertical dotted line represents the time when the compressive deformation was applied

austenite reverse transformation was accelerated significantly by compressive deformation at intercritical temperature (700 °C). This is presumably because lattice defects introduced in ferrite (tempered martensite) by compressive deformation act as nucleation sites for reverse transformation to austenite.

Figure 4 shows SEM images of the specimens which were water quenched after isothermal holding for 1181 s at 700 °C (a) without compressive deformation and (b) with 60% compressive deformation. The regions indicated by M_F correspond to fresh martensite that formed during water quenching from 700 °C. It can be seen that the morphology and size of fresh martensite (i.e., prior austenite just before quenching) are completely different between Fig. 4a and b. In Fig. 4a, the fresh martensite exhibits film-like shapes and exits between pre-existing martensite laths. In contrast, most of the fresh martensites in Fig. 4b have globular shapes and their grain sizes are relatively small (less than 1 μm). Because the regions of fresh martensite correspond to austenite reversely transformed at 700 °C, it can be said that compressive deformation changed the morphology and size of reversely transformed austenite. Based on the above-mentioned results, in can be concluded that compressive deformation at intercritical temperature accelerates reverse transformation and forms fine globular-shaped austenite.

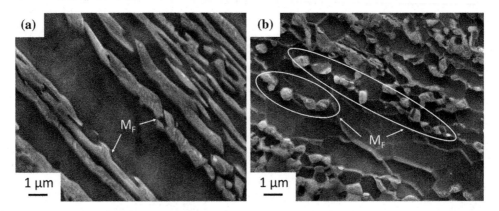

Fig. 4 SEM images of the specimens water quenched after isothermal holding for 1181 s at 700 °C **a** without compressive deformation and **b** with 60% compressive deformation. The regions indicated by M_F correspond to fresh martensite that formed during water quenching form 700 °C

Conclusions

In this study, austenite reverse transformation behaviors during thermo-mechanical processing in a medium manganese steel (Fe-5Mn-2Si-0.1C) were investigated by in situ neutron diffraction experiment and microstructure observation. The conclusions obtained are summarized below:

- When the specimens were isothermally held at 700 °C for 10 ks without compressive deformation, the fraction of reversely transformed austenite was much less than the equilibrium fraction. On the other hand, the fraction of reversely transformed austenite increased rapidly after the compressive deformation at 700 °C and reached to the equilibrium amount after subsequent isothermal holding at 700 °C. This result implied that the compressive deformation accelerated reverse transformation from (tempered) martensite to austenite during subsequent isothermal holding at 700 °C. This was presumably because lattice defects introduced in ferrite (tempered martensite) by compressive deformation acted as nucleation sites for reverse austenite transformation.
- The austenite reversely transformed during isothermal holding at 700 °C exhibited film-like shapes and existed between pre-existing martensite laths. In contrast, when the compressive deformation was applied during isothermal holding at 700 °C, most of the reversely transformed austenite had globular shapes with grain sizes less than 1 μm. The results suggested that the morphology and size of reversely transformed austenite were changed significantly by the compressive deformation.

Acknowledgements The neutron experiment at the Materials and Life Science Experimental Facility of the J-PARC was performed under a user program (Proposal No. 2016E0003, 2017E0001, 2017A0129). This study was financially supported by the Elements Strategy Initiative for Structural Materials (ESISM) through the Ministry of Education, Culture, Sports, Science and Technology (MEXT), Japan.

References

1. Lee YK et al (2015) Current opinion in medium manganese steel. Mater Sci Technol 31(7):843–856
2. Arlazarov A et al (2012) Evolution of microstructure and mechanical properties of medium Mn steels during double annealing. Mater Sci Eng A 542:31–39
3. Cai ZH et al (2015) Austenite stability and deformation behavior in a cold-rolled transformation-induced plasticity steel with medium manganese content. Acta Mater 84:229–236
4. Han J et al (2014) The effects of the initial martensite microstructure on the microstructure and tensile properties of intercritically annealed Fe–9Mn–0.05C steel. Acta Mater 78:369–377
5. Hanamura T et al (2011) Excellent total mechanical-properties-balance of 5% Mn, 30000 MPa% steel. ISIJ Int 51(4):685–687
6. Shi J et al (2010) Enhanced work-hardening behavior and mechanical properties in ultrafine-grained steels with large-fractioned metastable austenite. Scripta Mater 63(8):815–818
7. Xu HF et al (2012) Heat treatment effects on the microstructure and mechanical properties of a medium manganese steel (0.2C-5Mn). Mater Sci Eng A 532:435–442
8. Hu J et al (2015) The determining role of reversed austenite in enhancing toughness of a novel ultra-low carbon medium manganese high strength steel. Scripta Mater 104:87–90
9. Zhao C et al (2014) Effect of annealing temperature and time on microstructure evolution of 0.2C-5Mn steel during intercritical annealing process. Mater Sci Technol 30(7):791–799
10. Huang H et al (1994) Retained austenite in low carbon, manganese steel after intercritical heat treatment. Mater Sci Technol 10:621–626
11. Zhao C et al (2014) Austenite thermal stabilization through the concentration of manganese and carbon in the 0.2C-5Mn steel. ISIJ Int 54(12):2875–2880

Part V
Size Effects in Martensitic Transformations

Nanoscale Phase Field Modeling and Simulations of Martensitic Phase Transformations and Twinning at Finite Strains

Anup Basak and Valery I. Levitas

Abstract

A thermodynamically consistent phase field approach to martensitic phase transformations for a system with austenite and two martensitic variants has been developed. The model considers two order parameters, describing austenite ↔ martensite and variant ↔ variant transformations, respectively. The coexistence of three phases at a single material point are consistently penalized. Twinning in the nanoscale sample was studied for two different kinematic models (KMs) for the transformation stretch tensor U_t. In KM-I, U_t is taken as a linear combination of the Bain strains, and in KM-II, U_t is an exponential of the logarithm of the Bain stretch tensors. For these two KMs and for an additional model based on simple shear, analytical solutions for elastic stresses within a variant-variant boundary in an infinite twinned sample are presented. The results can be easily generalized for an arbitrary number of variants. They are crucial for further development of phase field approaches to multivariant martensitic transformations coupled to mechanics.

Keywords

Phase field • Martensitic transformation • Twinning
Interfacial stress • Finite strain

A. Basak (✉)
Department of Aerospace Engineering, Iowa State University, Ames, IA 50011, USA
e-mail: abasak@iastate.edu

V. I. Levitas (✉)
Departments of Aerospace Engineering, Mechanical Engineering, and Material Science and Engineering, Iowa State University, Ames, IA 50011, USA
e-mail: vlevitas@iastate.edu

V. I. Levitas
Ames Laboratory, Division of Materials Science and Engineering, Ames, IA 50011, USA

Introduction

Phase field approaches have been widely used for studying martensitic phase transformations [1–9]. The central idea in all phase field approaches is to consider order parameters, which describe the phases. The evolution of these order parameters is governed by a system of Ginzburg-Landau equations.

Various descriptions of the order parameters have been used in different models. N order parameters η_i, each describing A ↔ M_i transformation, have been used in [1, 2, 8]. The symbols A, M, and M_i denote austenite, martensite, and the i-th variant of martensite, respectively. In [1, 2], the description of M_i ↔ M_j transformation is not fully satisfactory, since the order parameters η_i and η_j vary independently in the order parameter space, and the interfacial energy and width depend on stress and temperature in an uncontrollable way. These shortcomings were overcome within the approach considered in [8], where the coexistence of a third phase between two other phases was controlled consistently. However, the interpolation function for the description of A ↔ M transformation cannot accommodate material parameters, and this is too restrictive. Also, penalization of energy due to the coexistence of three phases at a single material point (triple junction) is not satisfactory. Other multiphase models, such as those considered by Tůma et al. [9] and Steinbach [10] do not describe the instability criteria for transformations and are more suitable for microscale modeling. In another approach, $N + 1$ hyperspherical order parameters were considered for a system with N variants, where the order parameter η_0 describes A ↔ M transformations and another N order parameters η_i describe M_i ↔ M_j transformations, where all the η_i satisfy a nonlinear constraint [3, 6]. While within this model the triple junctions can be consistently penalized, due to the nonlinear constraint, instability conditions cannot be consistently formulated for a system with more than two variants. We have recently developed a general consistent model for

© The Minerals, Metals & Materials Society 2018
A. P. Stebner and G. B. Olson (eds.), *Proceedings of the International Conference on Martensitic Transformations: Chicago*, The Minerals, Metals & Materials Series, https://doi.org/10.1007/978-3-319-76968-4_25

multivariant martensitic transformations, which does not have all these drawbacks. A full description of the general model is beyond the scope of this paper and will be presented elsewhere. Here, we instead consider only two variants in using the model. The governing equations are thus simplified without suppressing the essential features.

Based on this model, we have studied variant-variant transformation (twinning) in a nanoscale sample for a two-variant system. The microstructure evolution can be completely captured by two order parameters η_0 ($\eta_0 = 0$ in A and $= 1$ in M) and η_i ($\eta_i = 1$ in M_i and $= 0$ in M_j), which describe A $\leftrightarrow$ M and $M_i \leftrightarrow M_j$ transformations, respectively. We have mainly considered two different kinematic models (KMs) for the transformation stretch tensor U_t: in KM-I, U_t is a linear combination of the Bain strains [7], and in KM-II, U_t is an exponential of the logarithm of the Bain stretch tensors [9]. Note that the deformation rule given by KM-II yields an isochoric variant $\leftrightarrow$ variant transformation process for all intermediate values of the order parameter; however, for KM-I the deformation is isochoric for $\eta_i = 1$ (within variants) only. The twinning solutions for these two models, obtained using the finite element (FE) method, are compared. Using these two transformation rules, the analytical solution of the elastic stresses within an infinite twinned sample, consisting of two variants of NiAl (cubic A and tetragonal M_i) separated by an interface, has been obtained. The solution for a simple shear-based kinematic model from crystallographic theory has also been obtained.

Notation: We denote the multiplication and the inner product between two second-order tensors as $(\mathbf{A} \cdot \mathbf{B})_{ij} = A_{ik}B_{kj}$ and $\mathbf{A} : \mathbf{B} = A_{ij}B_{ji}$, respectively, where A_{ij} and B_{ij} are the components of the tensors in an orthonormal Cartesian basis $\{e_1, e_2, e_3\}$. The repeated indices denote Einstein's summation. The trace, determinant, inverse, and transpose of $\mathbf{A}$ are denoted by $tr\mathbf{A}$, $det\mathbf{A}$, $\mathbf{A}^{-1}$, and $\mathbf{A}^T$, respectively. The reference, stress-free intermediate, and current configurations are denoted by Ω_0, Ω_t, and Ω, respectively. Gradient operator and Laplacian are denoted by ∇_0 and ∇_0^2, respectively, in Ω_0, and by ∇ and ∇^2, respectively, in Ω. $\mathbf{I}$ denotes the identity tensor.

System of Phase Field and Elasticity Equations

Kinematics: We decompose the total deformation gradient $\mathbf{F}$ into [7] $\mathbf{F} = \mathbf{F}_e \cdot \mathbf{F}_t = \mathbf{V}_e \cdot \mathbf{R} \cdot \mathbf{U}_t$, where $\mathbf{F}_e$ and $\mathbf{F}_t$, respectively, are the elastic and transformational parts of $\mathbf{F}$, $\mathbf{U}_t$ is the right transformation stretch tensor (symmetric), $\mathbf{V}_e$ is the left elastic stretch tensor (symmetric), and $\mathbf{R}$ is the lattice rotation tensor; the suffixes e and t stand for elastic and transformational parts, respectively. We also denote $J = det\mathbf{F}$, $J_t = det\mathbf{F}_t$, and $J_e = det\mathbf{F}_e$. We consider Eulerian

total and elastic strains as $\mathbf{b} = 0.5(\mathbf{F} \cdot \mathbf{F}^T - \mathbf{I}) = 0.5(\mathbf{V}^2 - \mathbf{I})$ and $\mathbf{b}_e = 0.5(\mathbf{F}_e \cdot \mathbf{F}_e^T - \mathbf{I}) = 0.5(\mathbf{V}_e^2 - \mathbf{I})$. The total deformation gradient satisfies the compatibility condition [11] $\nabla_0 \times \mathbf{F} = \mathbf{0}$, where $(\nabla_0 \times \mathbf{F})_{ij} = e_{ikl}\frac{\partial F_{jl}}{\partial r_{0k}}$, e_{ikl} is the third-order permutation tensor, and $\mathbf{r}_0$ denotes the position vector of a point in Ω_0. The transformation rules for our kinematic models are given by

$$\mathbf{U}_t^I = \mathbf{I} + \varphi\varepsilon_{tj} + \varphi\phi_i(\varepsilon_{ti} - \varepsilon_{tj}) \text{ and}$$
$$\mathbf{U}_t^{II} = \exp[\varphi \ln \mathbf{U}_{tj} + \varphi\phi_i(\ln \mathbf{U}_{ti} - \ln \mathbf{U}_{tj})], \tag{1}$$

where $\varphi = \eta_0^2(3 - 2\eta_0)$ and $\phi_i = \eta_i^2(3 - 2\eta_i)$ are the interpolation functions and the superscripts denote corresponding kinematic model numbers. It can be easily proven that $det\mathbf{U}_t^{II}$ is a constant for variant-variant transformations (using $\varphi = 1$), and hence the entire transformation path is isochoric; this is not, however, the case for $\mathbf{U}_t^I$.

Free energy: We consider the Helmholtz free energy per unit mass of the body in the following form:

$$\psi(\mathbf{F}, \eta_0, \eta_i, \theta, \nabla\eta_0, \nabla\eta_i) = \frac{J_t}{\rho_0}\psi_e(\mathbf{F}_e, \eta_0, \eta_i, \theta) + J\overset{\smile}{\psi}{}^\theta(\eta_0, \eta_i, \theta)$$
$$+ \tilde{\psi}^\theta(\eta_0, \eta_i, \theta) + \psi_p(\eta_0, \eta_i)$$
$$+ J\psi^\nabla(\nabla\eta_0, \nabla\eta_i), \quad \text{where}$$

$$\tag{2}$$

$$\psi_e = \frac{\lambda}{2}(tr\mathbf{b}_e)^2 + \mu|\mathbf{b}_e|^2; \quad \overset{\smile}{\psi}{}^\theta = A\eta_0^2(1 - \eta_0)^2 + \bar{A}\varphi\eta_i^2(1 - \eta_i)^2;$$

$$\rho_0\tilde{\psi}^\theta = \psi_0^\theta(\theta) + \Delta\psi^\theta(\theta)\varphi; \quad \psi^\nabla = \frac{\beta_{0M}|\nabla\eta_0|^2 + \beta_{ij}\varphi|\nabla\eta_i|^2}{2\rho_0};$$

$$\psi^p = K_{0ij}\eta_0^2\eta_i^2(1 - \eta_i)^2(1 - \varphi). \tag{3}$$

Here, ψ_e is the strain energy density per unit volume of Ω_t; $\overset{\smile}{\psi}{}^\theta$ is the barrier energy; $\rho_0\tilde{\psi}^\theta$ is the thermal energy; ψ^∇ is the gradient energy; ψ^p penalizes the triple junctions between A and two variants; ρ_0 is the mass density in Ω_0; λ and μ are the Lamé constants; A_{0M} and $\bar{A}$ are the barrier heights for A $\leftrightarrow$ M and $M_i \leftrightarrow M_j$ transformations, respectively; ψ_0^θ and $\Delta\psi^\theta$ are the thermal energy of A and the difference in thermal energy between A and M, respectively; β_{0M} and β_{ij} are the energy coefficients of A-M and M_i-M_j interfaces, respectively; K_{0ij} is the penalization coefficient. The material properties at each particle are determined using $M = (1 - \varphi)M_0 + \varphi(M_j + (M_i - M_j)\phi_i)$, where M_0 and M_i are the properties of A and M_i, respectively.

Mechanical equilibrium equation and stresses: The mechanical equilibrium equation is given by [7] $\nabla \cdot \boldsymbol{\sigma} = \mathbf{0}$ in Ω, where $\boldsymbol{\sigma} = \boldsymbol{\sigma}_e + \boldsymbol{\sigma}_{st}$, where $\boldsymbol{\sigma}$ is the total Cauchy stress tensor composed of the elastic $\boldsymbol{\sigma}_e$ and structural parts $\boldsymbol{\sigma}_{st}$ (also see [6]):

$$\sigma_e = \frac{1}{J_e} \mathbf{V}_e^2 \cdot (\lambda\, tr(\mathbf{b}_e)\mathbf{I} + 2\mu\mathbf{b}_e),$$

$$\sigma_{st} = \rho_0(\overset{\smile}{\psi}{}^{\theta} + \psi^{\nabla})\mathbf{I} - \beta_{0M}\nabla\eta_0 \otimes \nabla\eta_0 - \varphi\beta_{ij}\nabla\eta_i \otimes \nabla\eta_j. \tag{4}$$

Ginzburg-Landau equations: Using a consistent thermodynamic framework (see, for example, [7]), we have obtained the following Ginzburg-Landau equations for the order parameters:

$$\dot\eta_0 = L_{0M}X_0 \quad \text{and} \quad \dot\eta_i = L_{ij}X_i, \quad \text{where} \tag{5}$$

$$
\begin{aligned}
X_0 =& (\mathbf{P}_e^T \cdot \mathbf{F} - J_t\psi_e\mathbf{I}) : \mathbf{U}_t^{-1} \cdot \frac{\partial \mathbf{U}_t}{\partial\eta_0} - J_t\frac{\partial\psi_e}{\partial\eta_0}\Big|_{\mathbf{F}_e} - \rho_0\Delta\psi^\theta\frac{\partial\varphi}{\partial\eta_0} \\
&- \rho_0\bar{A}J\frac{\partial\varphi}{\partial\eta_0}\eta_i^2(1-\eta_i)^2 - \rho_0K_{0ij}\left(2\eta_0(1-\varphi) - \eta_0^2\frac{\partial\varphi}{\partial\eta_0}\right)\eta_i^2(1-\eta_i)^2 \\
&- \rho_0AJ(2\eta_0 - 6\eta_0^2 + 4\eta_0^3) - \frac{J\beta_{ij}}{2}|\nabla\eta_i|^2\frac{\partial\varphi}{\partial\eta_0} + \nabla_0\cdot(\beta_{0M}J\mathbf{F}^{-1}\cdot\nabla\eta_0),
\end{aligned}
$$

$$
\begin{aligned}
X_i =& (\mathbf{P}_e^T \cdot \mathbf{F} - J_t\psi_e\mathbf{I}) : \mathbf{U}_t^{-1} \cdot \frac{\partial \mathbf{U}_t}{\partial\eta_i} - J_t\frac{\partial\psi_e}{\partial\eta_i}\Big|_{\mathbf{F}_e} - \rho_0J\bar{A}\varphi(2\eta_i - 6\eta_i^2 + 4\eta_i^3) \\
&- \rho_0K_{0ij}\eta_0^2(2\eta_i - 6\eta_i^2 + 4\eta_i^3)(1-\varphi) + \nabla_0\cdot(J\varphi\beta_{ij}\mathbf{F}^{-1}\cdot\nabla\eta_i),
\end{aligned} \tag{6}
$$

and L_{0M} and L_{ij} are the kinetic coefficients. A detailed derivation will be presented elsewhere. The order parameters satisfy the following conditions at all external surfaces: $\nabla_0\eta_0 \cdot \mathbf{n}_0 = \nabla_0\eta_i \cdot \mathbf{n}_0 = 0$, where $\mathbf{n}_0$ is the unit normal to the surface in Ω_0.

Results and Discussion

In this section we present the FE solutions of twinning in martensite (variant-variant transformation) in a 2D sample for both KM-I and KM-II, and present the analytical solution for elastic stresses within an interface between two variants in an infinite twinned sample. The following material parameters for NiAl (cubic **A** and tetragonal **M**) have been used: $A_{0M} = 1.62$ GPa, $\bar{A} = 2.40$ GPa, $\beta_{0M} = 3.6 \times 10^{-10}$ N, $\beta_{ij} = 7.5 \times 10^{-11}$ N, $L_{0M} = L_{ij} = 2600$ (Pa-s)$^{-1}$, $\lambda = 74.6$ GPa, $\mu = 72$ GPa, $\theta_e = 215$ K, $\theta = 200$ K, $\Delta\psi^\theta = -1.4679$ MPa. The Bain tensors are $diag(\beta, \alpha, \alpha)$ and $diag(\alpha, \beta, \alpha)$, where $\alpha = 0.922$ and $\beta = 1.215$ are the constant parameters [7].

Twinning: We consider the 2D sample shown in Ω_0 in Fig. 1. Initially η_0 and η_i are randomly distributed between 0 and 1 (Fig. 1a). We apply displacements **u** at all the boundaries, obtained using $\bar{\mathbf{u}} = (\bar{\mathbf{F}} - \bar{\mathbf{I}}) \cdot \mathbf{r}_0$, where $\bar{\mathbf{F}}$ is the average deformation gradient obtained using the equation for **A**-twinned **M** interface of crystallographic theory (see [9, 12]). We have taken $K_{0ij} = 0$. We have solved coupled mechanical equilibrium equations and Eq. (5) considering the generalized plane strain approach. A nonlinear FE code has been written in deal.ii [13], a finite element open source framework. The stationary solutions for both KM-I and KM-II are shown in Fig. 1b and c, respectively, in Ω. The volume fraction of $\mathbf{M}_j$ is close to 30%, in agreement with crystallographic theory [12].

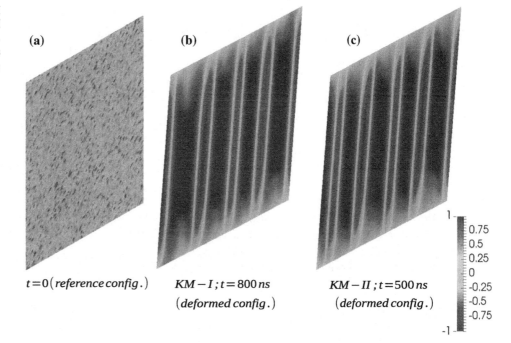

Fig. 1 Twin solutions for KM-I and KM-II obtained using FEM in terms of $\eta_{eq} = 2\eta_0(\eta_i - 0.5)$. Obviously, the dark red indicates $\mathbf{M}_i$ and the dark blue indicates $\mathbf{M}_j$

(a) $t = 0\,(reference\,config.)$

(b) $KM - I; t = 800\,ns$ (deformed config.)

(c) $KM - II; t = 500\,ns$ (deformed config.)

0.75
0.5
0.25
0
-0.25
-0.5
-0.75

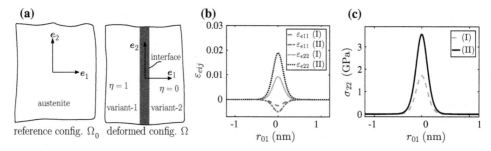

Fig. 2 **a** Schematic of the infinite sample in Ω_0 and Ω. Plots for solutions along $r_{02} = 0$ for KM-I and II: **b** ε_{e11} and ε_{e22}; **c** σ_{22}

Analytical solutions for elastic stresses and strains within variant-variant boundary in an infinite sample: For an infinite A sample that deforms to a twinned sample consisting of variants M_i and M_j and a diffuse interface, we analyze the elastic stresses within the variant-variant boundary, with the structural stress neglected here. Recently, the authors have explained the origin of large elastic stress within the $M_i - M_j$ boundary when U_t is considered from KM-I or KM-II in [14]. For these two models, U_t is not compatible with vanishing elastic strains within the variant-variant boundary. For the Bain tensors rotated by $\pi/4$ about the coordinate axes as shown in Fig. 2a, we obtain the solutions for the transformation stretches and elastic stresses as

$$
\begin{aligned}
U_{t11} &= U_{t22} = 0.5(\alpha + \beta), \\
U_{t12} &= 0.5(\beta - \alpha)(1 - 2\phi_i) \text{ KM} - \text{I};
\end{aligned}
\tag{7}
$$

$$
\begin{aligned}
U_{t11} &= U_{t22} = 0.5(\alpha_i^\phi \beta^{1-\phi_i} + \alpha^{1-\phi_i} \beta_i^\phi), \\
U_{t12} &= 0.5(\alpha_i^\phi \beta^{1-\phi_i} - \alpha^{1-\phi_i} \beta_i^\phi) \text{ KM} - \text{II}; \\
\sigma_{e11} &= \sigma_{12} = 0; \quad \sigma_{e22} = \frac{2\mu(\lambda' + \mu)V_{e22}(V_{e22}^2 - 1)}{2(\lambda' + \mu) - \lambda' V_{e22}^2} \quad \text{for KM} - \text{I \& II},
\end{aligned}
\tag{8}
$$

where the plane stress condition was considered, and $\lambda' = 2\lambda\mu/(\lambda + 2\mu)$. Plots for all the stresses and strains are shown in Fig. 2. Although, the variant-variant transformation for KM-II is isochoric, the elastic stress σ_{e22} is much larger within the interface for KM-II compared to that of KM-I. The maximum value for KM-II is more than twice that for KM-I. While there is no precise knowledge about these stresses, it was expected that they would be negligible, similar to the sharp interface approach. These stresses can cause artificial plastic deformation within the twin interface if dislocational plasticity is included. From this point of view, model KM-I is preferable over KM-II. Note that, for a simple shear model for the transformation deformation gradient, it has been shown that the elastic stresses are identically zero within the entire sample. However, this model cannot be generalized if more than two variants are present.

In summary, a thermodynamically consistent multiphase phase field approach to martensitic transformations with large strain and interfacial stresses was considered. This model consistently penalizes the triple junctions between austenite and any two variants. Twinning in this model was studied for two different kinematic models related to the transformation stretch tensor. Finally, analytical solutions for the elastic stresses within a variant-variant boundary in an infinite twinned sample were presented. This result is important in supporting further development of multiphase phase field approaches. A comprehensive and more general treatment of the model, along with detailed computational procedures, will be presented elsewhere.

Acknowledgements The authors gratefully acknowledge the supports from NSF (CMMI-1536925 and DMR-1434613), ARO (W911NF-12-1-0340), and Iowa State University (Schafer 2050 Challenge Professorship and Vance Coffman Faculty Chair Professorship). Simulations were performed at Extreme Science and Engineering Discovery Environment (XSEDE), allocations TG-MSS140033 and MSS170015.

References

1. Jin YM, Artemev A, Khachaturyan AG (2001) Acta Mater 49:2309–2320
2. Chen LQ (2002) Annu Rev Mater Res 32:113–140
3. Levitas VI, Preston DL (2002) Phys Rev B 66:134207
4. Clayton JD, Knap J (2011) Physica D 240:841–858
5. Hildebrand FE, Miehe C (2012) Philos Mag 92:1–41
6. Levitas VI, Roy AM, Preston DL (2013) Phys Rev B 88:054113
7. Levitas VI (2014) J Mech Phys Solids 70:154–189
8. Levitas VI, Roy AM (2015) Phys Rev B 91:174109
9. Tůma K, Stupkiewicz S, Petryk H (2016) J Mech Phys Solids 95:284–307

10. Steinbach I (2009) Model Simul Mater Sci Eng 17:073001
11. Jog CS (2007) Foundations and applications of mechanics. Volume I: continuum mechanics. Narosa, New Delhi
12. Bhattacharya K (2004) Microstructure of martensite: why it forms and how it gives rise to the shape-memory effect. Oxford University Press, Oxford
13. Bangerth W, Davydov D, Heister T, Heltai L, Kanschat G, Kronbichler M, Maier M, Turcksin B, Wells D (2016) J Numer Math 24 version 8.4
14. Basak A, Levitas VI (2017) Acta Mater 139:174–187

Phase Field Study of Lattice Instability and Nanostructure Evolution in Silicon During Phase Transformation Under Complex Loading

H. Babaei and V. I. Levitas

Abstract

An advanced phase-field approach (PFA) to study martensitic phase transformations is developed for finite strains, particularly taking into account crystal lattice instability conditions under complex triaxial compression-tension loading obtained using molecular dynamics (MD) simulations. Calibration of novel phase-field instability criteria with those from MD simulations for Si I $\leftrightarrow$ Si II phase transformations leads to unexpected interpolation functions for transformation strain and elastic constants. A finite element algorithm and a numerical procedure are developed and implemented using code deal.II. The effect of stress state on lattice instability and nanostructure evolution is studied. Within a specific stress range for which direct and reverse transformation instability stresses coincide, a unique homogeneous, hysteresis-free, and dissipation-free transformation is observed. For such a transformation, a continuum of intermediate phases exists along the transformation path, all in indifferent thermodynamic equilibrium. The absence of interfaces results in the absence of internal stresses and minimizes damage. All these properties are optimal for various PT-related applications.

Keywords

Phase field approach • Martensitic phase transformation • Lattice instability conditions

H. Babaei (✉)
Department of Aerospace Engineering,
Iowa State University, Ames, IA 50011, USA
e-mail: hbabaei@iastate.edu

V. I. Levitas
Ames Laboratory, Division of Materials Science & Engineering,
Departments of Aerospace Engineering, and Mechanical Engineering, and Material Science & Engineering,
Iowa State University, Ames, IA 50011-2274, USA
e-mail: vlevitas@iastate.edu

Introduction

A phase-field approach to martensitic phase transformations (PTs) has been broadly used for simulations [1–5]. In this paper, the phase-field theory of martensitic PTs is advanced to take into account crystal lattice instability conditions of atomistic simulations, in particular, molecular dynamics simulations (MD) of cubic-to-tetragonal Si I $\leftrightarrow$ Si II PTs [6]. Theory has been developed for the case of large elastic and transformational strain as well as for anisotropic and different elastic properties of the phases. Coupled phase-field and elasticity equations have been solved using the means of the finite element method. The impact of triaxial stress states on nucleation as well as propagation of martensitic nanostructures have been studied. A novel homogeneous hysteresis- and dissipation-free first-order PT with stable intermediate phases, revealed in MD simulations, was observed after calibration of PFA with MD.

Complete System of Equations

Kinematics. The deformation gradient $\boldsymbol{F}$ is multiplicatively decomposed into elastic $\boldsymbol{F}_e$ and transformational $\boldsymbol{U}_t$ parts as $\boldsymbol{F} = \boldsymbol{F}_e \cdot \boldsymbol{U}_t$. The transformation deformation gradient is expressed as

$$\boldsymbol{U}_t(\eta) = \boldsymbol{I} + \boldsymbol{\varepsilon}_t \circ \boldsymbol{\varphi}(\boldsymbol{a}_\varepsilon, \boldsymbol{w}_\varepsilon, \eta);$$
$$\boldsymbol{\varphi} := [\boldsymbol{a}_\varepsilon \eta^2 + (10\boldsymbol{\iota} - 3\boldsymbol{a}_\varepsilon + \boldsymbol{w}_\varepsilon)\eta^3 + (3\boldsymbol{a}_\varepsilon - 2\boldsymbol{w}_\varepsilon - 15\boldsymbol{\iota})\eta^4$$
$$+ (6\boldsymbol{\iota} - \boldsymbol{a}_\varepsilon + \boldsymbol{w}_\varepsilon)\eta^5].$$

$$(1)$$

Here, $\boldsymbol{\varepsilon}_t$ is the transformation strain of the product phase P_1; $\boldsymbol{\varphi}$ and consequently, $\boldsymbol{a}_\varepsilon, \boldsymbol{w}_\varepsilon$, and $\boldsymbol{\iota}$ are matrices (not second-rank tensors) having the same non-zero components and symmetry as $\boldsymbol{\varepsilon}_t$; non-zero components of $\boldsymbol{\iota}$ are equal to unity; $\boldsymbol{\varepsilon}_t \circ \boldsymbol{\varphi} := \{\varepsilon_t^{ij} \varphi^{ij}\}$ is the Hadamard product with no summation over i and j and η is an order parameter that

A. P. Stebner and G. B. Olson (eds.), *Proceedings of the International Conference on Martensitic Transformations: Chicago*, The Minerals, Metals & Materials Series, https://doi.org/10.1007/978-3-319-76968-4_26

describes phase transformation from parent phase P_0 ($\eta = 0$) to product phase P_1 ($\eta = 1$).

The Helmholtz free energy per unit mass can be expressed as [7]

$$\bar{\psi}(\boldsymbol{F}, \eta, \theta, \nabla_0 \eta) = J_t \psi^e + \psi^\theta + \psi^\nabla, \qquad (2)$$

where $J_t = det \boldsymbol{U}_t$ and the elastic energy ψ^e, the thermal energy ψ^θ and the gradient energy ψ^∇ are:

$$\psi^e = \frac{1}{2} \boldsymbol{E}_e : \boldsymbol{C}(\eta) : \boldsymbol{E}_e; \quad \boldsymbol{E}_e = \frac{1}{2}(\boldsymbol{F}_e^T \cdot \boldsymbol{F}_e - \boldsymbol{I});$$
$$\boldsymbol{C}(\eta) = \boldsymbol{C}_0 + (\boldsymbol{C}_1 - \boldsymbol{C}_0)\phi_e(\eta);$$
$$\psi^\theta = A\eta^2(1-\eta)^2 + \Delta\psi^\theta(3\eta^2 - 2\eta^3); \quad \psi^\nabla = \frac{\beta}{2}|\nabla_0\eta|^2.$$
$$(3)$$

Here, A is the double-well barrier between phases; $\Delta\psi^\theta$ is the difference between thermal free energies of phases; and β is the coefficient of the gradient energy. The elastic constants $\boldsymbol{C}$ are interpolated between values in both phases using $\phi_e(\eta) = \eta^3(10 - 15\eta + 6\eta^2)$, so that nonlinear in stress terms related to ψ^e do not contribute to the lattice instability condition (6) that, according to MD simulations in [6], is linear in stresses.

The Ginzburg-Landau equation for the PT in the reference configuration is

$$\dot{\eta} = LX$$
$$= L\left(\boldsymbol{P}^T \cdot \boldsymbol{F}_e : \frac{\partial \boldsymbol{U}_t}{\partial \eta} - J_t \frac{\partial \psi^e}{\partial \eta}\Big|_{\boldsymbol{E}_e} - J_t \psi^e \boldsymbol{U}_t^{-1} : \frac{\partial \boldsymbol{U}_t}{\partial \eta} - \frac{\partial \psi^\theta}{\partial \eta} + \beta\nabla_0^2\eta\right),$$
$$(4)$$

where L is the kinetic coefficient and $\boldsymbol{P} = J_t \boldsymbol{F}_e \cdot \frac{\partial \psi^e}{\partial \boldsymbol{E}_e} \cdot \boldsymbol{U}_t^{-1} = J_t \boldsymbol{F}_e \cdot \boldsymbol{C} : \boldsymbol{E}_e \cdot \boldsymbol{U}_t^{-1}$ is the first Piola-Kirchhof stress. The momentum balance equation is $\nabla_0 \cdot \boldsymbol{P} = \boldsymbol{0}$ and the boundary condition for the order parameter is either $\boldsymbol{n}_0 \cdot \nabla_0\eta = 0$ or periodic.

The lattice instability condition for the state $\hat{\eta} = 0$ or 1 derived in [8] is

$$\frac{\partial X(\boldsymbol{P}, \boldsymbol{F}_e, \hat{\eta})}{\partial \eta} = \boldsymbol{P}^T \cdot \boldsymbol{F}_e : \frac{\partial^2 \boldsymbol{U}_t}{\partial \eta^2} - J_t \frac{\partial^2 \psi^e}{\partial \eta^2}\Big|_{\boldsymbol{E}_e} - J_t \psi^e \boldsymbol{U}_t^{-1} : \frac{\partial^2 \boldsymbol{U}_t}{\partial \eta^2}$$
$$- \frac{\partial^2 \psi^\theta}{\partial \eta^2} \geq 0.$$
$$(5)$$

For cubic-to-tetragonal PT under action of the Cauchy stresses σ_i, $i = 1, 2, 3$, normal to the cubic faces, after substituting all terms and neglecting nonlinear in stress term ψ^e, the criteria of direct and reverse PTs transform to

$$P_0 \to P_1 : \quad (\sigma_1 + \sigma_2)\varepsilon_{t1}a_{\varepsilon1} + \sigma_3\varepsilon_{t3}a_{\varepsilon3} \geq \frac{1}{J_e}(A + 3\Delta\psi^\theta);$$

$$P_1 \to P_0 : \quad (\sigma_1 + \sigma_2)\frac{\varepsilon_{t1}w_{\varepsilon1}}{1 + \varepsilon_{t1}} + \frac{\sigma_3\varepsilon_{t3}w_{\varepsilon3}}{1 + \varepsilon_{t3}} \geq \frac{1}{J}(A - 3\Delta\psi^\theta),$$
$$(6)$$

where $J_e = det\boldsymbol{F}_e$ and $J = det\boldsymbol{F}$.

Simulation Results

The following material parameters are used: $L = 2600$ (Pa.s)$^{-1}$, $\beta = 2.59 \times 10^{-10}$ N, $\Delta\psi^\theta = 4.8$ GPaK^{-1}, $C_0^{11} = C_0^{22} = C_0^{33} = 167.5$ GPa, $C_0^{44} = C_0^{55} = C_0^{66} = 80.1$ GPa, $C_1^{11} = C_1^{22} = 174.76$ GPa, $C_1^{33} = 136.68$ GPa, $C_1^{44} = C_1^{55} = 60.24$ GPa, $C_1^{66} = 42.22$ GPa, $C_1^{12} = 102$ GPa, $C_1^{13} = C_1^{23} = 68$ GPa.

PFA calibration with MD. Lattice instability conditions for direct and reverse PTs between cubic Si I and tetragonal Si II were obtained using MD simulations for various combinations of all six stress components in [6]:

$$P_0 \to P_1 : \begin{cases} 0.36(\sigma_1 + \sigma_2) - \sigma_3 \geq 12.29 \text{ GPa} & if \quad -\sigma_3 \geq 6.23 \text{ GPa} \\ 0.19(\sigma_1 + \sigma_2) - \sigma_3 \geq 9.45 \text{ GPa} & otherwise \end{cases}$$
$$P_1 \to P_0 : \quad 0.19(\sigma_1 + \sigma_2) - \sigma_3 \leq 9.45 \text{ GPa}.$$
$$(7)$$

They are shown in Fig. 1a. Components of the transformation strain were obtained as $\varepsilon_t = (0.1753; 0.1753; -0.447)$. Adjusting the PFA instability criteria, Eq. (6), with that of MD, Eq. (7), the following transformation strain interpolation constants a_ε, w_ε and double-well barrier A are obtained: *if* $-\sigma_3 \geq 6.23$ GPa $\to A = 4.88$ GPa, $a_{\varepsilon1} = 3.39$, $a_{\varepsilon3} = 3.69$, $w_{\varepsilon1} = -2.39$, $w_{\varepsilon3} = -2.30$, *otherwise* $\to A = -5.61$ GPa, $a_{\varepsilon1} = 1.13$, $a_{\varepsilon3} = 2.32$, $w_{\varepsilon1} = -3.81$, $w_{\varepsilon3} = -3.67$. Jacobian determinants, J and J_e have been approximated as 0.64 and 0.76, respectively, consistent with their magnitude in our PFA simulations of Si I $\leftrightarrow$ Si II PT.

The stress-strain curve for a single element under uniaxial compressive strain and biaxial lateral tensile stress is shown in Fig. 1b, where it can be clearly seen that the direct and reverse PFA instability stresses are in satisfactory agreement with those found using MD simulations.

Problem formulation. A square thin plate with dimensions $20 \times 20 \times 1$ nm is considered. Periodic boundary conditions for both displacement and order parameter are considered for the top and bottom external faces. Constant stresses $\sigma_1 = \sigma_2$ are prescribed at lateral surfaces. The initial condition for the order parameter is considered to consist of randomly distributed values within the range 0–0.01. A C++ program library called deal.II [9] has been used to develop an FEM

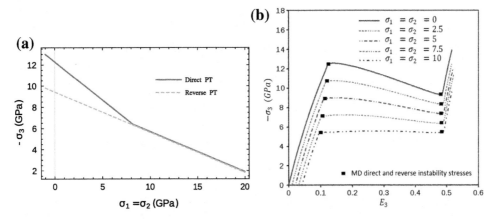

Fig. 1 **a** Crystal lattice instability stresses for direct and reverse PTs, obtained using MD simulations. **b** True compressive stress σ_3 versus Lagrangian strain E_3 for different lateral tensile stresses, applied on a single element and obtained with PFA; dots are instability stresses obtained using MD simulations in [6]

algorithm and numerical procedure. All size and time parameters are normalized to 1 nm and 1 ps, respectively. Results are shown in the current deformed configuration.

Effect of stress state on microstructure evolution. The impact of external stresses on the PT conditions and nanostructure evolution was investigated using three different lateral tensile stresses ($\sigma_1 = \sigma_2 = 0, 5$, and 10 GPa), and nanostructure evolution for all cases is shown in Fig. 2. It can be observed that when no lateral stress is applied, PT processes include nucleation and formation of distinct martensitic bands along with propagation of these bands. Stress-strain curves for all three stress cases are plotted in Fig. 3. After reaching the instability stress for direct PT and during the formation of bands, the stress drops from instability stress to approximately Maxwell (phase equilibrium) stress because of minimization of elastic and surface energies.

For $\sigma_1 = \sigma_2 = 5$, while observing nucleation and propagation of martensitic bands, it can be seen that phase interfaces are not as distinct and straight as for the previous case, and a much smaller part of the band reached complete martensite ($\eta = 1$). Interfaces are widened and there is an intermediate phase between bands. This can be explained by decreased stress hysteresis that in turn results in an increase of the interface width.

The third stress state, $\sigma_1 = \sigma_2 = 10$ GPa, corresponds to the merged region of instability lines for direct and reverse PT, corresponding the absence of stress hysteresis. As shown in Fig. 2, despite considering a heterogeneous initial condition for the order parameter, the system goes through a unique homogeneous hysteresis- and dissipation-free first-order PT with no nucleation or band formation. A similar homogeneous transformation was observed in MD simulation of Si I $\leftrightarrow$ Si II PT for this stress state [6]. The absence of stress hysteresis leads to the disappearance of the energy barrier between phases, in turn making interface energy tend to zero and the interface width tend to infinity. During the PT the

Fig. 2 Evolution of martensitic nanostructure for Si I $\leftrightarrow$ Si II PT under different lateral tensile stresses

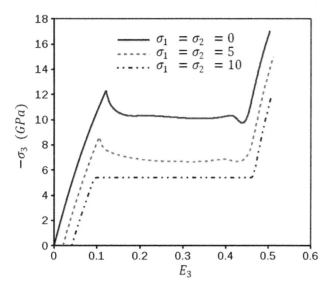

unique homogeneous, hysteresis-free and dissipation-free transformation has been obtained using simulation. An absence of interface minimizes internal stresses and consequent damage. All these properties are optimal for various PT-related applications.Support of NSF (CMMI-1536925 and DMR-1434613), ARO (W911NF-17-1-0225), XSEDE (TG-MSS140033), and ISU (Schafer 2050 Challenge Professorship and Vance Coffman Faculty Chair Professorship) is gratefully acknowledged.

Fig. 3 Global true compressive stress σ_3 versus Lagrangian strain E_3 for three different lateral tensile stresses

system passes through an infinite number (continuum) of intermediate phases, all having the same Gibbs energy and in indifferent equilibrium with product and parent phases. These intermediate phases, possibly with important and interesting properties, can be arrested and studied through fixing compressive strain.

To summarize, a PFA for martensitic PTs has been advanced to take into account crystal lattice instability conditions obtained by MD simulations. The novel instability criterion derived within the PFA has been calibrated with MD simulations for cubic-to-tetragonal Si I $\leftrightarrow$ Si II PTs. The effect of different stress states on the PT process, including nucleation and propagation of martensitic nanostructure, is studied using FEM. For the stress range within which direct and reverse transformation instability stresses coincide, a

References

1. Jin YM, Artemev A, Khachaturyan AG (2001) Three-dimensional phase field model of low-symmetry martensitic transformation in polycrystal: simulation of ζ'_2 martensite in AuCd alloys. Acta Mater 49:2309–2320
2. Chen LQ (2002) Phase-field models for microstructure evolution. Annu Rev Mater Res 32:113–140
3. Seol DJ, Hu SY, Li YL, Chen LQ, Oh KH (2003) Cubic to tetragonal martensitic transformation in a thin film elastically constrained by a substrate. Int J Mater Res 9:221–226
4. Levitas VI, Lee DW (2007) Athermal resistance to an interface motion in phase field theory of microstructure evolution. Phys Rev Lett 99:245701
5. Idesman AV, Cho J-Y, Levitas VI (2008) Finite element modeling of dynamics of martensitic phase transitions. Appl Phys Lett 93:043102
6. Levitas VI, Chen H, Xiong L (2017) Triaxial-stress-induced homogeneous hysteresis-free first-order phase transformations with stable intermediate phases. Phys Rev Lett 118:025701
7. Levitas VI (2014) Phase field approach to martensitic phase transformations with large strains and interface stresses. J Mech Phys Solids 70:154
8. Levitas VI (2013) Phase-field theory for martensitic phase transformations. Int J Plast 49:85
9. Bangerth W, Hartmann R, Kanschat G (2007) deal.II—a general purpose object oriented finite element library. ACM Trans Math Softw 33(4):1–27

Investigation of the Precipitation Processes in NiTi Filaments

Ondřej Tyc, Jan Pilch, Petr Šittner, and Petr Haušild

Abstract

Precipitation in Ni-rich NiTi alloy is an influential process affecting material properties such as fatigue life, tensile strength or thermomechanical response. In this work, a nonconventional method of pulse heating by electric current was employed to recover cold-worked NiTi filaments (50 µm in diameter) and set the initial microstructure without precipitation. The impact of recovery processes and Ni-rich precipitates on mechanical properties can be separated to some extent, owing to the ultrafast electric current annealing. Afterwards, the NiTi filaments were subjected to an aging at 350–520 °C for 2–120 min. The aging at appropriate temperature allows a fluent and predictable adjustment of superelastic response together with an increase in tensile strength of approx. 20%. Small angle neutron scattering experiments were also performed to determine the mean size of Ni-rich precipitates. Less than 10 nm precipitates are created at 350 °C and 2 h aging, which yields a stable superelastic response and higher tensile strength. On the other hand, the filaments annealed at 520 °C show significant instability during superelastic cycling, a decrease in tensile strength and a precipitate mean size that significantly increases up to 500 nm.

O. Tyc (✉) · P. Haušild
Faculty of Nuclear Sciences and Physical Engineering of the CTU, Department of Materials, Trojanova 13, 120 00 Praha 2, Czech Republic
e-mail: tycondre@fjfi.cvut.cz

P. Haušild
e-mail: petr.hausild@fjfi.cvut.cz

O. Tyc · J. Pilch · P. Šittner
Institute of Physics of the Czech Academy of Sciences, Na Slovance 2, 182 21 Praha 8, Czech Republic
e-mail: pilch@fzu.cz

P. Šittner
e-mail: sittner@fzu.cz

Keywords

NiTi alloy • Pulse heating • Precipitation hardening Superelasticity

Introduction

Near-equiatomic intermetallic NiTi alloy, well-known under the commercial name Nitinol, shows superelastic and shape memory properties. Owing to the unique functional behaviour, Nitinol has been utilized in a wide range of branches such as medicine, aerospace and automotive [1–3]. Reversible martensitic transformation from a B2 austenite to a B19′ martensite provides recoverable strain up to 10% [4]. However, cyclic martensitic transformation is also accompanied by irreversible strain [5].

Precipitation hardening is widely used to improve tensile strength and inhibit plastic strain of metallic alloys. Three precipitation stages (Ni_4Ti_3, Ni_3Ti_2 and Ni_3Ti) are known in a Ni-rich NiTi alloy. Ni is preferentially depleted from the matrix, which decreases transformation stresses in a superelastic cycle [6]. From the application point of view, Ni_4Ti_3 precipitates are the most important. These precipitates have rhombohedral crystal structure ($a = 0.6704$ nm, $\alpha = 113.85°$) and a lenticular shape, and they nucleate at relatively low temperatures [4, 7]. Small precipitates are coherent with the matrix and create strain fields in their neighbourhood [8], which increases yield stress and tensile strength [9] and promotes R-phase transformation [6, 10].

Since precipitation is mostly accompanied by other diffusional processes (recovery and recrystallization) during a conventional heat treatment, a unique method of pulse heat treatment [11, 12] was used in this work for recovery of cold-worked filaments, leading to a nanograin microstructure (approx. 100–200 nm). This type of heat treatment takes only milliseconds. Therefore, precipitation is suppressed, which enables to some extent the ability to separate the influence of recovery processes and precipitation on mechanical properties. In this work, NiTi filaments were subjected to heat treatment, followed by superelastic cycling and tensile test to rupture. Moreover, small angle neutron scattering experiments were performed to determine size and volume fraction of precipitates. The aim of this work is to contribute to a better understanding of the influence of heat treatment parameters and precipitation hardening on superelastic behaviour of Ni-rich NiTi filaments.

Experiments and Methods

Ni-rich superelastic filaments (Ti-50.9 at.% Ni) 50 μm in diameter and 35% cold work were used. These filaments were manufactured by Fort Wayne Metals company and tensile strength was 1900 MPa in the cold-worked state. Clamping to stainless steel capillaries on both sides of the filament was needed for easier manipulation and gripping to a testing rig. The length of the sample between the capillaries was 50 mm.

At first, the wire was pulse-heated by electric current (50 W mm^{-3}/50 ms) to trigger recovery processes. Further pulse heating leads to progressive increase in permanent strain without relevant change of transformation stresses. Then, isothermal aging followed in a furnace at various temperatures (350, 395, 460 and 520 °C) and times (2, 5, 15, 30, 60 and 120 min) to create Ni-rich precipitates.

Testing of mechanical properties can be seen in Fig. 1. The main evaluated quantities are: forward transformation stress and transformation strain in the first superelastic cycle and tensile strength, which correlates with a stability of superelastic cycles. The graph represents the initial superelastic response after the first step of heat treatment by electric

current, and the other samples were compared to this state. The tests were performed at 40 °C.

Small angle neutron scattering (SANS) experiments were performed in the PSI facility in Switzerland. The SANS method can determine size and volume fraction of heterogeneities in a whole volume of a tested sample. Thirteen wires of a diameter 1.7 mm and length 30 mm were stacked together in a sample holder in this experiment. However, other methods must be employed to determine type (crystal structure or chemical composition) of these heterogeneities. The samples were annealed at temperatures 350, 400, 450, 500, 550, 600, 650 and 900 °C for 2 h. The wavelength of the neutron source was $5-10 \times 10^{-10}$ m and the sample to ^{3}He detector was set within 1–6 m to cover the size distribution of heterogeneities at different temperatures. A 2D record of scattered neutron intensity was then integrated by azimuthal angle and transferred to a 1D curve from which size and volume fraction of scattering heterogeneities are determined.

Results and Discussion

As mentioned before, formation of Ni-rich precipitates decreases transformation stresses, and small coherent Ni$_4$Ti$_3$ precipitates create strain fields that increase tensile strength and stability of superelastic cycles with respect to accumulation of permanent strain. These phenomena are demonstrated in Fig. 2, which shows behaviour of filaments aged in the furnace at 350 °C for 2–120 min. A gradual decrease in transformation stresses and an increase in tensile strength in time compared to the initial state is evidence of the formation of Ni$_4$Ti$_3$ precipitates at that temperature. Furthermore, R-phase transformation is more pronounced with increasing time. However, precipitation is also responsible for the shortening of transformation strain. Surprisingly, ductility remains approx. 13% regardless of precipitates creation. More detailed information is stated in the following sections.

Forward Transformation Stress

The initial value of forward transformation stress was 685 MPa at 40 °C (see Figs. 2 and 3). The decrease in transformation stress is more pronounced at higher temperatures, and the maximal decrease was about 250 MPa after aging at 460 °C. On the other hand, the decrease in transformation stress is smaller at 520 °C, indicating slower precipitation kinetics above 500 °C. The decrease is almost exponential within the tested temperature range and relatively short time is needed to activate precipitation processes. Changes in functional properties were detectable even after 2 min aging.

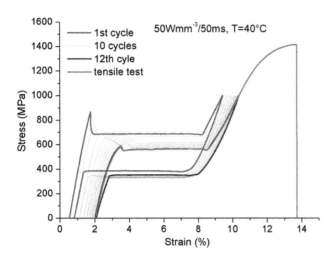

Fig. 1 Testing of mechanical properties, 1st superelastic cycle after heat treatment at 0.001 s^{-1} strain rate (green), 10 cycles at 0.01 s^{-1} strain rate (grey), 12th cycle (blue) and tensile test to rupture (red) again at 0.001 s^{-1} strain rate

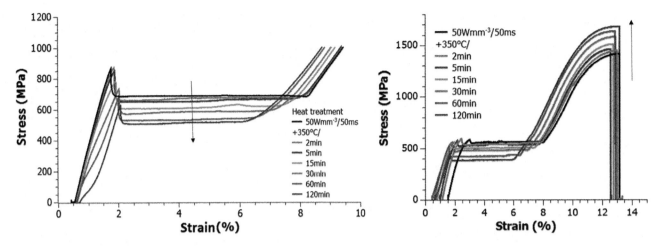

Fig. 2 Influence of aging time on transformation stress (left) and tensile strength (annealing temperature 350 °C)

Transformation Strain

The next important result is the shortening of transformation strain after the aging. The situation is similar to the transformation stress decrease. The fastest decrease in transformation strain was measured at 460 °C again. The initial value of transformation strain was 6.8% and then was reduced up to 4.5% as shown in Fig. 4. The first reason is a decrease in superelastic NiTi matrix content (precipitates are not superelastic). Maybe there is a change in a character of martensite bend front propagation and variants of created martensite in strain fields of the precipitates.

Tensile Strength

There is a significant difference between low-temperature and high-temperature aging (see Fig. 5). On one hand, the increase in tensile strength was measured at 350 and 395 °C.

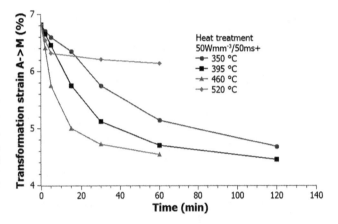

Fig. 4 Evolution of transformation strain after aging at different temperatures and time

On the other hand, tensile strength was decreased after aging at 520 °C. Small precipitates are created at low temperatures, as shown in a section below, and a gradual increase in tensile strength was observed from 1430 to 1680 MPa. Then, a steep increase followed by a decrease was measured at 460 °C, which could be attributed to coarsening of precipitates after 15 min aging. Finally, only a decrease in tensile strength was observed at 520 °C, enhanced by grain growth and maybe semi-coherent Ni_4Ti_3 or even Ni_3Ti_2 precipitates were created as no sign of R-phase transformation has been observed in the superelastic tests.

Small Angle Neutron Scattering

As mentioned in the description of experiments, a great advantage of the SANS method is the ability to determine a mean size of heterogeneities in a large volume of the NiTi matrix, unlike TEM microscopy. At first, 8 nm precipitates

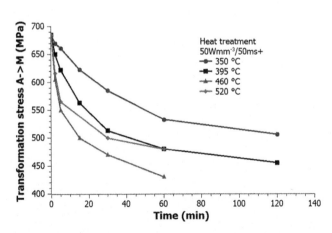

Fig. 3 Evolution of transformation stresses after aging at different temperatures and time

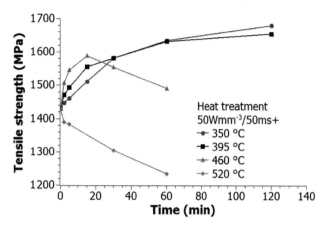

Fig. 5 Evolution of tensile strength after aging at different temperatures and time

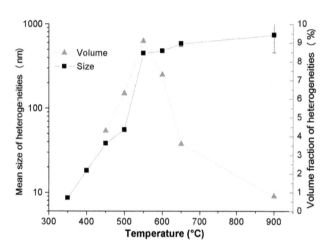

Fig. 6 Mean size and volume fraction of heterogeneities in NiTi matrix determined by SANS method

are created at 350 °C (Fig. 6) and effectively increase tensile strength as stated in the previous section. Size of precipitates gradually increases with temperature. However, there is a huge gap between 500 and 550 °C from 50 nm to almost

500 nm. That phenomenon corresponds to the decrease in tensile strength above 500 °C. Then, the size remains almost constant, but volume fraction decreases as the solubility of excess Ni in the matrix increases. Finally, only inclusions of TiC remain at 900 °C. Those inclusions are always present in the NiTi matrix, and the mean size is approx. 700 nm with larger scatter than precipitates, which corresponds to SEM observation of a transverse cut as shown in Fig. 7.

Conclusions

Heat treatment of 50 μm superelastic NiTi filaments was performed at various temperatures and times to determine the influence of precipitation on mechanical properties. Pulse heat treatment by electric current was employed to separate the influence of recovery processes and precipitation. A small angle neutron scattering experiment was also performed to determine size and volume fraction of precipitates. The results of these experiments can be summarized in the following points:

- Formation of small precipitates (approx. 10 nm) at 350 °C yields the best results with respect to an increase in tensile strength (1430 MPa → 1680 MPa) within the tested temperature range.
- No significant change in ductility (approx. 13%) introduced by precipitates was observed.
- A sharp increase in size of precipitates was observed in SANS experiment between 500 and 550 °C corresponding to the decrease in tensile strength and stability of superelastic cycles.
- A decrease in forward transformation stress of 250 MPa was achieved by precipitation processes.
- Precipitation can significantly reduce transformation strain.

Fig. 7 a SEM microscopy of a transverse cut of the wire annealed 350 °C/60 min; **b** detail in a higher resolution. Presence of structural heterogeneities (TiC carbides and Ti_2Ni intermatallic particles)

(a)

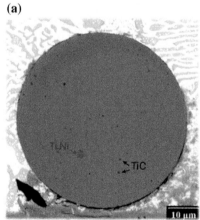

(b)

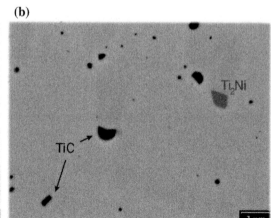

Acknowledgements We kindly acknowledge the financial support from the student grant SGS ČVUT no. SGS16/249/OHK4/3T/14 and the Czech science foundation through project no. 16-20264S.

References

1. Mahtabi MJ, Shamsaei N, Mitchell MR (2015) Fatigue of Nitinol: the state-of-the-art and ongoing challenges. J Mech Behav Biomed Mater 50:228–254
2. Hartl DJ, Lagoudas DC (2007) Aerospace applications of shape memory alloys. J Aerospace Engineering 221:535–552. https://doi.org/10.1243/09544100JAERO211
3. Rahim M, Frenzel J, Frotscher M, Pfetzing-Micklich J, Steegmüller R, Wohlschlögel M, Mughrabi H, Eggeler G (2013) Impurity levels and fatigue lives of pseudoelastic NiTi shape memory alloys. Acta Mater 61:3667–3686
4. Otsuka K, Wayman CM (1998) Shape memory materials. University Press, Cambridge
5. Sedmák P, Sittner P, Pilch J, Curfs C (2015) Instability of cyclic superelastic deformation of NiTi investigated by synchrotron X-ray diffraction. Acta Mater 94:257–270
6. Wang X, Kustov S, Li K, Schryvers D, Verlinden B, Van Humbeeck J (2015) Effect of nanoprecipitates on the transformation behavior and functional properties of a Ti–50.8 at.% Ni alloy with micron-sized grains. Acta Mater 82:224–233
7. Bojda O, Eggeler G, Dlouhý A (2005) Precipitation of Ni_4Ti_3-variants in a polycrystalline Ni-rich NiTi shape memory alloy. Scripta Mater 53:99–104
8. Tirry W, Schryvers D (2005) Quantitative determination of strain fields around Ni_4Ti_3 precipitates in NiTi. Acta Mater 53:1041–1049
9. Pelton AR, Dicello J, Miyazaki (2000) Optimisation of processing and properties of medical grade Nitinol wire. Min Invas Ther Allied Technol 9:107–118
10. Otsuka K, Ren X (2005) Physical metallurgy of Ti–Ni-based shape memory alloys. Prog Mater Sci 50:511–678
11. Delville R, Malard J, Pilch J, Sittner P, Schryvers D (2010) Microstructure changes during non-conventional heat treatment of thin Ni–Ti wires by pulsed electric current studied by transmission electron microscopy. Acta Mater 58:4503–4515
12. Pilch J, Heller L, Sittner P (2009) Final thermomechanical treatment of thin NiTi filaments for textile applications by electric current. Esomat 2009. EDP Sciences. https://doi.org/10.1051/esomat/20090524

Effect of Microstructure on the Low-Cycle Fatigue Properties of a Fe–15Mn–10Cr–8Ni–4Si Austenitic Alloy

Ilya Nikulin, Takahiro Sawaguchi, and Yuuji Kimura

Abstract

In the present study, the effect of the initial austenitic structure on the low-cycle fatigue (LCF) properties and ε-martensitic transformation (ε-MT) was studied in the Fe–15Mn–10Cr–8Ni–4Si seismic damping alloy under an axial strain control mode with total strain amplitude, $\Delta\varepsilon_t/2$, of 0.01. The microstructures with various grain size and texture conditions were produced by warm rolling followed by annealing at temperatures that ranged from 600 to 900 °C. It was found that the increase in the austenitic grain size observed in the studied temperature interval generally does not affect martensitic transformation and the LCF resistance of the studied alloy. However, strong texture and substructure remaining in the alloy after low temperature annealing at T ≤ 700 °C inhibit strain-induced phase transformation and reduce fatigue resistance of the studied alloy. As a result, the alloy annealed at T ≥ 800 °C shows higher fatigue resistance than that one annealed at T ≤ 700 °C.

Keywords

High-Mn steels • Low-cycle fatigue • Microstructure

Introduction

High-Mn austenitic alloys are attractive engineering materials because of a good combination of mechanical properties, high damping capacity, and reasonable cost. These advantages make it possible to consider high-Mn alloys as the construction materials for civil engineering applications in earthquake regions. In these applications, the structural components are often undergoing the cyclic loading and overloading that are major causes of failure. Therefore, the low-cycle fatigue (LCF) resistance should be taken into account at the development and evaluation of high-Mn alloys.

The present study examines the effect of the annealed microstructure on the low-cycle fatigue (LCF) resistance, fatigued microstructure, and strain-induced ε-martensitic transformation (ε-MT) of the Fe–15Mn–10Cr–8Ni–4Si seismic damping alloy [1]. After conventional thermomechanical processing, including forging and rolling at 1000 °C followed by annealing at 1000 °C for 1 h, the alloy shows fully annealed coarse grained (average austenite grain size, Dav = 95 μm) austenitic microstructure, excellent LCF resistance in a wide range of the strain amplitudes, and low yield stress, $\sigma_{0.2}$, of 230 MPa [2]. It has been shown that high fatigue resistance of the Fe–15Mn–10Cr–8Ni–4Si alloy is associated with the moderate kinetics of the strain-induced ε-MT accompanied by the planar glide of partial dislocation in austenite. It is known that the microstructure variables such as austenite grain size, texture, and dislocation structure influence the phase transformation behavior and consequently properties of high-Mn alloys [3–5]. The grain size, dislocation structure, and texture can be controlled by selection of the appropriate annealing temperature [6]. In the present study, the effect of a various annealed microstructure on the LCF behaviour and fatigue microstructure of the Fe–15Mn–10Cr–8Ni–4Si alloy subjected to symmetrical uniaxial tension–compression loading at a total strain control mode is investigated. The importance of the homogeneity of

I. Nikulin · T. Sawaguchi (✉) · Y. Kimura
National Institute for Materials Science, 1-2-1 Sengen, Tsukuba, Ibaraki 305-0047, Japan
e-mail: sawaguchi.takahiro@nims.go.jp

I. Nikulin
e-mail: nikulin.ilya@nims.go.jp

Y. Kimura
e-mail: kimura.yuuji@nims.go.jp

© The Minerals, Metals & Materials Society 2018
A. P. Stebner and G. B. Olson (eds.), *Proceedings of the International Conference on Martensitic Transformations: Chicago*, The Minerals, Metals & Materials Series, https://doi.org/10.1007/978-3-319-76968-4_28

the martensitic transformation on the LCF resistance of the Fe–15Mn–10Cr–8Ni–4Si alloy is discussed.

Materials and Experimental Procedure

The Fe–15Mn–10Cr–8Ni–4Si austenitic alloy with a chemical composition of Fe–14.9%Mn–10.3%Cr–8Ni–3.7%Si–0.004%C (in wt%) was produced by induction furnace melting. The alloy was rolled at 800 °C with a rolling reduction of 50%, annealed at 600, 700, 800, and 900 °C for 1 h and water cooled.

Smooth cylindrical samples with a gauge diameter of 8 mm and a gauge length of 12.5 mm were machined for fatigue tests so that the specimen longitudinal axis was aligned along the former rolling direction. LCF tests were carried out with a symmetric tension-compression loading under an axial strain control mode with the total strain amplitude, $\Delta\varepsilon_t/2$ of 0.01. The mean strain was maintained at zero (minimum to maximum strain ratio, $R_\varepsilon = -1$) in all tests. A strain-gage clip-on extensometer, "Shimadzu dynastrain", was used to control the total strain range, $\Delta\varepsilon_t$. The strain rate was fixed at 0.004 s^{-1}. The fatigue life, N_f, of the samples was defined in accordance with the ASTM E606/E606 M-12 standard as the number of cycles at which the maximum stress decreases by approximately 25% compared to the one at mid-life.

The microstructure before and after LCF tests was examined by electron-backscattering diffraction (EBSD) and X-ray diffraction (XRD) analysis. The sections containing the loading direction (LD) and transversal direction (TD) were used for observations of as-annealed and fatigued microstructures. The details of the microstructural observations and samples preparation were described in a previous report [2].

Results and Discussion

Figure 1 shows EBSD (100) inverse pole figure (IPF) maps, the corresponding phase-grain boundary maps, and kernel average misorientation (KAM) maps of the microstructures evolved in the alloy at 700 and 800 °C. These microstructures are characteristic microstructures developed by annealing at T ≤ 700 and T ≥ 800 °C, respectively. Fully austenitic structure was revealed (Fig. 1b, e) in the alloy at all annealing conditions. The coarse, elongated grains and roughly uniaxial fine grains were observed in the alloy annealed at T ≤ 700 °C (Fig. 1a, b, and Table 1). A high fraction of low angular boundaries (LABs), V_{LABs} (Table 1), associated with substructure was found inside coarse grains. The KAM map (Fig. 1c), qualitatively indicating the strain distribution in terms of dislocation density [7], shows

homogenous strain distribution within the fine grains (Fig. 1c). Large strain localization due to substructure was observed in the coarse grains. Thus, the fine and coarse grains were considered as recrystallized and un-recrystallized grains [8]. The volume fraction of recrystallized fine grain, V_{RexG}, was 0.4 and 0.5 after annealing at 600 and 700 °C (Table 1), respectively. The disappearance of the un-recrystallized grain and coarsening of the recrystallized grains were observed in alloy annealed at T 800 °C (Fig. 1d, e and Table 1). No strain localization was found in the alloy annealed at T ≥ 800 °C (Fig. 1f).

Figure 2 shows the evolution of the orientation distribution function (ODF) in $\varphi_2 = 45°$ section with the annealed temperatures. Some ideal texture components for face-centered cubic (FCC) materials are shown in Fig. 2a. Pronounced texture was observed in the alloy annealed at T ≤ 700 °C. EBSD results showed that un-recrystallized grains are responsible for the strong texture components of the alloy annealed at T ≤ 700 °C. The annealing at T 800 °C produces much weaker texture in the fully recrystallized alloy (Fig. 2c, d). In the alloy annealed at T 800 °C, the high orientation densities are placed around Goss, Goss/Brass, and Brass components (i.e., α-fiber) (Fig. 2a–c). Strong orientation density was also observed around Cu and/or Goss Twin components after annealing at T ≤ 700 °C. In general, the intensity of the above-listed texture components decreases as the annealing temperature increases from 600 to 800 °C. Completely different texture was observed in the alloy annealed above rolling temperature (Fig. 2d). Relatively high orientation densities were observed around Cube and Goss/Brass components and also around {221}<463>, {113}<131>, and {112}<351> components.

Thus, three types of the microstructure were developed in the studied alloy by annealing at temperatures from 600 to 900 °C. Strongly textured, partly recrystallized microstructure is formed at T ≤ 700 °C since the recrystallization is inhibited at temperatures below the hot rolling temperature [6]. A weak-textured, fully recrystallized microstructure is formed at 800 °C. Finally, the fully recrystallized microstructure with no texture was developed at 900 °C.

The effect of the annealing temperature on the fatigue life, N_f, of the alloy subjected to LCF at $\Delta\varepsilon_t/2$ of 0.01 is shown in Table 1. As can be seen, the alloy annealed at T ≤ 700 °C shows lower fatigue resistance than the alloy annealed at T ≥ 800 °C. The highest N_f was observed in the alloy annealed at 800 °C, and no further improvement in fatigue resistance was observed at 900 °C, in spite of the noticeable difference in grain size (Table 1). It is interesting to note that the alloy with partly recrystallized microstructure and low fatigue life shows higher tensile 0.2% yield strength, $\sigma_{0.2}$ ($\sigma_{0.2}$ was obtained from the first quarter of the first hysteresis loop), while the fully recrystallized alloy demonstrates

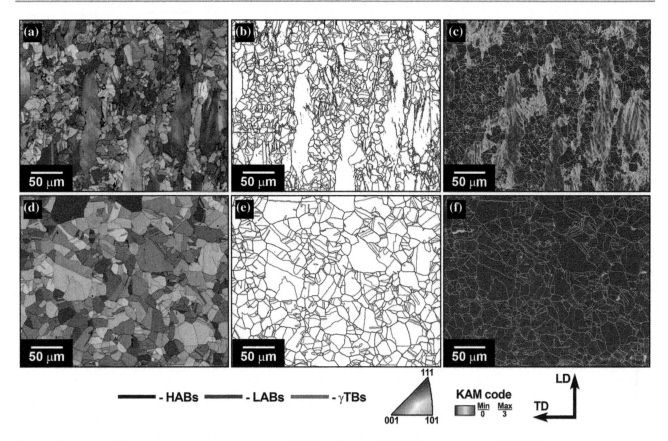

Fig. 1 EBSD maps of the microstructure annealed for 1 h at 700 °C (**a**, **b**, **c**) and 800 °C (**d**, **e**, **f**): **a**, **d** the (100) IPF, **b**, **e** phase-grain boundary and **c**, **f** KAM maps. LABs, HABs and TBs are the low angular, high angular and twin boundaries, respectively

Table 1 Microstructural and LCF parameters of the alloy annealed at 600–900 °C for 1 h

	600 °C	700 °C	800 °C	900 °C
As-annealed				
LABs fraction, V_{LABs}	0.46	0.44	–	–
RexG fraction, V_{RexG}	0.40	0.50	1	1
D_{RexG} (μm)[a]	10.3/8.4	11.6/9.5	18.9/17.4	33.0/29.6
D_{UnRexG} (μm)[a]	24.8/12.4	36.6/11.4	–	–
Low cycle fatigue				
N_f, cycles	7550	9700	14250	13830
0.2% yield strength, $\sigma_{0.2}$ (MPa)	450	400	280	240
ε-martensite fraction (%)	65	67	86	78

[a]Numerator and denominator are grain sizes measured in the longitudinal and transverse directions, respectively

reduced $\sigma_{0.2}$ (Table 1). This is opposite to the early observed effect of the alloying elements on the $\sigma_{0.2}$ and the fatigue resistance of Fe–15Mn–10Cr–8Ni–xSi and Fe–30Mn–(6-x) Si–xAl alloys [9, 10], in which high fatigue resistance was observed in the alloys with higher yield strength. This contradictory correlation of $\sigma_{0.2}$ and N_f may be due to the different nature of hardening/softening mechanisms introduced by thermomechanical processing and by alloying, respectively.

The effect of the annealing temperature on the variation of the stress amplitude with the number of cycles is shown in a semi-logarithmic scale in Fig. 3a. Figure 3b shows evolution of the cyclic hardening degree calculated in accordance with [11]. In general, an increase in annealing temperature decreases the cyclic stress amplitude and increases cyclic hardening degree (Fig. 3). Large cyclic stress amplitudes and weak cyclic hardening cause a relatively stable cyclic stress response during LCF deformation were observed in

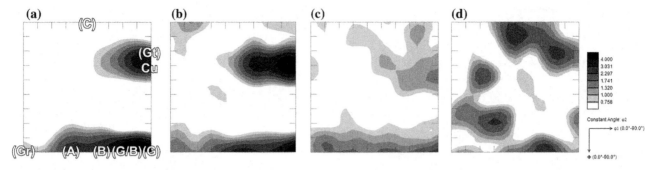

Fig. 2 Evolution of the $\varphi_2 = 45°$ section of the orientation distribution function (ODF) with annealing temperature: **a** 600 °C; **b** 700 °C; **c** 800 °C; and **d** 900 °C. Cube (C), Copper (Cu), Goss twin (Gt), Goss (G), Goss/Brass (G/B), Brass (B), (A), and Rotated Goss (Gr) components are indicated

Fig. 3 Changes in stress amplitude (**a**) and hardening degree (**b**) with increasing number of cycles during the LCF test at $\Delta\varepsilon_t/2$ of 0.01

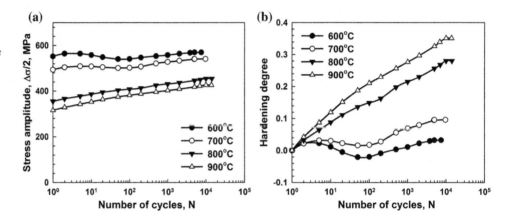

the alloy annealed at T ≤ 700 °C. The alloy annealed at T ≥ 800 °C shows lower initial stress amplitudes and pronounced cyclic hardening in comparison with the alloy annealed at T ≤ 700 °C (Fig. 3b).

The effect of the LCF deformation on the annealed microstructure was studied in samples deformed at total strain amplitude, $\Delta\varepsilon_t/2$, of 0.01. Figure 4 shows EBSD IPF maps, the corresponding phase-grain boundary maps, and KAM maps of the microstructure evolved during LCF deformation in the alloy annealed at 700 and 800 °C. The variation of the ε-martensite fraction with the annealing temperature measured by XRD analysis in fatigue fractured samples is listed in Table 1. In general, the alloy annealed at T ≥ 800 °C demonstrates higher fatigue life and fraction of ε-martensite as compared to the alloy annealed at T 700 °C. It is interesting to note that the highest fatigue life was observed in the alloy annealed at T = 800 °C and this alloy also shows the highest fraction of ε-martensite. On the other hand, the alloy annealed at T = 600 °C shows lowest fatigue life and fraction of ε-martensite. It was found that in the alloy annealed at T ≤ 700 °C, the martensitic transformation is more complete in fine recrystallized grains than in areas of the coarse un-recrystallized grains (Fig. 4a, b),

despite the earlier established fact that coarse grains are more likely to undergo martensitic transformation in iron-based high-Mn shape memory alloys [3]. In other words, in the alloy annealed at T ≤ 700 °C, the martensitic transformation is more likely to appear in the fine recrystallized grains with low KAM (Fig. 4b, c). At the same time, homogeneous martensitic transformation was observed in the alloy annealed at T ≥ 800 °C, which is associated with higher uniformity (i.e., uniform KAM distribution) of the annealed austenitic structure developed at higher temperatures (Fig. 1). It should be noted that inhomogeneity in KAM observed in annealed austenite of the alloy annealed at 700 °C (Fig. 1c) is inherited by the martensite developed by LCF deformation (Fig. 4c).

Figure 5 shows ODF sections at $\varphi_2 = 45°$ of the retained austenite of the alloy subjected to LCF at $\Delta\varepsilon_t/2 = 0.01$. As can be seen, the main texture components observed in the as-annealed alloy (Fig. 1) remain during LCF deformation, except for the alloy annealed at 900 °C (Fig. 5). Noticeable increase in the intensity of the texture along the α-fiber was found after LCF deformation in the alloy annealed at T 700 °C (Figs. 2a, b and 5a, b), suggesting contribution of the dislocation slip to the texture evolution by rotation of grain

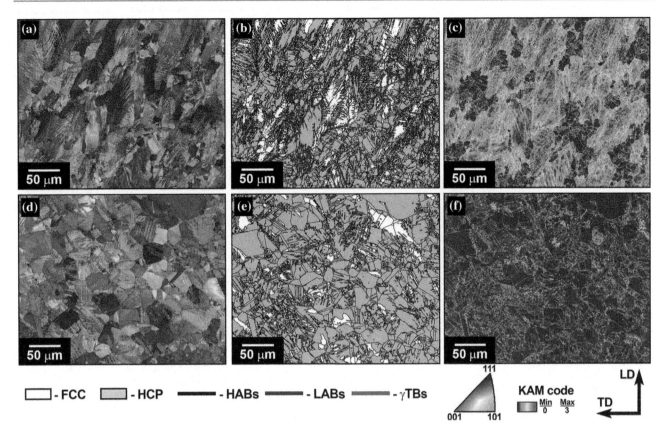

Fig. 4 EBSD maps of the alloy subjected to LCF at $\Delta\varepsilon_t/2 = 0.01$ following to annealing at 700 °C (**a**, **b**, **c**) and 800 °C (**d**, **e**, **f**): **a**, **d** the (100) IPF, **b**, **e** phase, and **c**, **f** KAM maps. FCC and HCP are the austenite and ε-martensite indicated on phase maps. LABs, HABs and TBs are the low angular, high angular, and twin boundaries, respectively

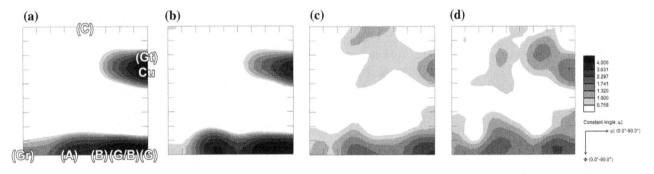

Fig. 5 Effect of the LCF deformation on the evolution of the $\varphi_2 = 45°$ sections of the orientation distribution function (ODF) with annealing temperature: **a** 600 °C; **b** 700 °C; **c** 800 °C; and **d** 900 °C

orientations. On the other hand, weak change in intensity of the texture components formed in the parent austenite of the alloy annealed at $T \geq 800$ °C was observed (Figs. 2c, d and 5c, d). In addition, a clear intensity near the A orientation can be recognized in the alloy annealed at $T \geq 800$ °C.

Figure 6 shows ODF at φ_2 of 0° and 30° of the martensite developed at LCF deformation. High and low intensity of the ε-martensite texture was observed in the alloy annealed at $T \leq 700$ °C and $T \geq 800$ °C, respectively. High intensity of ODF around {01-13}<0-332>, {11-25} <-1-121>, and {11-20}<1-101> was observed on the φ_2 of 0° and 30° sections. The additional peak of intensity around {01-10}<2-1-1 2> was observed in the alloy annealed at $T \geq 800$ °C, suggesting selective development of ε martensite during strain-induced martensitic transformation of the alloy annealed at $T \leq 700$ °C and $T \geq 800$ °C.

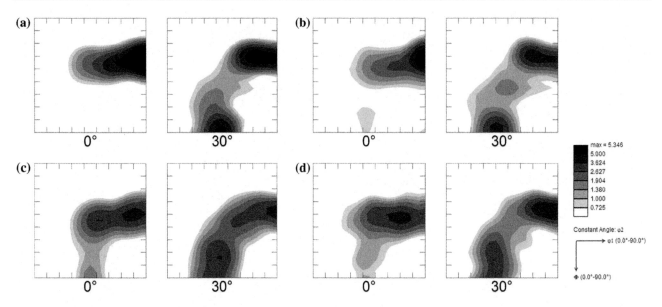

Fig. 6 Evolution of the φ_2 of 0 and 30° sections of the orientation distribution function (ODF) of the ε-martensite induced by LCF deformation at of $\Delta\varepsilon_t/2 = 0.01$ in the alloy annealed temperature: **a** 600 °C; **b** 700 °C; **c** 800 °C; and **d** 900 °C

Summary

In the examined strain interval, the alloy annealed at T ≥ 800 °C shows higher fatigue resistance than that one annealed at T ≤ 700 °C. Microstructural observations of the as-annealed and fatigue fractured samples showed that the homogeneity of the as-annealed microstructure affects the martensitic phase transformation, fatigue resistance, and fatigue behaviour of the Fe–15Mn–10Cr–8Ni–4Si austenitic seismic damping alloy. The appearance of the strong deformation texture and/or substructure in the austenitic matrix reduces fatigue resistance of the studied alloy annealed at T ≤ 700 °C by the development of the ε-martensite in limited orientations and increased inhomogeneity of the ε-MT. Uniform ε-martensitic transformation observed in the alloy annealed at T ≥ 800 °C is required to increase LCF resistance.

Acknowledgements The authors are grateful to the Materials Manufacturing and Engineering Station, NIMS, for materials processing.

References

1. Sawaguchi T, Maruyama T, Otsuka H, Kushibe A, Inoue Y, Tsuzaki K (2016) Design concept and applications of Fe–Mn–Si-based alloys—from shape-memory to seismic response control. Mater Trans 57:283–293

2. Nikulin I, Sawaguchi T, Kushibe A, Inoue Y, Otsuka H, Tsuzaki K (2016) Effect of strain amplitude on the low-cycle fatigue behavior of a new Fe–15Mn–10Cr–8Ni–4Si seismic damping alloy. Int J Fatigue 88:132–141

3. Jiang BH, Sun L, Li R, Hsu TY (1995) Influence of austenite grain size on γ-ε martensitic transformation temperature in Fe–Mn–Si–Cr alloys. Scr Met Mat 33:63–68

4. Hilkhuijsen P, Geijselaers HJM, Bor TC, Perdahcioglu ES, Boogaard AH, Akkerman R (2013) Strain direction dependency of martensitic transformation in austenitic stainless steels: the effect of g-texture. Mater Sci Eng, A 573:100–105

5. Gu Q, Van Humbeeck J, Delaey L (1994) A review on the martensitic transformation and shape memory effect in Fe–Mn–Si alloys. J Phys IV France 04:C3-135–C3-144

6. Humphreys FJ, Hatherly M (2005) Recrystallization and related annealing phenomena. Elsevier, Oxford

7. Wright SI, Nowell MM, Field DP (2011) A review of strain analysis using electron backscatter diffraction. Microsc Microanal 17:316–329

8. Mannan P, Saleh AA, Gazder AA, Casillas G, Pereloma EV (2016) Microstructure and micro-texture evolution during the dynamic recrystallisation of a Ni–30Fe–Nb–C model alloy. J Alloys Compd 689:250–265

9. Nikulin I, Sawaguchi T, Ogawa K, Tsuzaki K (2016) Effect of γ to ε martensitic transformation on low-cycle fatigue behaviour and fatigue microstructure of Fe–15Mn–10Cr–8Ni–xSi austenitic alloys. Acta Mater 105:207–218

10. Nikulin I, Sawaguchi T, Tsuzaki K (2013) Effect of alloying composition on low-cycle fatigue properties and microstructure of Fe–30Mn–(6-x)Si–xAl TRIP/TWIP alloys. Mater Sci Eng, A 587:192–200

11. Skelton RP (1987) High temperature fatigue: properties and prediction. Elsevier, London/New York

Martensitic Transformations of Ni–Mn–X Heusler Alloys with X = Ga, In and Sn

Peter Entel, Markus E. Gruner, Mehmet Acet, Asli Çahır, Raymundo Arroyave, Thien Duong, Anjana Talapatra, and Ibrahim Karaman

Abstract

Martensitic transformations of rapidly quenched and less rapidly cooled Heusler alloys of type Ni–Mn–X with X = Ga, In and Sn are investigated by ab initio calculations. For the rapidly cooled alloys, we obtain the magnetocaloric properties near the magnetocaloric transition. For the less rapidly quenched alloys these magnetocaloric properties start to change considerably. This shows that none of the Heulser alloys is in thermal equilibrium. Instead, each alloy transforms during temper-annealing into a dual-phase composite alloy. The two phases are identified to be cubic Ni–Mn–X and tetragonal NiMn.

Keywords

Magnetocaloric effect • Heusler alloys • First order phase transitions • Segregation

P. Entel (✉) · M. E. Gruner · M. Acet
Faculty of Physics and CENIDE, University of Duisburg-Essen, 47048 Duisburg, Germany
e-mail: entel@thp.uni-due.de

M. E. Gruner
e-mail: markus.gruner@uni-due.de

M. Acet
e-mail: mehmet.acet@uni-due.de

A. Çahır
Metalurji Ve Malzeme Mühendisliği Bölümü, Muğla Üniversitesi, 4800 Muğla, Turkey
e-mail: cahir@mu.edu.tr

R. Arroyave · T. Duong · A. Talapatra · I. Karaman
Department of Materials Science & Engineering, Texas A&M University, College Station, TX 77843, USA
e-mail: raymundo@fastmail.fm

T. Duong
e-mail: teryduong84@tamu.edu

A. Talapatra
e-mail: anjanatalapatra@tamu.edu

I. Karaman
e-mail: ikaraman@tamu.edu

Introduction

We have investigated the magnetocaloric properties of Ni–Mn–(Ga, In, Sn) alloys by means of first-principles ab initio calculations and Monte Carlo simulations. These alloys show giant magnetocaloric effects, especially if a few percent Co and Cr are added to the alloys [1, 2]. The extra magnetic ions occupy regular sites of the Heusler lattice and can enhance the magnetization jump and adiabatic temperature change at the magnetocaloric transition. On the other hand, we have systematically investigated the segregation tendencies which occur in these alloys near the martensitic transformation, see [3–7].

Other remarkable features like the magnetic shape-memory effect have recently been highlighted by Planes et al. [8].

$$\Delta S_{mag}(T, H) = \mu_0 \int_0^H dH' \left(\frac{\partial M}{\partial T}\right)_{H'},$$
$$\Delta T_{ad}(T, H) = -T \frac{\Delta S_{mag}(T, H)}{C(T, H)}. \tag{1}$$

Figure 1 shows the magnetocaloric effect of $Ni_{45}Co_5Mn_{32}Cr_5In_{13}$ with an adiabatic temperature change of nearly 10 K. This is a record value for Heusler alloys and is close to $\Delta T_{ad} = 12.9$ K of Fe–Rh [9, 10] and 9.2 K for $Fe_{49}Rh_{51}$ on the first application of a magnetic field of $\Delta \mu_0 H = 1.9$ T, which remains as high as 6.2 K during the cycling in alternated fields of the same magnitude [11]. It has been shown that the cooperative contribution of electronic, magnetic and vibrational degrees of freedom is the main origin of the giant magnetocaloric effect in FeRh [12]. The alloy $Ni_{45}Co_5Mn_{32}Cr_5In_{13}$ is stable with respect to decomposition into all individual elements.

© The Minerals, Metals & Materials Society 2018
A. P. Stebner and G. B. Olson (eds.), *Proceedings of the International Conference on Martensitic Transformations: Chicago*,
The Minerals, Metals & Materials Series, https://doi.org/10.1007/978-3-319-76968-4_29

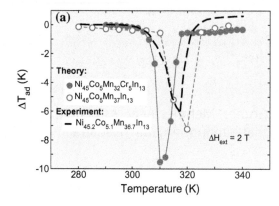 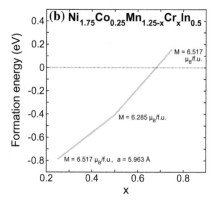

Fig. 1 **a** Adiabatic temperature change of Ni–Mn–In alloys doped with Co (open circles) compared with experimental data (dashed line) and doped with Co and Cr (blue circles). In the latter case, the adiabatic temperature change approaches 10 K in an external field of 2 T, which is close to the magnetocaloric effect shown by Fe–Rh. **b** Formation (mixing) energy of $Ni_7CoMn_5CrIn_2$ calculated as difference between the total energy of the 16-atom $L2_1$ supercell with ferromagnetic order and all partial total energies of the pure elements (with stability up to $x = 6$ at.% Cr)

$$E_{mixing} = E_{Ni_7Co_1Mn_5Cr_1In_2} - \frac{7}{16}E_{Ni} - \frac{1}{16}E_{Co} - \frac{5}{16}E_{Mn} \\ - \frac{1}{16}E_{Cr} - \frac{2}{16}E_{In}. \tag{2}$$

shown in Fig. 1b [13]. However, this is not the correct segregation, as the alloy will transform during temper-annealing to a dual-phase composite.

It seems that Co and Cr can have a cooperative, beneficial effect on the magnetostructural transition, which enhances the size of the magnetic jump [2].

Dependence on the Cooling Rate After Heat Treatment

In this section we briefly discuss the magnetization jump of Ni–Mn–In alloys as a function of the heat treatment, which yields different degrees of order in the alloys. The influence of annealing on the magnetization curves is shown in Fig. 2 [14].

For high annealing temperature A_Q, for example, $A_Q = 1173$ K, and rapid quenching of $Ni_{50.2}Mn_{33.4}In_{16.4}$, the form of the magnetization curve is very steep, but with decreasing annealing temperature, the forward and reverse martensite transformation temperatures are smoothed, for example, in very low field of 10 mT [14]. In order to retain states with different degrees of LRO (long-range order), the alloys were subjected to a 30 min annealing treatment at three different temperatures, followed by quenching into ice water. Another piece of the same alloy was slowly cooled from 1173 K (labeled AQ 300 K) for comparison with the the quenched samples.

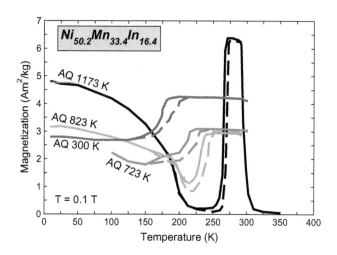

Fig. 2 Temperature dependence of magnetization in a field of 0.1 T of $Ni_{50.2}Mn_{33.4}In_{16.4}$ subject to four different heat treatments labeled AQ 1173 K–AQ 300 K, which yield different degrees of order. The slowly cooled sample labeled AQ 300 K exhibits the largest degree of order. Figure from [14], copyright Elsevier (2012)

Similar magnetization curves for ordered and less ordered $Ni_{50}Mn_{34.5}In_{15.5}$ alloys were obtained by Barandiaran et al. in [15, 16]. Rectangular-like magnetization curves have also been observed for $Ni_{50}Mn_{35}Sn_{15}$ alloys together with a splitting of ZFC (zero-field cooled), FC (field cooled), and FH (field heated) curves [17]. As already mentioned in a series of publications, a sytematic adjustment of composition which fixes the magnetic ordering and giant magnetocaloric effect was discussed [3, 4]. It appears that for the Ni–Mn–Sn alloy system, this compositional tuning has dramatic effects on the microstructural development. It was concluded that the martensitic transformation occurs only for those compositions where the single phase $L2_1$ has been retained in a metastable state of cooling [3]. The off-stoichiometric Ni–Mn–Sn Heusler alloys, which undergo a martensitic transformation, are metastable in a temperature range around $T = 773$ K.

There are other experiments in which the thermal stability, for example, of Ni–Mn–Ga alloys in the temperature range 620–770 K was investigated [18]. If the ratio e/a is increased by substitution of Ga by Mn, very stable alloys are achieved. Other interesting aspects like large precipitates, "self-patterning", and the influence of vacancies on

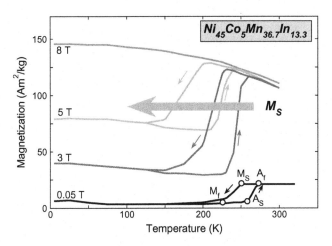

Fig. 3 Magnetization of $Ni_{45}Co_5Mn_{36.7}In_{13.3}$ across the magnetostructural transition in various fields showing the shift of martensitic transformation with field to lower temperatures. From H_{Af}, H_{Ms} and $H_0 = (H_{Af} + H_{Ms})/2$ the magnetic field-temperature phase diagram can be evaluated showing the kinetic arrest phenomenon. Figure adapted from [22], copyright AIP (2008); see also [23]

modulated and non-modulated phases were investigated [19–21].

We conclude this chapter by citing a result for the rapidly quenched alloys, namely the shift of the isofield magnetization curves in an external field and the kinetic arrest phenomenon [22–24]. Figure 3 shows the shift of the magnetization curves of $Ni_{45}Co_5Mn_{36.7}In_{12.3}$ with increasing magnetic field to lower temperatures [22]. This was predicted by Ghatak et al. using a degenerated e_g-band model for coexisting magnetic and martensitic transformation [25], yielding

$$\frac{\Delta T_{MS}}{T_{MS}^0} = -\left(\frac{\mu_B H}{k_B T_{MS}^0 (1 - U_{\text{eff}} \rho(\varepsilon_F))}\right)^2. \quad (3)$$

On the other hand, T_{MS} is arrested at low temperatures because the chemical driving force (the difference in energy between austenite and martensite) cannot overcome the additional energy barrier due to the displacement fields arising from the atomic disorder.

On the Nature of Decomposition in Heusler Alloys

The structural metastability of $Ni_{50}Mn_{45}Ga_5$ alloys is an inherent property of the $Ni_{50}Mn_{50-x}X_x$ alloys with X as Ga, In and Sn [3–7]. The alloys transform during temper-annealing to a dual-phase composite alloy, cubic $L2_1$ Heusler and $L1_0$ $Ni_{50}Mn_{50}$. This may lead to time-dependent effects. For example, supercells with non-stoichiometric composition show strong atomic relaxations because of the

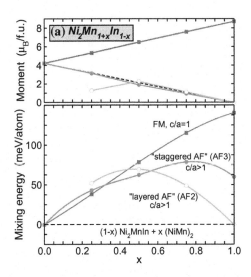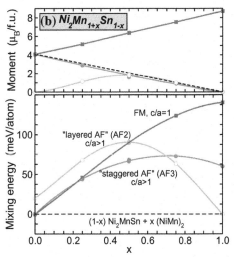

Fig. 4 Mixing energies and magnetic moments of **a** $Ni_2Mn_{1+x}In_{1-x}$ and **b** $Ni_2Mn_{1+x}Sn_{1-x}$ for two different antiferromagnetic spin orderings, AF2 and AF3. The mixing energy is calculated by assuming that over the whole concentration range the decomposition of the alloy will lead

to a dual-phase composite alloy like $L2_1$ $Ni_{50}Mn_{25}In_{25}$ ans $L1_0$ $Ni_{50}Mn_{50}$. The Heusler precipitates are ferromagnetic while the antiferromagnetic matrix is assumed to have AF2 or AF3 ordering

Table 1 Energy difference between B2 and L2$_1$ structures (fully relaxed 432-atoms supercell, $4 \times 4 \times 4$ k-points)

System	$\Delta E = E_{B2} - E_{L2_1}$
Ni$_2$MnAl	163 meV/f.u. = 41 meV/atom
Ni$_2$MnGa	198 meV/f.u. = 50 meV/atom
Ni$_2$MnIn	125 meV/f.u. = 32 meV/atom
Ni$_2$MnSn	359 meV/f.u. = 90 meV/atom

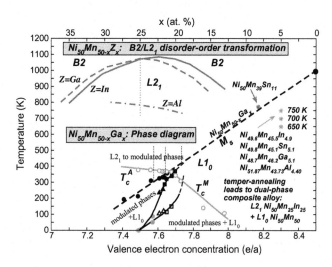

Fig. 5 B2/L2$_1$ disorder-order transformation of Ni$_{50}$Mn$_{50-x}$Z$_x$ as a function of x (upper axis) and e/a (lower axis) for $Z = $ Al, Ga, In [26]. The crosses mark the alloy systems for which segregation has been investigated [5–7]. They are all very close to the martensitic transformation. The phase diagram of Ni$_{50}$Mn$_{50-x}$Ga$_x$ has been added to the plot

non-cubic environment of the individual atoms. This directly probes the effect of disorder and may cause the onset of segregation for alloys close to the martensitic transformation (this may also trigger the L2$_1$-B2 transformation of Ni–Mn–In [26], Ni–Mn–Al [27], and Ni–Mn–Ga).

First results for the mixing energies of Ni–Mn–In and Ni–Mn–Sn alloys are shown in Fig. 4. This is different from the B2-L2$_1$ order transition (compare Table 1).

The final phase diagram is shown in Fig. 5. It is obvious that martensitic instabilities and segregation are close to each other in composition and are interrelated.

The final phase diagram showing martensite and Curie temperatures of Ni–Mn–Ga, B2/L2$_1$ ordering temperatures measured so far as well as position of alloys undergoing decomposition in Fig. 5 illustrates that the various instabilities are interrelated.

Acknowledgements We acknowledge support by the DFG priority program 1599 "ferroic cooling". Calculations were carried out on the magnitude supercomputer (DFG grant INST 20876/208-1) at the CCSS at University of Duisburg-Essen.

References

1. Liu J, Gottschall KP, Skokov KP, Moore JD, Gutfleisch O (2012) Nat Mater 11:620
2. Sokolovskiy VV, Entel P, Buchelnikov VD, Gruner ME (2015) Phys Rev B 91:220409(R)
3. Yuhasz WM, Schlagrel DI, Xing Q, Fennis KW, McCallum RW, Lograsso TA (2009) J Appl Phys 105:07A921
4. Yuhasz WM, Schlagel DL, XCing Q, McCallum RW, Lograsso TA (2010) J Alloys Comp 492:681
5. Çahr A, Acet M, Farle M (2016) Sci Rep 6:28931
6. Krenke T, Çahr A, Scheibel F, Acet M, Farle M (2016) J Appl Phys 120:243904
7. Çahr A, Acet M, Wiedwald U, Krenke T, Farle M (2017) Acta Mater 127:117
8. Planes A, Mañosa L, Acet M (2009) J Phys Condens Matter 21:233201
9. Kouvel JD, Hartelius CC (1912) J Appl Phys 32:1343
10. Nikitin SA, Myalikgulyev G, Tissin AM, Annaorazov MP, Asatryan KA, Tyurin AL (1990) Phys Lett 148:363
11. Chirkova A, Skokov KP, Schultz L, Baranov NV, Gutfleisch O, Woodcock T (2016) Acta Mat 106:15
12. Wolloch M, Gruner ME, Keune W et al (2016) Phys Rev B 94:174435
13. Pavlukhina O, Sokolovskiy V, Buchelnikov V, Entel P (2016) MSF 845:138
14. Recarte V, Pérez-Landazábal JI, Sánchez-Alarcos V (2012) J Alloys Comp 536:S5308
15. Barandiaran JM, Chernenko VA, Cesari E, Salas D, Lapzpita P, Gutierrez J, Oru I (2013) Appl Phys Lett 102:071904
16. Barandiaran JM, Chernenko VA, Cesari E, Salas D, Gutierrez J, Lapzpita P (2013) J Phys Condens Matter 25:484005
17. Cong DY, Roth S, Schultz L (2012) Acta Mater 60:5335
18. Cesari E, Font J, Muntashell J, Ochin P, Pons J, Santamarta R (2008) Scipta Mater 58:259
19. Aseguinoaza JR, Colub V, Saluk OY, Muntifering B, Knowlton WB, Müllner P, Barandiaran JM, Cernenko VA (2016) Acta Mater 111:163
20. Merida D, Garćia JA, Sánchez-Aloarca V, Perez-Landazábal JI, Recarte V, Plazola F (2015) J Alloys Comp 639:180
21. Pérez-Sierra AM, Pons J, Santamarta R, Vernaut P, Ochin P (2015) Act Mater 93:164
22. Ito W, Itto K, Umetsu RY, Kainuma R, Koyama K, Watanabe W, Fujita A, Okawa K, Ishida K, Kanomata T (2008) Appl Phys Lett 92:021908
23. Monroe JA, Raymond JE, Xu X, Nagasako M, Kainuma R, Chumlyakov YI, Arroyave R, Karaman I (2015) Acta Mat 101:107
24. Gottschall T, Skokov KP, Frincu B, Gutfleisch O (2015) Appl Phys Lett 106:021901
25. Ghatak SK, Ray DK (1985) Phys Rev B 31:3064
26. Miyamoto T, Ito W, Umetsu RY, Kainuma R, Kanomata T, Ishida K (2010) Scripta Mater 62:151
27. Kainuma R, Geijima F, Sutou Y, Ohnuma I, Aoko K, Ishida K (2000) JIM 41:943

Reversible Negative Thermal Expansion Response and Phase Transformation Behavior of a Ti-Rich Ti$_{54}$Ni$_{46}$ Alloy Prepared by Rapid Solidification

Zhong-Xun Zhao, Xiao Ma, Cai-You Zeng, Shanshan Cao, Chang-Bo Ke, and Xin-Ping Zhang

Abstract

In this study, a Ti-rich Ti–Ni alloy (Ti$_{54}$Ni$_{46}$) was prepared by rapid solidification technique through vacuum suction casting into a water-cooled copper mold. The microstructure, thermal expansion, and phase transformation behavior of the alloy were studied systematically. The results show that the rapidly solidified Ti$_{54}$Ni$_{46}$ alloy exhibits negative thermal expansion (NTE) response in both vertical and horizontal measuring directions upon heating and cooling. The discrepancy in the NTE response between the two mutually perpendicular directions of the alloy is small, indicating an implicit anisotropic NTE behavior. A one-to-one correspondence exists between the characteristic temperatures of phase transformation and NTE, as well as between their changes during thermal cycling. It is conclusive that the NTE strains generated upon heating and cooling originate from the volume changes accompanying the forward and reverse martensitic transformations in Ti$_{54}$Ni$_{46}$ alloy. Characteristic temperatures of both phase transformation and NTE of the alloy rapidly shift to lower temperatures due to the multiplication of dislocations during the initial approximately 20 thermal cycles, and then tend to be relatively unchanged in subsequent thermal cycling as the transformation-induced defects reach saturation. The absolute values of the coefficient of thermal expansion of the NTE stage upon heating and cooling decrease rapidly during the initial approximately 20 thermal cycles, and thereafter become relatively stable with the increase of thermal cycle number, which is mainly attributed to the decrease of the effective fraction of the B19′ martensite participating in the forward and reverse martensitic transformations.

Keywords

Rapid solidification • Ti-rich Ti–Ni alloy • Thermal cycling • Negative thermal expansion • Phase transformation

Introduction

Development of precision instruments such as specialized measurement apparatuses and control devices needs to use materials with low thermal expansion properties. Fe–Ni Invar alloys are the most common low-expansion materials and have been widely used in many engineering applications, but it has proved difficult to achieve near-zero thermal expansion (NZTE), that is, a coefficient of thermal expansion (CTE) value less than $1.0 \times 10^{-6} \, \text{K}^{-1}$, in Invar alloys [1]. Recently, it was reported that NZTE can be realized in Ti–Nb–Ta–Zr-based alloys, namely Gum Metal, via severe plastic deformation processing such as cold rolling and wire drawing, despite the fact that the so-induced NZTE response is highly anisotropic and its stability is yet to be studied [2]. At present, the most feasible way to prepare NZTE materials is to fabricate composites by combining substances that possess positive thermal expansion and NTE properties by means of CTE compensation. Heretofore, most of the currently available NTE materials are inorganic non-metallic materials, so that their applications are hindered by the undesired high brittleness and poor conductivity. It is of great interest that NTE behavior is found to exist in only a small number of metallic materials, mostly shape memory alloys (SMAs), and their NTE response is stronger than the inorganic counterparts [3].

So far, research on metallic NTE materials mainly focused on some well-studied near equiatomic Ti–Ni SMAs, and their NTE response is usually regarded as the deformation strain associated with the unique two-way shape memory effect in the alloys. By contrast, our latest work showed that as-fabricated Ti-rich and equiatomic Ti–Ni

Z.-X. Zhao · X. Ma · C.-Y. Zeng · S. Cao · C.-B. Ke
X.-P. Zhang (✉)
School of Materials Science and Engineering, South China University of Technology, Guangzhou, 510641, China
e-mail: mexzhang@scut.edu.cn

© The Minerals, Metals & Materials Society 2018
A. P. Stebner and G. B. Olson (eds.), *Proceedings of the International Conference on Martensitic Transformations: Chicago*, The Minerals, Metals & Materials Series, https://doi.org/10.1007/978-3-319-76968-4_30

alloys, obtained by conventional melting without subsequent functional training or mechanical processing treatments, exhibit an apparent NTE response that originates from the volume change accompanying the thermoelastic martensitic phase transformation (PT) in the alloys [4]. Since NTE materials are normally working under cyclic thermal loading conditions, it is imperative to study the evolution of thermal expansion and PT behavior of Ti–Ni NTE materials during thermal cycling. In addition, it has been well documented that the rapidly solidified Ti–Ni alloy possesses better properties, owing to a fine-grained and strongly textured microstructure with high solubility and less segregation as compared to the microstructure of alloys fabricated by conventional casting processes [5, 6]. Therefore, in this study, a Ti-rich alloy ($Ti_{54}Ni_{46}$) was prepared by rapid solidification technique through vacuum suction casting into a water-cooled copper mold. The microstructure, thermal expansion, and PT behavior of the alloy were studied systematically with the purpose of revealing the influence of thermal cycling on NTE response and PT behavior of the Ti-rich Ti–Ni alloy.

Experiments

Sponge titanium (purity > 99.7%) and electrolytic nickel (purity > 99.9%) with a nominal Ti content of 54.0 at.% were used to prepare the $Ti_{54}Ni_{46}$ master alloy by a non-self-consumable vacuum arc melting process in a water-cooled copper crucible. The melting process was repeated six times for each master alloy ingot to ensure the homogeneity in composition and microstructure of the alloy. Afterwards, the molten master alloy was sucked into a water-cooled copper mold under vacuum condition and rapidly solidified into $Ti_{54}Ni_{46}$ alloy cylindrical ingot with a dimension of 6 × 30 (diameter × height, mm). Then, the cylindrical ingots were sealed in a quartz tube under Ar atmosphere (99.99% purity) and subjected to solution treatment at 1223 K for 8 h followed by quenching in water. The $Ti_{54}Ni_{46}$ alloy samples used for microstructural characterization and thermal expansion measurement, with the dimension of 4.0 × 4.0 × 4.5 (length × width × height, mm), were cut from the center part of the cylindrical ingots by using wire electric discharge machining, as shown in Fig. 1. Phase constituents of the alloy were characterized by X-ray diffraction (XRD, X'pert Pro M-Philips) with CuKα radiation at a scanning step of 0.02°. Microstructure analysis was performed in horizontal and vertical cross-sections of the sample, as depicted in Fig. 1, by using a scanning electron microscope (SEM, ProX-Phenom, FEI) equipped with an energy dispersive spectrometer (EDS). Thermal expansion behavior of $Ti_{54}Ni_{46}$ alloy was characterized by a thermo-mechanical analyzer (TMA, 402F3-Netzsch)

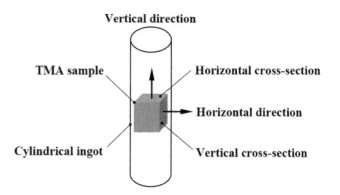

Fig. 1 Schematic illustration of the sampling positions for microstructure characterization and directions for thermal expansion measurement in a cuboid-shape sample cut from the rapidly solidified cylindrical ingot

between 253 and 423 K with a heating/cooling rate of 3 K/min, and CTEs along the vertical direction (VD) and horizontal direction (HD) of samples, as denoted in Fig. 1, were measured respectively. Phase transformation behavior was analyzed by a differential scanning calorimeter (DSC, Q200-TA) with a heating/cooling rate of 10 K/min between 253 and 423 K.

Results and Discussion

XRD result of the rapidly solidified $Ti_{54}Ni_{46}$ alloy and backscattered electron (BSE) images in horizontal and vertical cross-sections are shown in Fig. 2. It is clear that the $Ti_{54}Ni_{46}$ alloy mainly consists of two phases, identified by XRD analysis shown in Fig. 2a. These phases are Ti_2Ni phase (dark-gray region) and B19′ martensite phase (light-gray region), as shown in Fig. 2b, c. During rapid solidification, there is a large temperature gradient between the center and the cylindrical surface in the horizontal cross-section of the cylindrical ingot, which results in preferential grain growth morphological features as shown in Fig. 2b. In contrast, grains with relatively homogeneous size and morphology distribute uniformly in the vertical cross-section, as depicted in Fig. 2c. It is worth noting that the average grain size of the suction cast $Ti_{54}Ni_{46}$ alloy is only one-fourth of that of the conventional cast alloy, as the grain growth rate is restrained by the rapid cooling in the suction cast process [5, 6].

Thermal expansion curves of the rapidly solidified $Ti_{54}Ni_{46}$ alloy measured along HD and VD during heating and cooling are illustrated in Fig. 3. Upon heating, the strain (dl/l_0), in which l_0 is the initial sample length at 253 K, of the alloy increases at first and then decreases before further increasing in both HD and VD, indicating that the alloy possesses NTE response within a certain temperature range.

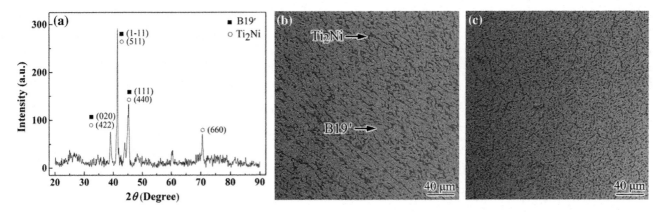

Fig. 2 XRD pattern (**a**) and backscattered electron images showing the microstructures in horizontal (**b**) and vertical (**c**) cross-sections of the rapidly solidified $Ti_{54}Ni_{46}$ alloy

Fig. 3 Thermal expansion curves (**a**) of the $Ti_{54}Ni_{46}$ alloy measured along HD and VD upon heating and cooling, and the enlarged view (**b**) of the region marked in (**a**)

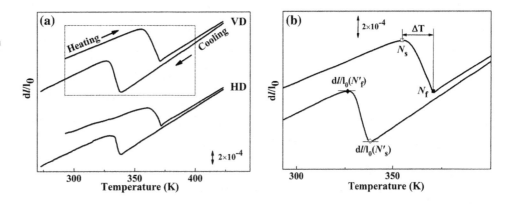

The critical temperatures at which the strain of the alloy begins and finishes decreasing are defined as the starting and finishing temperatures of NTE upon heating (i.e., N_s and N_f) respectively, as depicted in Fig. 3b. Thus, the temperature range of NTE behavior (ΔT) and the total strain produced in the NTE stage ($\Delta l/l_0$) during heating can be defined as follows:

$$\Delta T = N_f - N_s \qquad (1)$$

$$\Delta l/l_0 = dl/l_0(N_f) - dl/l_0(N_s) \qquad (2)$$

where $dl/l_0(N_s)$ and $dl/l_0(N_f)$ are the strains corresponding to N_s and N_f respectively. Then, the CTE of the NTE stage upon heating (α_N) can be obtained by

$$\alpha_N = \frac{dl/l_0(N_f) - dl/l_0(N_s)}{N_f - N_s} \qquad (3)$$

On the other hand, with decreasing temperature, the starting and finishing temperatures of NTE upon cooling

(i.e., N_s' and N_f') are the critical temperatures that correspond to when the strain of the alloy begins and finishes increasing, respectively. The temperature range ($\Delta T'$) and the CTE (α_N') of the NTE stage during cooling are defined in a similar way as described above for the case of heating. As can be seen in Fig. 3a, rapidly solidified $Ti_{54}Ni_{46}$ alloy exhibits NTE behavior along both measuring directions. Moreover, there is a small discrepancy in the NTE response between VD and HD; that is, the anisotropy of NTE behavior of Ti-rich $Ti_{54}Ni_{46}$ alloy is implicit, in contrast to the case of the equiatomic $Ti_{50}Ni_{50}$ alloy where there is an apparent anisotropic NTE response as reported in our previous work [4]. Therefore, for simplicity, we only present the thermal expansion measurement data along VD during thermal cycling in the following.

To clarify the influence of thermal cycling on NTE response of the rapidly solidified $Ti_{54}Ni_{46}$ alloy, thermal expansion along VD of the alloy was measured for 100 thermal cycles under the same thermal cycling conditions; for clarity, the heating and cooling parts of the thermal expansion curves are demonstrated in Fig. 4a, b,

Fig. 4 Thermal expansion curves along VD of the Ti$_{54}$Ni$_{46}$ alloy upon **a** heating and **b** cooling; and **c** DSC curves and **d** characteristic temperatures of PT and thermal expansion during thermal cycling

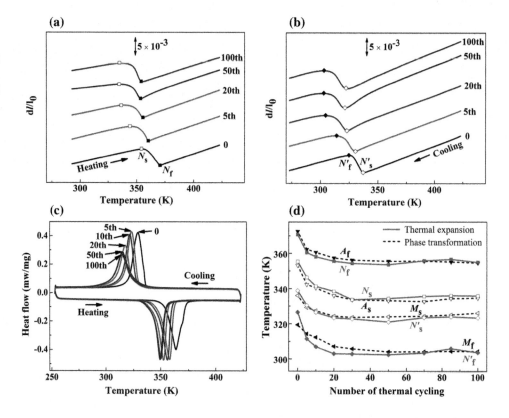

respectively. Clearly, the Ti$_{54}$Ni$_{46}$ alloy still possesses NTE response in both heating and cooling stages during thermal cycling, and there is no apparent change in the NTE strains. However, the starting and finishing temperatures of NTE upon heating and cooling, that is, N_s, N_f, N_s', and N_f', all shift to lower temperatures with increasing thermal cycle number. Figure 4c shows DSC curves of the Ti$_{54}$Ni$_{46}$ alloy during thermal cycling, in which the starting and finishing temperatures of austenitic transformation (A_s and A_f) and that of martensitic transformation (M_s and M_f) are obtained through tangential method. These critical PT temperatures, as well as the characteristic NTE temperatures mentioned above, are plotted against thermal cycle number as shown in Fig. 4d. Apparently, a one-to-one correspondence exists between the characteristic temperatures of PT and NTE, as well as between their changes during thermal cycling, which provides further evidence supporting the previously proposed mechanism that the NTE strain originates from the volume change accompanying the martensitic transformation in Ti–Ni alloys [4]. In addition, the characteristic temperatures of both PT and NTE of the alloy rapidly drop to lower temperatures after undergoing an initial approximately 20 thermal cycles, and then tend to be relatively unchanged in subsequent thermal cycling. This is because dislocations are generally generated at the parent-martensite interface during PT, resulting in the increase of defect density, hence the shift of PT temperature during the first couple of thermal cycles.

Then, after a certain number of thermal cycles, the PT behavior and the corresponding NTE response can be stabilized due to the saturation of the transformation-induced defects.

Figure 5 shows the changes of CTE values of the Ti$_{54}$Ni$_{46}$ alloy and transformation enthalpies upon heating and cooling with thermal cycle number. It can be seen clearly that the absolute value of α_N decreases rapidly from 29.2×10^{-6} K^{-1} to 20.7×10^{-6} K^{-1} during the initial 20 thermal cycles; similarly, α_N' drops from 34.8×10^{-6} K^{-1} to 20.7×10^{-6} K^{-1}, as shown in Fig. 5a, meaning that the decreasing rate of CTE in the initial 20 thermal cycles is -29.2% upon heating and -40.5% upon cooling, respectively. However, the decreasing rate between the 21st and 100th thermal cycles is reduced to 11.7% and 4.8%, respectively, which indicates that the NTE effect, in terms of the CTE value, of the alloy is suppressed significantly in the initial stage of thermal cycling process and becomes relatively stable in subsequent thermal cycling. Further, both transformation enthalpies of the Ti$_{54}$Ni$_{46}$ alloy upon heating and cooling decrease rapidly before approximately 40 cycles and thereafter become relatively stable with the increase of thermal cycle number, as shown in Fig. 5b. Notably, previous studies indicated that the decrease of enthalpy during thermal cycling of Ti–Ni alloys can be attributed to the increasing fraction of residual B19′ martensite phase during thermal cycling [7, 8]. As described above, the NTE strain of the Ti–Ni alloy is mainly attributed

Fig. 5 Changes of **a** CTE values of the Ti$_{54}$Ni$_{46}$ alloy and **b** transformation enthalpies upon heating and cooling with thermal cycle number

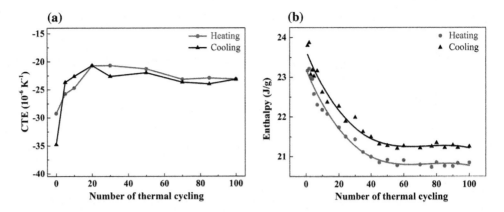

to the transformation strain that originates from the volume change accompanying the B2 $\rightleftharpoons$ B19′ PTs. Since the accumulated residual B19′ martensite is stabilized and cannot transform into austenite unless an additional driving force is supplied, the effective fraction of the B19′ martensite that participates in the forward and reverse martensitic transformations is decreasing, which results in the weakening of the transformation-induced NTE effect, as depicted in Fig. 5a.

effective fraction of the B19′ martensite participating in the forward and reverse martensitic transformations.

Acknowledgements This work was supported by the National Natural Science Foundation of China under Grant Nos. 51571092 and 51401081, and Key Project Program of Guangdong Provincial Natural Science Foundation under Grant No. S2013020012805.

Conclusions

(1) Rapidly solidified Ti$_{54}$Ni$_{46}$ alloy prepared by suction cast process exhibits NTE response in both vertical and horizontal measuring directions upon heating and cooling. The discrepancy in NTE response between the two mutually perpendicular directions of the alloy sample is small, indicating an implicit anisotropic NTE behavior.

(2) A one-to-one correspondence exists between the characteristic temperatures of PT and NTE, as well as between their changes during thermal cycling. The NTE strains generated upon heating and cooling originate from the volume changes accompanying the forward and reverse martensitic transformations in the Ti$_{54}$Ni$_{46}$ alloy.

(3) Characteristic temperatures of both PT and NTE of the alloy rapidly shift to lower temperatures due to the multiplication of dislocations during the initial approximately 20 thermal cycles, and then tend to be relatively unchanged in subsequent thermal cycling as the transformation-induced defects reach saturation.

(4) The absolute values of CTE of the NTE stage upon heating and cooling decrease rapidly during the initial approximately 20 thermal cycles and thereafter become relatively stable with the increase of thermal cycle number, which is mainly attributed to the decrease of the

References

1. Shiga M (1996) Invar alloys. Curr Opin Solid State Mater Sci 1 (3):340–348
2. Kim HY, Wei L, Kobayashi S, Tahara M, Miyazaki S (2013) Nanodomain structure and its effect on abnormal thermal expansion behavior of a Ti–23Nb–2Zr–0.7Ta–1.2O alloy. Acta Mater 61 (13):4874–4886
3. Monroe JA, Gehring D, Karaman I, Arroyave R, Brown DW, Clausen B (2016) Tailored thermal expansion alloys. Acta Mater 102:333–341
4. Zhao ZX, Ma X, Cao SS, Ke CB, Zhang XP (2017) Anisotropic negative thermal expansion behavior of the as-fabricated Ti-rich and equiatomic Ti–Ni alloys induced by preferential grain orientation. Shape Mem Superelasticity. https://doi.org/10.1007/s40830-017-0145-9 (in press)
5. Li YY, Yao XY, Cao SS, Ma X, Ke CB, Zhang XP (2017) Rapidly solidified and optimally constraint-aged Ni$_{51}$Ti$_{49}$ shape memory alloy aiming at making a purpose-designed bio-actuator. Mater Design 118:99–106
6. Li YY, Cao SS, Ma X, Ke CB, Zhang XP (2017) Influence of strongly textured microstructure on the all-round shape memory effect of rapidly solidified Ni$_{51}$Ti$_{49}$ alloy. Mater Sci Eng, A 705:273–281
7. McCormick PG, Liu Y (1994) Thermodynamic analysis of the martensitic transformation in NiTi—II. Effect of transformation cycling. Acta Metall Mater 42(7):2407–2413
8. Soejima Y, Motomura S, Mitsuhara M, Inamura T, Nishida M (2016) In situ scanning electron microscopy study of the thermoelastic martensitic transformation in Ti–Ni shape memory alloy. Acta Mater 103:352–360

Part VII
Engineering Applications and Devices

Film and Foil-Based Shape Memory Alloy Microactuators for Fluid Handling

Hinnerk Ossmer, Marcel Gueltig, Christoph Wessendorf, Manfred Kohl, and Christof Megnin

Abstract

In this contribution, the potential of film and foil-based shape memory alloy (SMA) microactuators for fluid handling applications is explored. SMAs provide a high work density and allow for compact and robust actuator designs based on the one-way shape memory effect. Compared to more commonly used wires, actuators fabricated from thin film or foil material allow for more complex designs having several degrees of freedom and enable shorter switching times in the range of 10 ms. In order to commercialize such actuators, the "*memetis GmbH*" was founded as a high-tech spin-off of the Karlsruhe Institute of Technology (KIT) in Germany. *Memetis* is focused on miniature fluid handling products and combines rapid prototyping and rapid manufacturing techniques such as 3D printing, laser cutting and CNC milling of polymer housings for customer-specific device development. A normally open microvalve is presented here as an example, which is actuated by a novel fatigue-free TiNiCu film actuator.

Keywords

Shape memory alloy • Fatigue-free • Miniature valve • Fluid handling

Introduction

In many technical fields, an ongoing trend for miniaturization is observed, leading to smaller devices with an increasing number of functional features. Well-known examples are laptop computers and smart phones, but also scientific and medical handheld devices like analytic instruments and insulin pumps. In these devices, compact actuators are required to create a small-scale movement, e.g., in optical image stabilization, microvalves or micropumps. However, not all actuation technologies are equally applicable for miniaturization. Although electromagnetic drives are very common, they require a number of components and their forces show an unfavorable scaling behavior upon miniaturization. Therefore, they are barely built smaller than 5 mm. Piezoelectric drives, on the other hand, may be realized in a very compact layer-based way, provide fast switching frequencies up to the kHz range and a high accuracy, but their strokes—which are in the order of 10^{-4} of the actuator size—are too low for many applications. Furthermore, widely used standard PZT actuators contain lead.

In cases where a compact actuator with high force and stroke is required while moderate switching times in the order of 10 ms are sufficient, shape memory alloy (SMA) based actuators may represent an excellent alternative [1]. Besides having a high work density in the order of $10^7 \, Jg^{-1}$, these actuators are compact, silent and robust and even provide self-sensing capability. However, while SMAs have been successfully applied in stents and other passive elements in the medical field for several decades, their use as active elements (actuators) is still restricted to niche applications. The two main technical challenges in the past have been the relatively slow operation due to the thermal operation principle as well as a limited lifetime due to structural and functional fatigue. The switching time may be decreased by increasing the surface-to-volume ratio of SMA actuators, e.g., by using thinner wires, in order to facilitate heat transfer. Products based on hair-thin SMA wires are already

H. Ossmer (✉) · M. Gueltig · C. Wessendorf · C. Megnin
Memetis GmbH, Hermann-von-Helmholtz-Platz 1,
76344 Eggenstein-Leopoldshafen, Germany
e-mail: hinnerk.ossmer@memetis.com; hinnerk.ossmer@kit.edu

H. Ossmer · M. Gueltig · C. Wessendorf · M. Kohl · C. Megnin
Karlsruhe Institute of Technology, Institute of Microstructure Technology, Hermann-von-Helmholtz-Platz 1,
76344 Eggenstein-Leopoldshafen, Germany

© The Minerals, Metals & Materials Society 2018
A. P. Stebner and G. B. Olson (eds.), *Proceedings of the International Conference on Martensitic Transformations: Chicago*,
The Minerals, Metals & Materials Series, https://doi.org/10.1007/978-3-319-76968-4_31

at the market, e.g., autofocus units for mobile phones [2] and microvalves [3]. However, wires provide a one-dimensional motion and the force decreases with decreasing diameter. As an alternative, actuators may be fabricated from thin SMA sheet material like films or foils. Planar SMA actuators exhibit a number of specific advantages including compactness and high forces in combination with fast response times due to their high surface-to-volume ratio. Furthermore, complex actuator shapes having several degrees of freedom and different actuation modes like tension, bending and rotation may be realized in a single element.

The second challenge faced by engineers of SMA actuators is the limited lifetime of the commonly used NiTi-based alloys (Cu- or Fe-based SMA exhibit even stronger fatigue effects and are therefore not commercially used): Due to microstructural changes within the material during the first 10^2 operation cycles, SMA actuators may show strong training behavior, sometimes even leading to crack formation and catastrophic failure. The understanding of these fatigue phenomena has tremendously increased in recent years. It could be shown that fatigue is closely related to crystallographic compatibility between austenite (A) and martensite (M), the phases involved in the shape memory effect [4]. On this basis, TiNiCu-based thin film materials have been developed, showing stable shape memory behavior for over 10^7 cycles [5].

Film and foil-based microactuators have been investigated by the group of Professor Manfred Kohl at the Institute of Microstructure Technology (IMT) of the Karlsruhe Institute of Technology (KIT) for several decades [6, 7]. In order to develop these actuators into products and push their market entry, the "*memetis GmbH*" was founded as a KIT spin-off in spring 2017. Memetis is focused on the development of miniature fluid handling products for the fast-growing life science market, including valves, pumps and dosing units. State-of-the-art SMA film and foil materials are combined with rapid prototyping and rapid manufacturing techniques such as CNC milling and 3D printing of polymer housings for customer-specific device development.

In the following, the shape memory effect is briefly introduced. Then, a normally open microvalve is presented as an application example. The microvalve is actuated by a novel fatigue-free TiNiCu film actuator and the housing is fabricated by milling and stereolithographic 3D-printing.

SMA Films and Foils

Due to a first-order phase transformation between the high-temperature austenite (A) and low-temperature martensite (M) phase, SMAs exhibit unique functional behavior referred to as (one-way) shape memory effect. In contrast to the highly symmetric, cubic austenite, the martensite lattice

has several crystallographic variants. These variants may be reoriented—or "detwinned" (M^+)—by low mechanical stresses, leading to a macroscopic, apparently plastic deformations in the order of several percent. However, the deformation may be reversed by heating the SMA up until it transforms to the cubic austenite phase—the material appears to remember its initial shape. Upon cooling until retransformation into martensite, a mixture of crystallographic M variants results—a "twinned" state without macroscopic deformation. In practical applications, a reset mechanism like a spring is used to deform SMA actuators in the M phase and electrical heating is applied for restoring the shape in A phase.

In order to use the shape memory effect for actuation, SMAs have to be selected with phase transformation temperatures above the intended temperature of application. In this work, a binary TiNi foil and a ternary TiNiCu film alloy are selected for fabrication of microvalve actuators. Both alloys are obtained from third-party manufacturers and have a thickness of 20 μm. The TiNi foil is fabricated by cold-rolling, whereas the TiNiCu film is produced by DC magnetron sputtering [8]. Phase transformation temperatures are determined by differential scanning calorimetry (DSC), see Fig. 1. Peaks indicate the temperature intervals of endothermal $M \rightarrow A$ and exothermal $A \rightarrow M$ transformation. In the case of TiNiCu, the transformation peaks are sharp, having a hysteresis of 13 K and an austenite finish temperature of 60 °C. The binary TiNi alloy, on the other hand, exhibits wider peaks and a two-step transformation upon cooling, involving an intermediate R-phase ($A \rightarrow R \rightarrow M$). In order to increase device lifetime, only the $A \rightarrow R$

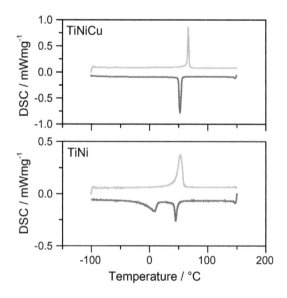

Fig. 1 Differential scanning calorimetry (DSC) measurement of sputter-deposited TiNiCu and cold-rolled TiNi samples with a thickness of 20 μm. The ternary alloy shows a direct transformation between austenite (A) and martensite (M), whereas an intermediate R-phase is observed in the binary alloy

transformation is used, which has a reduced transformation strain of 0.8%. In comparison, the transformation strain of the TiNiCu alloy is twice as large.

Microvalves

The SMA films and foils described above are used for fabrication of microfluidic components such as microvalves, which are suitable for both liquid and gas handling. Valves can be realized having normally open (NO), normally closed (NC) as well as bistable functional behavior [7, 9]. Microvalves based on TiNi foil actuators have been extensively described in previous publications, therefore a NO valve with novel fatigue-free TiNiCu film actuator shall be presented here as an application example.

The operation principle of the valve is schematically shown in Fig. 2a. The layer-based valve design consists of a fluidic part sealed by a membrane with a chamber having an inlet and an outlet, and an actuator part which is coupled to the fluidic part by a guided spherical plunger. The actuator part contains a cross-shaped TiNiCu bridge actuator mounted on a small circuit board for electrical contacting. Each of the crossing SMA beams has a width of 200 μm and a length of 7 mm. In the current-free state, the actuator may be deformed in out-of-plane direction by the applied fluid pressure acting on the membrane—the valve opens. As soon as a sufficiently high electrical current is applied to the

actuator, the crossed beams heat up and push the spherical plunger down onto the membrane, which in turn seals a valve seat within the fluidic chamber—the valve closes. In the present case, the electrical resistance of the actuator is about 1 Ω and the current for closing the valve is 400 mA. Figure 2b shows a valve prototype with a polymer housing fabricated by CNC milling. The cross-shaped actuator may be seen through the transparent top cover. As an alternative, full valve functionality is also obtained using a 3D-printed housing, which opens up new routes for rapid product development and customization. In the case that aggressive or reactive fluids shall be controlled by the valve, chemically inert materials like ethylene propylene diene monomer (EPDM) for the membrane and polyether ether ketone (PEEK) for the housing are chosen. Due to the media-separating membrane, the SMA actuator never contacts the fluid inside the valve chamber.

The actuator part is characterized separately using a tensile test machine equipped with a 5 N force sensor. Figure 3 shows the force-deflection characteristic of the actuator up to a maximum force of 0.5 N. In the heated state (A), the actuator exhibits linear elastic behavior, leading to a parabolic curve in the present case of out-of-plane deflection. Upon first mechanical loading of the twinned actuator in the cold state (M), a considerably flatter curve is obtained, which is related to the detwinning process. When the cold, detwinned actuator (M$^+$) is loaded a second time, the curve is linear elastic as in the heated case, but exhibiting a pre-deflection of 580 μm.

A similar force-deflection characteristic is obtained for the fluidic part at different applied nitrogen gas pressures between 0 and 2 bar. The fluid channels have a diameter of 0.75 mm. As Fig. 4 shows, the forces acting on the membrane increase with increasing applied pressure and with decreasing membrane deflection. However, when the

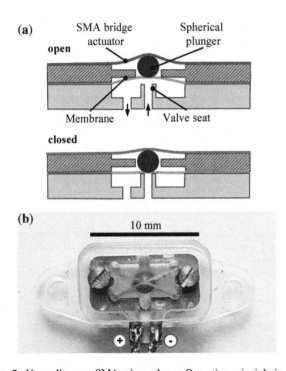

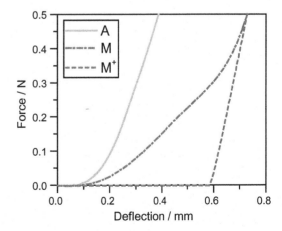

Fig. 2 Normally open SMA microvalve. **a** Operation principle in side view, **b** micrograph of the valve in top view

Fig. 3 Force-deflection characteristic of a cross-shaped TiNiCu bridge actuator

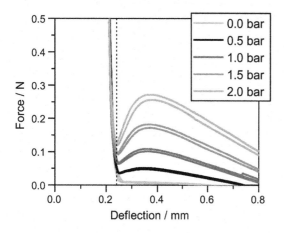

Fig. 4 Force-deflection characteristic of the fluidic part in a microvalve for various applied gas pressures

membrane comes close to the valve seat, a pressure drop is observed. This behavior is attributed to Bernoulli forces due to faster flow of the fluid during valve closing, as well as the smaller fraction of the membrane on which the pressure can act in an almost closed valve. The closed state is indicated by a dashed line in Fig. 4.

By adjusting the diameter of the spherical plunger, the pre-strain of the actuator is matched with respect to the valve seat in such a way that actuator forces are sufficient to close the valve. The actuator part and fluidic part are then assembled to form a valve. The valve performance is investigated using a computer-controlled test setup including a gas flow sensor, a power source and an electrical circuit for controlling the electrical input power. Flow rate characteristics of the valve at various applied pressure differences are shown in Fig. 5. As can be seen, the valve has a maximum flow rate of 5800 ml per minute in the open state at an applied pressure of 2 bar. In this case, the power input for

complete closing is about 85 mW and the valve opens again when the power is reduced to less than 40 mW. The response time for valve opening is about 100 ms. The time for opening depends on the applied electrical power and is up to an order of magnitude faster. As has been shown previously, fluid flow can be even controlled continuously by combining an SMA microvalve with a flow sensor and a power source providing a pulse width modulation (PWM) signal [7].

Conclusion

Microactuation is required in an increasing number of technical applications, and in many cases, conventional technologies like electromagnetic actuation reach their limits. In these cases, shape memory actuation represents a powerful alternative. As one of the first commercial suppliers, *memetis* offers products based on SMA films and foils, which combine great flexibility, robustness and high forces with fast switching times. Novel fatigue-free alloys enable exceptionally long product lifetimes. Besides many other fields, these actuators are attractive for fluid handling applications like switching and pumping in Life Sciences.

Acknowledgements The authors acknowledge funding by Exist Forschungstransfer, a program of ESF and BmWi.

References

1. Kohl, M (2013) Shape memory microactuators. Springer Science & Business Media
2. http://www.actuatorsolutions.de/products/miniature-camera/. Accessed 9 Aug 2017
3. http://www.takasago-fluidics.com/p/valve/s/smad_valve/SMV/49/. Accessed 9 Aug 2017
4. Chen X, Srivastava V, Dabade V, James RD (2013) Study of the cofactor conditions: Conditions of supercompatibility between phases. J Mech Phys Sol 61(12):2566–2587. https://doi.org/10. 1016/j.jmps.2013.08.004
5. Chluba C, Ge W, Lima de Miranda R, Strobel J, Kienle L, Quandt E, Wuttig M (2015) Ultralow-fatigue shape memory alloy films. Science 348(6238):1004–1007. https://doi.org/10.1126/science. 1261164
6. Kohl M, Krevet B, Just E (2002) SMA microgripper System. Sens Actuators A 97–98:646–652
7. Megnin C, Kohl M (2014) Shape memory alloy microvalves for a fluidic control system. J Micromech Microeng 24:025001. https:// doi.org/10.1088/0960-1317/24/2/025001
8. Lima de Miranda R, Zamponi C, Quandt E (2012) Micropatterned freestanding superelastic TiNi films. Adv Eng Mat 15(1–2):66–69. https://doi.org/10.1002/adem.201200197
9. Barth J, Megnin C, Kohl M (2012) A Bistable Shape Memory Alloy Microvalve With Magnetostatic Latches. J Mircomechanical Systems 21(1):76–84. https://doi.org/10.1109/JMEMS.2011.2174428

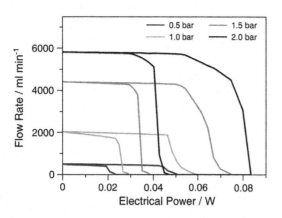

Fig. 5 Flow characteristics of normally open SMA microvalve for various gas pressure differences as a function of applied electrical power

Functional Stability of the Ni$_{51}$Ti$_{49}$ Two-Way Shape Memory Alloy as Artificial Anal Sphincter During Thermo-Mechanical Cycling

Yuan-Yuan Li, Cai-You Zeng, Shanshan Cao, Xiao Ma, and Xin-Ping Zhang

Abstract

Ni$_{51}$Ti$_{49}$ alloy strip with optimal two-way shape memory effect to be potentially used in a purpose-designed artificial anal sphincter (AAS) was prepared by means of rapid solidification process followed by constraint-aging treatment. The functional stability in terms of phase transformation behavior and deformation performance during thermo-mechanical cycling (TMC) was studied. Results show that the forward and reverse R-phase transformation temperatures of the alloy remain in the required operation temperature range of 35–55 °C with small shifts during TMC. The alloy strip exhibits stable deformation performance with steady recovery ratio during TMC, and its microstructure after TMC remains featured fine and stable Ni$_4$Ti$_3$ precipitates together with limited number of dislocations, indicating that the external force and cycling temperature have no influence on the size of Ni$_4$Ti$_3$ precipitates and coherent stress field. Such excellent stability of microstructure and corresponding functionalities are attributed to the stabilized Ni$_4$Ti$_3$ precipitates formed through optimal constraint-aging treatment and the small lattice distortion of R-phase and reverse R-phase transformations during TMC. Nevertheless, the Ni$_{51}$Ti$_{49}$ alloy strip has the maximum displacement of at least 10 mm within 35–55 °C and an irreversible displacement of 4 mm.

Keywords

NiTi shape memory alloy • Artificial anal sphincter Functional stability • Thermo-mechanical cycling

Introduction

As the most important member of shape memory alloys (SMAs), NiTi alloy has been extensively studied for several decades. With the extraordinary superelasticity (SE) and shape memory effect (SME), the NiTi SMA has been widely applied in many fields such as aerospace and aeronautical, biomedical engineering and so forth as actuators. Usually, the function of the NiTi SMA actuator is realized by using one-way shape memory effect (OWSME) or SE. Previously, an artificial anal sphincter (AAS) based on two-way shape memory effect (TWSME) of the NiTi alloy was proposed [1], following with several attempts to solve relevant biomedical problems [2, 3]. Recently, we fabricated the Ni$_{51}$Ti$_{49}$ alloy strip for the potential use as the AAS through rapid solidification process followed by constraint-aging treatment so as to enhance the recovery strain and narrow the temperature ranges of phase transformations [4, 5]. In our studies [4, 6], the purpose-designed AAS consists of two Ni$_{51}$Ti$_{49}$ alloy strips with optimal TWSME, as shown in Fig. 1. On cooling and heating, the forward and reverse R-phase transformations occur in the Ni$_{51}$Ti$_{49}$ alloy strips, which can realize the "close" and "open" functions of the AAS with the maximum gap of at least 20 mm in the required operation temperature range of 35–55 °C. Furthermore, we firstly evaluated and studied the functional stability of the Ni$_{51}$Ti$_{49}$ alloy for the AAS during high-cycle thermal cycling (TC) [6]. Results show that the alloy exhibits stable phase transformation behavior, deformability and deformation recoverability during long-term thermal cycling up to 30000 cycles without applying external stress. However, it should be noted that actually the AAS operates under around 60 mmHg pressure in human body [3]. Thus, it is imperative to study the functional stability of the Ni$_{51}$Ti$_{49}$ alloy, in terms of phase transformation behavior and deformation performance, under the condition of applying both external force and cycling temperature for long time, i.e., the long-term thermo-mechanical cycling (TMC).

Y. Y. Li · C. Y. Zeng · S. Cao · X. Ma · X. P. Zhang (✉)
School of Materials Science and Engineering, South China
University of Technology, Guangzhou, 510640, China
e-mail: mexzhang@scut.edu.cn

Y. Y. Li
e-mail: l.yuanyuan03@mail.scut.edu.cn

A. P. Stebner and G. B. Olson (eds.), *Proceedings of the International Conference on Martensitic Transformations: Chicago*,
The Minerals, Metals & Materials Series, https://doi.org/10.1007/978-3-319-76968-4_32

Fig. 1 Snapshot views of "close" (**a**) and "open" (**b**) states of the AAS prototype made of two $Ni_{51}Ti_{49}$ alloy strips

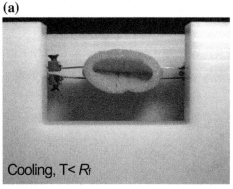

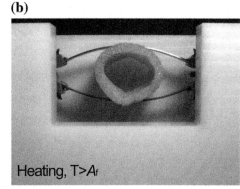

(a) Cooling, T< R_f

(b) Heating, T> A_f

During TC, the instability of functionalities of the NiTi alloy embodies the occurrence of R-phase, new precipitates and lattice distortion induced defects such as dislocations [7]. Comparatively, in the case of TMC, besides the abovementioned phenomena during TC, the addition of external stress can lead to stress induced martensitic (SIM) transformation, plastic deformation, twining and martensitic stabilization [8]. For example, if the external stress is applied at the temperature between A_f (i.e., finishing temperature of austenitic transformation) and M_d (the highest temperature of stress induced martensitic transformation), the NiTi alloy will undergo SIM transformation, which can result in large lattice distortion and consequent generation and multiplication of dislocations. Differently, the external stress can also directly result in plastic deformation, accompanying with slip and rapid dislocation multiplication in the matrix of the NiTi alloy at the temperature above M_d during TMC. Furthermore, it has been proved that multiple factors, such as processing method, stress level, cycling temperature and grain orientation, can largely affect the functional stability of the NiTi alloy during TMC [9]. Clearly, there is plenty of diversity of functionalities corresponding to different applications of the NiTi alloy. Importantly, the influence of the existing external stress in the AAS on the functional stability of the NiTi alloy during TMC has not been studied yet. Therefore, the present work aims at evaluating and studying the functional stability, that is, the shift of phase transformation temperature and the degradations of deformability and deformation recoverability. The influential mechanism was also studied through microstructure analysis.

(length × width × thickness, mm) was obtained through sucking the molten master alloy into a water-cooled thick copper mold under vacuum condition and solidifying at a high cooling rate of about 1×10^4 °C/s. Finally, the as-fabricated strip was subjected to solution treatment at 850 °C for 3 h followed by constraint-aging at 400 °C for 100 h in order to acquire optimal TWSME. The device for constraint was an arch shaped steel mold with a diameter of 66 mm [6].

Two $Ni_{51}Ti_{49}$ alloy strips fabricated through the same process were assembled on a specially designed AAS stage for TMC, as illustrated in Fig. 2. Both ends of two strips were linked to hinges by junctors, one hinge was fixed on the stage and the other was movable in a sliding groove. A cylindrical sponge with a diameter of 20 mm was placed in the gap between two strips, which can provide a maximum force of around 2 N during the "close" of the AAS, being equivalent to approximately 60 mmHg pressure. During TMC, both strips were heated by resistance heating films attached to surface of the strips and cooled by two fans on both sides [10]. The durations of heating and cooling were 1 min and 2 min respectively and the cycling temperature range was hence controlled within 25–80 °C. The temperature and displacement of the strip were synchronously detected by using thermo-couple and a laser displacement sensor (HG-C1050, SUNX) respectively. The displacement (D) refers to the distance between the surface of the central point of the upper strip and the axis of sliding grove, as shown in Fig. 2. Notably, the cycle number of TMC was set to be 30000, which was

Experiments

Electrolytic nickel (purity > 99.9%) and sponge titanium (purity > 99.7%) with a nominal atomic ratio of 51% Ni to 49% Ti were smelted repeatedly (six times) in a non-consumable vacuum arc melting furnace (WK-I mode) to fabricate homogeneous $Ni_{51}Ti_{49}$ ingot. Then, the $Ni_{51}Ti_{49}$ alloy strip with a dimension of 70 × 8 × 0.7

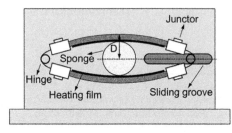

Fig. 2 Illustration of the $Ni_{51}Ti_{49}$ alloy strips assembled on the AAS stage for thermo-mechanical cycling test

calculated based on four times "open" and "close" of the AAS per day during 20 years of service. Notably, 2 mm of the so-obtained displacement was from the existence of hinges. In addition, the phase transformation behavior was analyzed by differential scanning calorimeter (DSC, Q200, TA) with a heating/cooling rate of 5 °C/min. The microstructure of the alloy strip was characterized by transmission electron microscope (TEM, 2100F, JEOL). Samples for TEM observation were prepared through twin-jet electro-polishing using mixed solution of 20 vol. % H$_2$SO$_4$ and 80 vol. % methanol at 5 °C.

Results and Discussion

Phase Transformation Behavior

DSC curves and corresponding phase transformation temperatures of the Ni$_{51}$Ti$_{49}$ alloy before and after TMC for 30000 cycles are comparatively displayed in Fig. 3 and Table 1, respectively. As indicated in Fig. 3, the starting and finishing temperatures of forward R-phase transformation and martensitic transformation are represented by R_s and R_f as well as M_s and M_f, respectively. Similarly, A_s and A_f represent the starting and finishing temperatures of austenitic transformation. Through partial DSC measurement, the onset and end points of the reverse R-phase transformation were captured and indicated by R_s' and R_f'. Clearly, the phase transformation path of the alloy, A-R-M (i.e., austenite to R-phase to martensite) on cooling and M-A (i.e., martensite to austenite) on heating, does not change after TMC for 30000 cycles. Besides, all phase transformation temperatures except M_s and M_f keep almost unchanged during TMC, as listed in Table 1. Apparently, M_s and M_f shift 5.5–3.2 °C toward higher temperature, respectively. The possible reasons of the small increase of martensitic transformation temperature include the slight Ni depletion in the matrix of the alloy and the decrease of coherent stress between the matrix and Ni$_4$Ti$_3$ precipitates [4]. However, such microstructure difference is too slight to be detected during TMC and the shifts of phase transformation temperatures are just probably caused by the sampling position difference. Nevertheless, it can be confirmed that the as-fabricated Ni$_{51}$Ti$_{49}$ alloy possesses stable forward and reverse R-phase transformations in the required operation temperature range of 35–55 °C.

Deformability and Deformation Recoverability

As a very important functionality of the AAS, the deformation performance, including deformability and deformation recoverability, can be evaluated by the displacement and corresponding recovery ratio of the alloy strip. Figure 4 shows the TMC curves of the displacement versus temperature of the alloy strip after undergoing different cycles and the subsequent TC curves for 30 cycles after removing external force, in which the latter was shown as a comparative investigation for clarifying the influence of the external force on the deformation behavior of the alloy strip. The recovery ratio (r) of the alloy strip during 35–55°C can be calculated by the following Eq. (1):

$$r = \frac{D_{55} - D_{35}}{D_{55} - 2} \times 100\% \qquad (1)$$

where D_{35} and D_{55} represent the displacements at 35 °C on cooling and 55 °C on heating respectively, as shown in Fig. 4b. It should be indicated that in Eq. (1) the unrecoverable displacement of 2 mm induced by the hinges in the AAS has been eliminated. As can be seen, the TMC curves are stable and highly repeatable with a recovery ratio of 53.0% in the first cycle and a slightly decreased recovery ratio of 48.4% in the 30000th cycle. This indicates that the Ni$_{51}$Ti$_{49}$ alloy can deform steadily with stable deformability

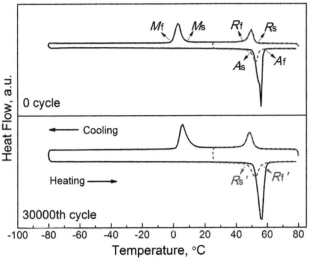

Fig. 3 DSC curves of the Ni$_{51}$Ti$_{49}$ alloy before test and after thermo-mechanical cycling for 30000 cycles

Table 1 Phase transformation temperatures of the Ni$_{51}$Ti$_{49}$ alloy before and after thermo-mechanical cycling	Cycle number	Phase transformation temperatures (°C)							
		R_s	R_f	R_s'	R_f'	M_s	M_f	A_s	A_f
	0	52.3	45.4	48.8	54.9	6.4	−0.9	51.3	57.2
	30000	52.3	44.3	47.9	56.1	11.9	2.3	51.4	58.7

Fig. 4 Thermo-mechanical cycling curves of the displacement versus temperature of the $Ni_{51}Ti_{49}$ alloy after undergoing different cycles (**a**) and subsequent thermal cycling curves for 30 cycles after removing external force (**b**)

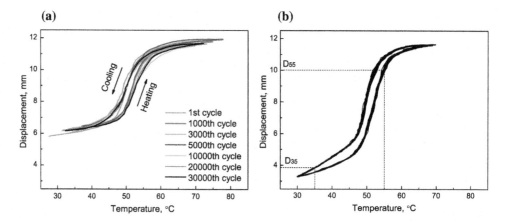

and recoverability after undergoing TMC for 30000 cycles. Further, after removing the external force, the alloy strip still deforms continuously and stably during TC, as shown in Fig. 4b. Thus, it can be demonstrated that the external force of 2 N has little influence on the stability of deformation performance of the $Ni_{51}Ti_{49}$ alloy during TMC.

Referring to the requirement of "close" and "open" functions of the AAS, the alloy strip should possess D_{35} around 2 mm and D_{55} of at least 10 mm. Although D_{55} of the strip is beyond 10 mm during TMC, there still remains an irreversible displacement (D_{35}-2) around 4 mm, as shown in Fig. 4a, resulting in the degradation of the well "close" function of the AAS. It should be pointed out that the irreversible displacement is not caused by the functional instability. Because the irreversible displacement and recovery ratio of the strip can recover to 1.8 mm and 77.4% respectively after removing external force, as shown in Fig. 4b. As a result, the external force leads to little degradation of displacement and recovery ratio after TMC for 30000 cycles, but prevents the alloy strip from recovering to the near straight shape on cooling.

Microstructure

Taking into account the results of phase transformation behavior and deformation performance, microstructure characterization was carried out to study the mechanism of functional stability of the $Ni_{51}Ti_{49}$ alloy during TMC. The bright field (BF) image, together with selected area electron diffraction (SAED) pattern, of the alloy after TMC for 30000 cycles, and the size distribution of 290 Ni_4Ti_3 precipitates determined by statistical analysis are shown in Fig. 5. As can be seen in Fig. 5a, lenticular Ni_4Ti_3 particles preferentially precipitate in the matrix of the $Ni_{51}Ti_{49}$ alloy, which consists of austenite (B2) and R-phase. Thus, the phase constituents of thermo-mechanically cycled $Ni_{51}Ti_{49}$ alloy are similar to the one only undergoing TC [6]. Moreover, the Ni_4Ti_3 precipitates in the $Ni_{51}Ti_{49}$ alloy after TMC has an average size (L) of 130.0 ± 48.3 nm, as shown in Fig. 5b. Comparing with the precipitate size of the as-fabricated $Ni_{51}Ti_{49}$ alloy, 131.4 ± 36.1 nm [4], the Ni_4Ti_3 precipitates are quite stable during TMC. Further, only a limited number of dislocations are observed at the interface between Ni_4Ti_3 precipitates and

Fig. 5 Bright field image along with SAED pattern of the $Ni_{51}Ti_{49}$ alloy for the AAS after thermo-mechanical cycling for 30000 cycles (**a**) and the size distribution of 290 Ni_4Ti_3 precipitates (**b**)

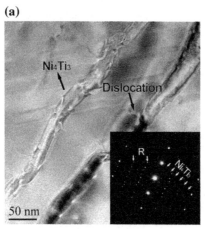

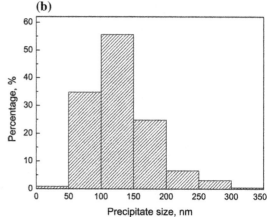

the matrix in thermo-mechanically cycled Ni$_{51}$Ti$_{49}$ alloy. Therefore, the excellent stability of Ni$_4$Ti$_3$ precipitates in particle size and the relatively stable microstructure indicate that the external force of 2 N has limited influence on the microstructural evolution of the Ni$_{51}$Ti$_{49}$ alloy for the AAS during TMC.

During TMC, the external force and cycling temperature play a dominant role in determining the functional stability of the Ni$_{51}$Ti$_{49}$ alloy. As the presence of the external force is approximately below 55 °C ($< A_f$), at which the cylindrical sponge is compressed, the SIM transformation does not take place in the alloy strip during TMC. In addition, the applied external force (2 N) is not sufficient to cause plastic deformation in the alloy strip. Therefore, the small number of dislocations in the alloy after TMC are caused by the alternate forward and reverse R-phase transformations accompanying with slight lattice distortion. Such a limited microstructural damage may not lead to obvious functional instability like the shift of phase transformation temperature and degradation of deformation performance. Besides, since the Ni$_4$Ti$_3$ precipitates are formed and stabilized during long time constraint-aging for 100 h, whose further growth can not be triggered and driven by TC especially TMC within 25–80 °C, the coherent stress field between the matrix and Ni$_4$Ti$_3$ precipitates remains steady during TMC. Therefore, neither the external force nor the TMC temperature is sufficient to bring about plastic deformation, large lattice distortion or the growth of Ni$_4$Ti$_3$ precipitates, hence the alloy strip possesses stable functionalities. As for the irreversible displacement, it can be explained by the fact that the residual stress in the Ni$_{51}$Ti$_{49}$ alloy after removing cylindrical sponge can not be released by heating during TC within 25–80 °C.

In our further work, the focus will be placed on clarifying the role of the irreversible displacement and further optimizing the deformability and deformation recoverability of the Ni$_{51}$Ti$_{49}$ alloy strip during TMC so as to meet optimally the requirement of the AAS.

Conclusion

The specially designed Ni$_{51}$Ti$_{49}$ alloy strip, aiming at making the artificial anal sphincter, exhibits excellent functional stability with small shifts of phase transformation temperatures and slight degradations of displacement and recovery ratio during thermo-mechanical cycling for 30000 cycles under a pre-determined external load condition. The

thermo-mechanically cycled alloy strip shows excellent stability of microstructure in terms of nearly unchanged size of Ni$_4$Ti$_3$ precipitates and limited number of dislocations. Thus, neither the external force nor the cycling temperature has obvious influence on the corresponding functionalities of the Ni$_{51}$Ti$_{49}$ alloy strip during thermo-mechanical cycling. However, the external load impedes the recovery of the currently fabricated alloy strip from fully open state to fully closed state on cooling during thermo-mechanical cycling and may lead to the degradation of the designed "close" function of the artificial anal sphincter.

Acknowledgements This research was supported by Key Project Program of Guangdong Provincial Natural Science Foundation under Grant No. S2013020012805 and the National Natural Science Foundation of China under Grant Nos. 51401081 and 51571092.

References

1. Luo Y, Takagi T, Matsuzawa K (2001) Design of an artificial sphincter using shape memory alloys. Int J Appl Electrom 14 (1):411–416
2. Luo Y, Higa M, Amae S, Yambe T, Okuyama T, Takagi T, Matsuki H (2005) The possibility of muscle tissue reconstruction using shape memory alloys. Organogenesis 2(1):2–5
3. Liu H, Luo Y, Higa M, Zhang X, Saijo Y, Shiraishi Y, Sekine K, Yambe T (2007) Biochemical evaluation of an artificial anal sphincter made from shape memory alloys. J Artif Organs 10 (4):223–227
4. Li YY, Yao XY, Cao SS, Ma X, Ke CB, Zhang XP (2017) Rapidly solidified and optimally constraint-aged Ni$_{51}$Ti$_{49}$ shape memory alloy aiming at making a purpose-designed bio-actuator. Mater Design 118:99–106
5. Li YY, Cao SS, Ma X, Ke CB, Zhang XP (2017) Influence of strongly textured microstructure on the all-round shape memory effect of rapidly solidified Ni$_{51}$Ti$_{49}$ alloy. Mater Sci Eng A 705:273–281
6. Li YY, Zeng CY, Ma X, Cao SS, Zhang XP (2017) Phase transformation behavior and reversible deformability of Ni$_{51}$Ti$_{49}$ two-way shape memory alloy and its functional stability as an artificial anal sphincter during high-cycle thermal cycling. J Mech Behav Biomed Mater (submitted)
7. Liu YN, Laeng J, Chin TV, Nam TH (2008) Partial thermal cycling of NiTi. J Alloy Compd 449(1–2):144–147
8. Suresh KS, Bhaumik SK, Suwas S (2013) Effect of thermal and thermo-mechanical cycling on the microstructure of Ni-rich NiTi shape memory alloys. Mater Lett 99(20):150–153
9. Saikrishna CN, Ramaiah KV, Prabhu SA, Bhaumik SK (2009) On stability of NiTi wire during thermo-mechanical cycling. Bull Mater Sci 32(3):343–352
10. Zhang XP, Zeng CY, Li YY, Zhao ZX, Li WY (2017) Testing system of thermal cycling functional stability and functional fatigue for shape memory alloys. China Patent 201710068496, 10 May 2017

Influence of Contact Friction on Force-Deflection of Orthodontic NiTi Archwire: A Computational Study

M. F. Razali and A. S. Mahmud

Abstract

The force response of NiTi archwire with respect to tooth movement in orthodontic leveling treatment depends largely on the sliding resistance of a bracket system. This study investigated the influence of contact friction between the wire and the bracket towards the force-deflection behavior of superelastic NiTi wire. A finite-element model of a three-bracket bending configuration was developed, and a user material subroutine was employed to predict the force response. The archwire was bent to a certain displacement representing the curvature of the wire when installed in a patient, and the coefficient of contact friction with the brackets was defined at a range of 0.1–0.5. This investigation revealed that the force plateau of NiTi archwire occurred at positive slope, with steeper gradient recorded on the model with a higher friction coefficient. This implies that lower contact friction is preferable in a bracket system to preserve the force plateau characteristic.

Keywords

Nickel-titanium archwire · Bracket friction
Force-deflection · Bending

Introduction

In fixed appliance therapy, the force-induced movement of a tooth is obtained from the force being released during the recovery of the deflected archwire. The early discovery of this therapy considered the usage of stainless steel archwire, before the orthodontist shifted to NiTi archwire for its

superelastic property. The superelasticity allows the NiTi archwire to deliver light and constant force over a large magnitude of bending activation—the suitable force characteristics to move a highly displaced tooth during the initial stage of orthodontic treatment. In today's market, the dental bracket is manufactured from several materials, ranging from stainless steel to ceramic and plastic. The development of brackets from ceramic and plastic is to meet aesthetic demands requested by the patients, as these materials promote a translucent look with a color of a tooth. Unfortunately, these types of brackets are reported to induce more sliding resistance upon the sliding of archwire along the bracket slot [1].

In orthodontic studies, the force delivery behaviors of NiTi archwire under bending loads are evaluated through the force-deflection curve. Although there are two force levels on the force-deflection curve, the unloading curve is in fact the portion of interest because it reflects the magnitude of force released by the archwire to the teeth [2]. Until today, the force-deflection curves were obtained over different bending models, including cantilever [3], three-point bending [4], and modified three-point bending [5]. It is reported that in cantilever and three-point bending tests, the force released by NiTi archwire is characterized by an unloading plateau [6]. However, the incorporation of a bracket attachment in the modified three-point bending model is found to alter the force plateau into a slope [7] due to the generation of contact friction between the wire and the bracket.

The contact friction in archwire sliding mechanics can be classified into classic friction and binding [8]. Classic friction refers to sliding resistance created by elastomer ligatures when it drives the archwire against the base of the bracket slot. Meanwhile, binding refers to friction developed when the archwire is bent, with the magnitude of friction increases as the curvature of the bend increases [9]. During tooth movement, along with sliding of the wire on the adjacent brackets, the deactivation force of NiTi archwire is partially used to overcome the binding friction developed due to the bend of the wire, hence lowering the magnitude of effective

M. F. Razali · A. S. Mahmud (✉)
Nanofabrication and Functional Materials Research Group,
School of Mechanical Engineering, Universiti Sains Malaysia,
Engineering Campus, 14300 Penang, Malaysia
e-mail: abdus@usm.my

M. F. Razali
e-mail: fauzinizam88@gmail.com

© The Minerals, Metals & Materials Society 2018
A. P. Stebner and G. B. Olson (eds.), *Proceedings of the International Conference on Martensitic Transformations: Chicago,*
The Minerals, Metals & Materials Series, https://doi.org/10.1007/978-3-319-76968-4_33

force to induce tooth movement [5]. Studies pertaining to binding friction have been carried over various combinations of bracket materials and NiTi wire sizes. It is reported that binding increases with an increase in the size of the archwire [10] and the friction coefficient (μ) of the meeting surfaces [11].

Although numerous studies have evaluated the binding magnitude with different archwire-bracket combinations, the influence of contact friction on the force-deflection behaviors of NiTi archwire during bending has remained unreported. Therefore, the objective of this work is to evaluate the force-deflection released by the superelastic NiTi wire upon couples with different bracket material. This investigation was conducted using computational technique by developing a three-dimensional finite-element model that simulated the bending of archwire at varied contact friction. This approach allows a direct control of the friction coefficient encountered for different wire-bracket combinations. A common combination of appliances for a standard configuration of leveling was considered for the evaluation of the force-deflection. The analysis focused on the magnitude and slope of the force during the deactivation course. This finding may assist orthodontists in selecting the best bracket material to comply with the light and constant force criteria suggested for effective tooth movement.

Methodology

Experimental Testing

The force-deflection of superelastic NiTi archwires in the bracket system was investigated by using a modified three-point bending test as described in [5]. The concept of this setup incorporates the effect of contact friction towards force-deflection behavior during bending. This setup considered three aligned brackets, with the central and adjacent brackets mounted on a movable indenter and fixed supports, respectively. Three brackets with a slot size of 0.46×2.80 mm were selected for their zero torque and angulation design. The pairing of 0.4 mm wire with the 0.46 mm-slot bracket provides sufficient clearance for the free sliding of the wire [12]. No ligature was installed in securing the wire specimen inside the bracket slot to avoid the unnecessary friction. As shown in Fig. 1, the interbracket distance (IBD) was set to 7.5 mm, and the specimen was deflected to 4.0 mm by moving the indenter vertically downwards at a rate of 1.0 mm/min. A heating chamber was used to maintain the testing environment at 36 °C. This bending test was repeated twice for consistency purposes, and the curves were directly compared with the numerical result for validation.

A sliding test was conducted to determine the static friction coefficient between NiTi wire and stainless steel

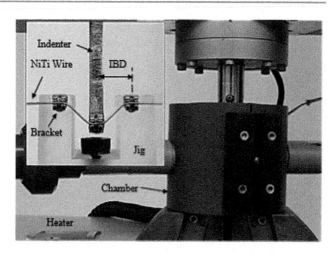

Fig. 1 Modified three-point bending setup equipped with a heating chamber

brackets, which was later used to define the frictional properties in the numerical model. The sliding test was carried out by using a Ducom TR-20 pin-on-disk tribometer. In order to allow greater clearance between the archwire and bracket slot, a 0.40×0.56 mm NiTi archwire and a 0.56 mm bracket slot were selected for the test.

A straightened wire specimen was glued to the surface of a movable sliding plate and aligned against the fixed bracket slot. The sliding test began with applying a 2 kg (19.62 N) dead weight on the loading pan, which subsequently caused the bracket to press the archwire. Then, the NiTi wire was reciprocally slid along the slot at a speed of 1 mm/sec for 4 mm displacement. The friction coefficient was obtained by dividing the friction data with the applied load. The sliding test was repeated three times with new archwires and brackets for each run.

Finite Element Modeling

The finite-element model and the force analysis were performed using a commercial finite-element analysis package of Abaqus/Standard version 6.12-1 in combination with UMAT/Nitinol subroutine. The subroutine has been developed based on the constitutive model of superelastic NiTi alloys by Auricchio and Taylor [13]. The material data that constitutes the mechanical properties and shape memory deformation behaviors of the specimen used in the subroutine are tabulated in Table 1. Each parameter in this table was measured and calculated from the uniaxial tensile stress-strain curve. Additionally, since the compression test on such a small wire specimen is impossible, the start of transformation stress in compression (σ_{SCL}) was set to be 1.2 times higher than the start of transformation stress in tension (σ_{SL}) [14].

Table 1 Mechanical properties and shape memory deformation behaviors of NiTi archwire measured from the uniaxial stress-strain curve

Parameter	Description	Value (unit)
E_A	Austenite elasticity	44 (GPa)
(ν_A)	Austenite Poisson's ratio	0.33
E_M	Martensite elasticity	23 (GPa)
(ν_M)	Martensite Poisson's ratio	0.33
(ε_L)	Transformation strain	0.06
$(\delta\sigma/\delta T)_L$	Stress rate during loading	6.7 (MPa/°C)
σ_{SL}	Start of transformation loading	377 (MPa)
σ_{EL}	End of transformation loading	430 (MPa)
T_0	Reference temperature	26 (°C)
$(\delta\sigma/\delta T)_U$	Stress rate during unloading	6.7 (MPa/°C)
σ_{SU}	Start of transformation unloading	200 (MPa)
σ_{EU}	End of transformation unloading	140 (MPa)
σ_{SCL}	Start of transformation stress in compression	452 (MPa)

A nonlinear finite-element model of modified three-point bending was developed to evaluate the force-deflection behavior of the NiTi archwire prior to the displacement of the middle bracket. The 30 mm length specimen was modeled by using 72,000 linear hexahedral elements with reduced integration (C3D8R). The bracket was modeled by using a bilinear rigid quadrilateral element (R3D4) with the actual slot dimensions of 0.46 × 2.80 mm. Figure 2 illustrates the assembly of the wire and the brackets, with a center-to-center distance between the brackets of 7.5 mm. Each bracket was assigned to its own reference point (RP), so that the boundary condition set to the reference point could be applied to the entire bracket. The middle bracket was set to be free to move only along the vertical axis and the displacement rate was controlled at 1.0 mm/min (Ux = Uz = 0). The adjacent brackets were fixed in all displacement directions (Ux = Uy = Uz = 0).

The contact between the wire and the rigid brackets was modeled using the finite sliding, surface-to-surface formulation. For validation with the experimental work, a friction coefficient of 0.27 (acquired from the sliding test) was defined at the possible contacted surfaces. Then, the simulation was expanded to different friction conditions by varying the coefficient values from 0.1 to 0.5. In detail, the coefficient values of 0.1–0.3 and 0.4–0.5 reflected the friction coefficients range of NiTi wire when in contact with stainless steel [15] and ceramic [16] brackets, respectively. All simulations were conducted at a constant temperature of 36 °C. The force-deflection result was attained by requesting the vertical reaction force (RF2) and displacement (U2) at the reference point of the middle bracket.

Results and Discussion

Figure 3 displays the experimental results of frictional force and friction coefficient established from the sliding test. The static friction was observed at the beginning of the wire movement as indicated by the maximum frictional peak. Beyond static friction are the peaks of kinetic friction that fluctuated at a slightly lower magnitude. In brief, the static friction coefficient was averaged at 0.27. This coefficient

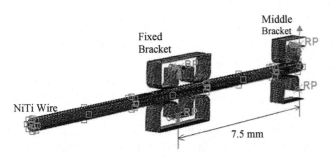

Fig. 2 Finite element model for the modified three-point bending test

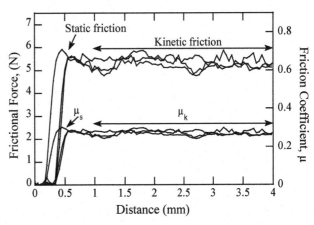

Fig. 3 Plot of frictional force and friction coefficient along the 4.0 mm sliding distance

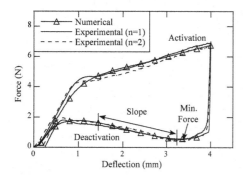

Fig. 4 Force-deflection curve of NiTi archwire bent in three-bracket configuration at 36 °C

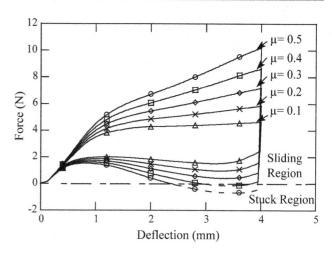

Fig. 5 Numerical results of the force-deflection curves at various friction coefficients

value falls within the expected coefficient range for stainless steel bracket coupling, as reported in [15]. It was also acknowledged that a small variation was registered in the coefficients of static friction (standard deviation of 0.20).

Figure 4 compares the force-deflection curve obtained from the numerical and experimental work. A relatively accurate agreement in force magnitude was observed between the experimental and the numerical curves, as indicated by the small discrepancy of force magnitude (0.2 N) at 4.0 mm deflection. Since the test was performed over the austenite finish temperature, the numerical results portrayed the exact superelastic expression represented by the full deflection recovery. As for the experimental result, a small magnitude of residual elongation was observed, indicating a small volume of the specimen being plastically deformed during the activation sequence.

In orthodontic practices, the actual force delivered to the periodontium is represented by the deactivation curve of the force-deflection. Since the deactivation curve exhibited over a slope trend, the evaluation of the effective force for tooth movement was focused on the magnitude of minimum force and the slope of the curve. As illustrated in Fig. 4, the minimum force of the archwire is measured at the valley of the deactivation curve, of which the deflection distance is 3.2 mm. Meanwhile, the slope of the force was measured from the best linear region along the deactivation curve, as indicated by the arrow in the figure. This gradient force behavior of superelastic NiTi wire during bending in the bracket system was consistently correlated to the linear increase of friction at the wire-bracket interface [17]. The valley at the onset of the deactivation curve indicates that a greater portion of the force was utilized to encounter the friction.

The force-deflection curves of the archwire at various friction coefficients are plotted in Fig. 5. At $\mu_s = 0.1$, the bending deformation behavior displayed a typical superelastic curve, such as indicated by the flat force plateaus on both wire activation and deactivation. The activation plateau corresponded to stress-induced martensitic transformation (SIMT), while the deactivation plateau at a lower force level denoted

the reversed transformation of the wire phase at the onset of stress removal. As the friction coefficient increased, the stress plateau level also increased proportionally. The activation plateau stress increased in positive slope, but the deactivation plateau increased in negative slope. This force slope is related to the variation of binding created at the contact location between the wire curvature and the bracket edges [9].

Additionally, it is also interesting to highlight that the increment in coefficient value delayed the deactivation curve to a lower force level. For the case in which friction coefficients were 0.4–0.5, the minimum force was plummeted beyond the zero force level. The zero force marks the end of the sliding mechanics of the archwire as the spring-back force was no longer capable to overcome the overpowered contact friction (binding) at the adjacent bracket slot. Therefore, the archwire was subsequently stuck at the onset of deactivation. If one were to translate this phenomenon from a clinical perspective, no movement of the tooth will be induced until the wire can be released from its stuck position.

For this simulation, one should note that the middle bracket incorporated in the model was set to return to its original position at the end of the deactivation mode. Therefore, the negative force magnitude shall be denoted as the minimum force required to surpass the contact friction, hence allowing the archwire to slide again for the remaining deflection. This similar zero force behavior was reported previously prior to the deactivation of the NiTi archwire from 6.0 mm bracket displacement [18].

The magnitude of the minimum force and the deactivation slope measured in Fig. 5 is plotted in Fig. 6. It was recognized that the increase of coefficient values from 0.1–0.5 has increased the slope of the deactivation curve from 0.31 N/mm to 1.10 N/mm, respectively. This slope rate signified that the high friction coefficient case would lead to superior force changes as the deflection recovers; a condition which is

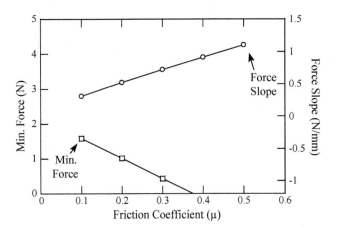

Fig. 6 Variation of minimum force and deactivation slope at various friction coefficients

unsuitable to induce tooth movement. It can also be seen that the minimum force decreased linearly from 1.60 to 0.42 N when the friction coefficient was increased from 0.1 to 0.3. Noted that the data points of minimum force for cases with a friction coefficient higher than 0.4 were not included in this figure due to their values being below zero. This observation signified that the spring-back potential of the archwire during the deactivation was used to overcome the high contact friction, thus hindering further recovery. In this regard, it is recommended that the friction coefficient between the wire and bracket should be limited to lower than 0.3 for effective tooth movement. Subsequently, the use of ceramic brackets is highly not recommended due to the fact the contact friction between ceramic and NiTi archwire can be higher than 0.5 due to its rougher surface morphology [11]. Thus, the orthodontist is strongly suggested to consider the stainless steel bracket ($\mu \leq 0.3$) for the leveling treatment, as this bracket promotes positive forces with a minimum force variation.

Conclusion

The main findings are summarized below:

1. The binding friction increased proportionally as the wire deflection magnitude increased.
2. The bending of NiTi archwire at friction coefficients higher than 0.4 reduced the deactivation force to zero value, and thus would inhibit further tooth movement.

Acknowledgements The authors are grateful for the financial support provided by Universiti Sains Malaysia under the grant RUI 1001/PMEKANIK/814244.

References

1. Williams CL, Khalaf K (2013) Frictional resistance of three types of ceramic brackets. J Oral Maxillofac Res 4:e3
2. Segner D, Ibe D (1995) Properties of superelastic wires and their relevance to orthodontic treatment. Eur J Orthod 17:395–402
3. Pesce RE et al (2014) Evaluation of rotational control and forces generated during first-order archwire deflections: a comparison of self-ligating and conventional brackets. Eur J Orthod 36:245–254
4. Gatto E et al (2013) Load-deflection characteristics of superelastic and thermal nickel-titanium wires. Eur J Orthod 35:115–123
5. Nucera R et al (2014) Influence of bracket-slot design on the forces released by superelastic nickel-titanium alignment wires in different deflection configurations. Angle Orthod 84:541–547
6. Bartzela TN, Senn C, Wichelhaus A (2007) Load-deflection characteristics of superelastic nickel-titanium wires. Angle Orthod 77:991–998
7. Badawi HM et al (2009) Three-dimensional orthodontic force measurements. Am J Orthod Dentofac Orthop 136:518–528
8. Burrow SJ (2009) Friction and resistance to sliding in orthodontics: a critical review. Am J Orthod Dentofacial Orthop 135:442–447
9. Hamdan A, Rock P (2008) The effect of different combinations of tip and torque on archwire/bracket friction. Eur J Orthod 30:508–514
10. Major PW et al (2014) Effect of wire size on maxillary arch force/couple systems for a simulated high canine malocclusion. J Orthod 41:285–291
11. Doshi UH, Bhad-Patil WA (2011) Static frictional force and surface roughness of various bracket and wire combinations. Am J Orthod Dentofac Orthop 139:74–79
12. Proffit WR, Fields Jr HW, Sarver DM (2014) Contemporary orthodontics. Elsevier Health Sciences
13. Auricchio F, Taylor RL (1997) Shape-memory alloys: modelling and numerical simulations of the finite-strain superelastic behavior. Comput Methods Appl Mech Eng 143:175–194
14. Auricchio F, Sacco E (1999) A temperature-dependent beam for shape-memory alloys: Constitutive modelling, finite-element implementation and numerical simulations. Comput Methods Appl Mech Eng 174:171–190
15. Whitley JQ, Kusy RP (2007) Influence of interbracket distances on the resistance to sliding of orthodontic appliances. Am J Orthod Dentofac Orthop 132:360–372
16. Kusy RP, Whitley John Q (1990) Coefficients of friction for arch wires in stainless steel and polycrystalline alumina bracket slots. I. The dry state. Am J Orthod Dentofac Orthop 98:300–312
17. Thalman TD (2008) Unloading behavior and potential binding of superelastic orthodontic leveling wires. M.Sc. thesis, Saint Louis University
18. Baccetti T et al (2009) Forces produced by different nonconventional bracket or ligature systems during alignment of apically displaced teeth. Angle Orthod 79:533–539

Features Cavitation Resistance of Multifunctional Coatings from Materials with a Shape Memory Effect

D. V. Dmitrenko, Zh. M. Blednova, and E. U. Balaev

Abstract

It is established on the basis of the analysis of structural components of criterion of cavitation resistance of multifunctional coatings from materials with a shape memory effect (SME) that the major factors influencing the cavitation resistant coatings are: tendency to strain hardening, reversible deformation, the adhesion strength, microhardness, uniformity of structure and roughness of a surface. Weights coefficients for assessment of influence of each of the specified components are defined. The paper studies the mechanism of enhancing cavitation resistance of coatings from materials with SME based on structural and phase transformations in the coating material, subject to local cavitation effects. Formation of surface compositions was carried out by the high-velocity oxy-fuel spraying in a protective environment. Tests on cavitation resistance of multilayer coatings of materials with SME performed according to standard methods of analysis of structural-phase state and functional-mechanical properties that they are allowed to recommend as a cavitation-resistant. Architecture layered surface composition is proposed, which provides increase of durability of products in the conditions of cavitation-abrasive environment.

Keywords

Cavitation • Surface • Multicomponent material Multifunctional coatings • SME

D. V. Dmitrenko (✉) · Zh.M. Blednova · E. U. Balaev
Department of dynamics and strength of machines,
Kuban State Technological University,
Moskovskaya 2, 350072 Krasnodar, Russia
e-mail: ddv-kk@yandex.ru

Zh.M. Blednova
e-mail: blednova@mail.ru

E. U. Balaev
e-mail: balaev1122@mail.ru

Introduction

Increasing the reliability and durability of machine parts and mechanisms is primarily due to the elimination of wear caused by various types of impacts, in particular, on their surface layers. One of the types of such impact on the surface of parts working in a fluid environment is cavitation erosion of the metal, which is caused by the mechanical action of fast moving fluid particles on the surface of the metal, sand particles of solids, suspensions, gas bubbles, etc. Such influence is characterized by a high rate of loading, its small duration, locality and cyclicity. At the same time, stresses appear in the surface layers, they can be compared with the ultimate strength of the material and are concentrated in volumes close to the dimensions of its structural components. Effective ways to improve the strength properties and resistance to corrosion-erosive effects are thermal and thermomechanical hardening treatments and the deposition of protective coatings on the metal surface [1].

Taking into account the crucial role of the surface layer in the accumulation of damage and destruction, the task of providing multifunctionality, as well as increasing reliability and resource, can be successfully solved at the final stage of processing by means of technologies based on the principles of layer-by-layer synthesis using combined, functionally oriented macro-, micro-, and nanotechnologies. [2, 3].

A promising direction in the implementation of layered synthesis technologies is the intellectualization of products using shape memory effect (SME) materials. They have a wide range of functional and mechanical capabilities: effects of thermomechanical memory and superelasticity, high strength and damping properties, thermomechanical reliability and durability, wear and corrosion resistance [4]. Works [5–7] show that multicomponent TiNi-based coatings with addition of alloying elements such as Zr in an amount up to 10 mass%, allow to create a composite material with a gradient of properties at the interface of layers, which gives new properties to coatings and significantly improves their performance. The use of materials with shape memory effect (SME) as surface layers [8] or in the composition of surface layered compositions [6] improves the durability under cyclic loading and reverse friction due to the adaptation of the functional layers to external influences.

The aim of this work is to identify the main factors that affect cavitation resistance of multilayer coatings, based on the analysis of the structural components of the criterion of cavitation resistance of multifunctional coatings from materials with shape memory effect (SME).

Experiment

Studies carried out by a number of authors [9] show that the main influence on the cavitation stability of the surface is caused by such factors as surface strength, surface hardness and the damping ability of the material which allows to dampen the energy of deformation and vibration oscillations that occur during cyclic cavitation process. Taking into consideration that materials of high hardness have a lower viscosity, it becomes necessary to create layered gradient structures that provide damping and stress relaxation. The relaxation functions in the surface composition can be performed by layers made of SME materials [8]. There are works of different researchers [9, 10] convincingly showing the positive effect of increasing the wear and cavitation resistance of the SME alloy with TiNi. The use of SME alloys for the manufacture of engineering products is limited

for costs reasons. Therefore, the development of layered surface compositions of SME materials will solve the energy saving problem; and the synergistic effect from the use of surface-gradient structures will provide multi-functionality and expand the field of practical use. The purpose of this work is to obtain new data on the cavitation resistance of a layered surface composition of engineering designation made of multicomponent SME materials with a gradient distribution of the properties of the "TiNi–TiNiZr–wear-resistant layer cBN-Co", which improves the functionality of products.

The formation technology of a surface composition with SME material is a complex multi-operation process involving the preparation of the surface and the deposited material (milling and mechanical activation), high-velocity oxygen-fuel spraying (HVOF), and subsequent thermo-mechanical (TM) treatment. Characteristics of the layers are presented in Table 1.

HVOF of preliminary mechanically activated powder, was carried out on the upgraded GLC-720 unit [11], equipped with a vacuum chamber, and with a device for moving the HVOF gas flame burner. The steel samples (Steel 1045) were coated with a composite coating consisting of: a $Ti_{49}Ni_{51}$ functional layers (550–600 μm), a layer of $Ti_{33}Ni_{49}Zr_{18}$ (1000–1050 μm), and a wear-resistant cBN-10%Co layer with the thickness of 200–210 μm (Fig. 1). On the basis of the studies [4], we determined optimal regimes of HVOF of the constituent composition layers on a steel substrate according to the criteria of porosity and adhesion strength of

Table 1 Layer characteristics

Layers	Material	Layer thickness (μm)	T (°C)	M_s (°C)	M_f (°C)	A_s (°C)	A_f (°C)	Microhardness (hPa)	Phase state
Base	Steel 1045	–	20					1,9–2,1	–
1	$Ti_{49}Ni_{51}$	600 ± 50	20	23	55	13	31	10,2–11,9	Austenite
2	$Ti_{33}Ni_{49}Zr_{18}$	1000 ± 50	20	186	249	215	298	2,1–2,9/9,5–12,7	Martensite/austenite
3	cBN-10% Co	200 ± 10	20					34,6–35,6	–

M_f is the temperature of martensitic transformation expiry
M_s is the temperature of martensitic transformation onset
A_s is temperature of austenitic transformation onset
A_f is the temperature of austenite transformation expiry

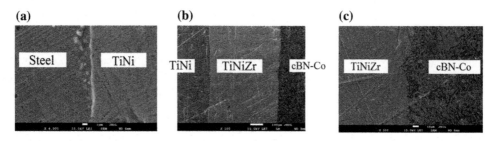

Fig. 1 Surface layer with a material with SME: **a** base steel–layer $Ti_{49}Ni_{51}$, **b** layers $Ti_{49}Ni_{51}$–$Ti_{33}Ni_{49}Zr_{18}$–cBN-Co, **c** layers $Ti_{33}Ni_{49}Zr_{18}$–cBN-Co, **(c)**

the surface layer and the substrate. HVOF is accompanied by large plastic deformations and the crystallization of particles occurs at high degrees of supercooling and is accompanied by the formation of crystalline nucleating centers whose size is determined by the degree of supercooling determined by the parameters of the process [8]. Complex thermo-mechanical processing was performed to impart pseudo-elasticity or shape memory properties to the surface layer and increase the adhesive strength, depending on the thickness of the composite layer. We carried out a complex thermo-mechanical treatment for the coatings, including surface plastic deformation (SPD) by rolling and ultrasonic treatment (UT).

The outer layer in contact with the working medium must have a high wear resistance with an increased bearing and damping ability, corrosion resistance with an optimum combination of strength and toughness. To meet these requirements, it is proposed to use a combination of layers of dissimilar materials. The outer layer is a hard and wear-resistant cNB-Co material, intermediate layers of $Ti_{33}Ni_{49}Zr_{18}$–$Ti_{49}Ni_{51}$, which are in austenitic-martensitic and austenite phase state, respectively. The purpose of the intermediate layer is to relax the stresses, damping the oscillations and blocking the cracks that appear in the outer layer and propagate deep into the material. An intermediate functional layer made of materials with an SME in the martensitic state, which has an increased elasticity, helps to reduce the stress concentration at the tip of the crack and inhibits or blocks its movement. The gradient of properties in the outer contact and functional layers of materials with SME is provided by a gradient of the temperatures of the phase transformations, which are set by the regulation of the chemical composition and thermomechanical processing.

In order to assess the quality of the composition functional layers, the sprayed samples were tested for cavitation resistance and fatigue strength. The micro-hardness test was performed on the Falcon-500 hardness tester. Cavitation tests were carried out in accordance with the American Standard ASTM G32-10 (Standard Test Method for Cavitation Erosion Using Vibratory Apparatus), using an ultrasonic vibrating unit and standard statistical processing of the experimental data [12, 13].

Ultrasonic waves were used to create the cavitation zone. The signal from the generator is fed to a magnetostrictive transducer mechanically connected to a waveguide, at the end of which an experimental sample of a cylindrical shape with a diameter of 15 mm is rigidly mounted, which provides an amplitude of oscillation of the sample end surface up to 50 ± 2 μm at a frequency of 20 kHz. The sample is immersed in a vessel with water. On the end surface of the sample, a zone with developed cavitation is formed. To assess the effect of corrosion processes on the magnitude of cavitation erosion, experiments were conducted in fresh- and sea-water.

Damage accumulation and destruction of the samples surface was carried out with the visual inspection and area damage assessment on a stereoscopic zoom microscope Lomo MSP-1. Monitoring of the samples profile change, subjected to the cavitation fracture, as well as the damage depth was performed on the contour-tracing apparatus ABRIS PM7, the weight loss was measured on a DEMCON DL213 balance with an accuracy of ± 0.001 g. The dependence of the mass loss on the time of cavitation effect was measured; and from these data, the kinetic curves of the sample material destruction were constructed, which are characterized by the presence of the initial segment, when the destruction is small, and by the section with the maximum quasi-constant velocity.

Experimental Results and Discussion

The results of X-ray diffraction analysis showed that at room temperature the initial phase state of $Ti_{33}Ni_{49}Zr_{18}$ layer, after the HVOF of mechanically activated powder in a protective atmosphere (argon), are a martensitic phases B19′ with monoclinic lattice, austenitic phases B2 with cubic lattice, intermetallic phases Ni_3Ti, Ti_2Ni, $NiZr$, $NiZr_2$, with cubic and hexagonal lattice and there is a small amount of titanium oxide (TiO)—less than 2%. The phase analysis of layer cNB-10% Co BN with a cubic lattice ($\approx 85,3$–$87,4\%$), Co with a hexagonal lattice ($\approx 10,3$–$11,2\%$), B_2CN with orthorhombic lattice ($\approx 2,3$–$3,5\%$), BC_2N with orthorhombic lattice [6].

The hardness value was measured by Vickers scale at a test load of 0.2 kg and the average of eight indentation values were reported. The results of the hardness are given in Fig. 2a.

The kinetics of material destruction process is described by the erosion curve (mass or volume loss in time), on which, as in the fatigue process, the stages of erosion formation and development are identified. The traditional criterion for the erosion resistance of a material is the loss of its mass over a certain period of time. Figure 2b presents the results of cavitation failure study of uncoated samples and with various coatings in fresh and sea water. In fresh water cavitation destruction of the surface is a classical curve of cavitation destruction: there is an incubation period (30–60 min) with a period of intensive weight loss in the future. The intensity of the mass loss of the hardened $Ti_{49}Ni_{51}$–$Ti_{33}Ni_{49}Zr_{18}$–cBN-10%Co coating have a longer incubation period (exceeding 120 min), which is associated with an increased micro-hardness of the coating surface.

The presence of an aggressive component in the medium (salty sea water) somewhat smoothes the mass loss curve for a material without coating and increases the intensity of mass loss. The value of the mass loss is comparable for materials of the layers when testing in fresh water and sea water, as these materials have high corrosion resistance.

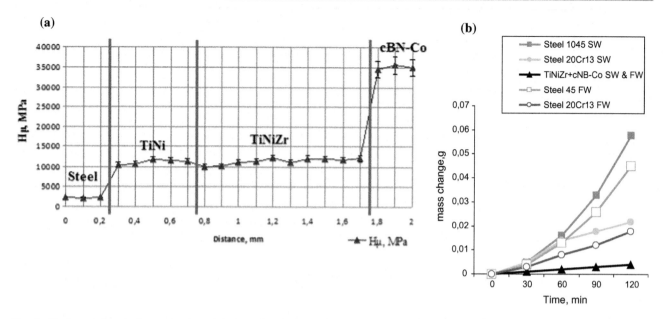

Fig. 2 Hardness values in the coating layer (**a**); dynamics of the sample mass change as a result of cavitation influences in see ware (SW) and fresh water (FW) (**b**)

In seawater, the dynamics of mass loss of the strengthened $Ti_{49}Ni_{51}–Ti_{33}Ni_{49}Zr_{18}–cBN-10\%Co$ coating does not change in comparison with the tests in fresh water. This suggests that the set of properties of the hardening coating, and namely its corrosion resistance, high microhardness and increased strength, combined with a damping ability, contribute to increasing resistance of multicomponent coatings to cavitation erosion in corrosive environments.

The surface without coating of Steel 1045, Steel 20Cr13 and the composite coating $Ti_{33}Ni_{49}Zr_{18}–cBN-10\%Co$ before and after cavitation for 30 and 120 min in the sea water are shown in Fig. 3. It can be seen that the surface micro-relief practically does not change, i.e. the coating is practically not subjected to cavitation erosion. The distribution was obtained by computer analysis of the measurement results of surface roughness in different areas of the samples. In the above

Fig. 3 Destruction of the surface by cavitation for 30 and 120 min in the sea water: **a** Steel 1045; **b** Steel 20Cr13; **c** composite coating $Ti_{33}Ni_{49}Zr_{18}–cBN10\%Co$

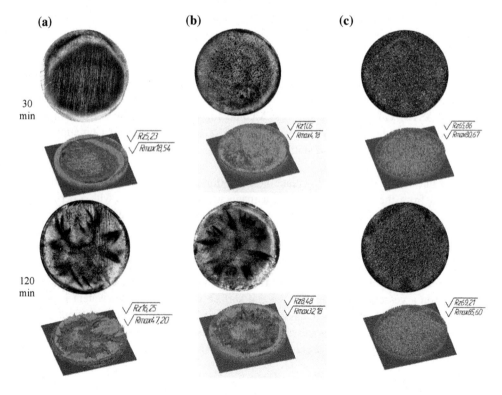

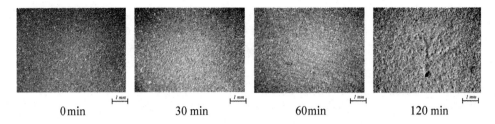

0 min	30 min	60min	120 min

Fig. 4 Cavitation destruction of the composite layer $Ti_{49}Ni_{51}$–$Ti_{33}Ni_{49}Zr_{18}$–cBN10%Co

diagrams the depth of erosion destruction of the surface was determined by the depth of asperities.

The composite coating in the process of cavitation for 120 min in the sea water is shown in Fig. 4. Cavitation damage of the composite layer is only visualized after 120 min of impact.

For a qualitative description of the cavitation fracture of a surface, it is expedient to investigate the parameters of the surface profile. The most typical factor to assess the degree of surface cavitation wear are the parameters of the roughness profile: Rz and Rmax.

Table 2 shows the dynamics of surface roughness variation on different samples during the cavitation effect in the sea water environment. The surface roughness of the samples with coatings is much higher than the initial roughness of the uncoated sample. This is due to the formation technology of surface coatings by the method of HVOF by the powder material particles, which forms a characteristic surface micro-relief.

Analysis of Table 2 and Fig. 3 shows that the change in the roughness of the samples with coatings is less than the change in the roughness of the sample without coating by 1,5–2 times. As we know, cavitation as a phenomenon, is associated with the discontinuity of the moving fluid, the formation of vacuum voids (caverns) in it, followed by their closing, accompanied by large hydrodynamic impacts. According to the ideas of a number of cavitation phenomena researchers [9–12], the cause of cavitation destruction of parts is the fatigue destruction of micro-volumes of a material due to repeated exposure of high-frequency pulses both as a detachment action (during formation and detachment of cavitation caverns) and shock action accompanying their clamping. In this case, the application site of the cavitation action pulses is so limited that the process of cavitation destruction should be considered from the point of the

cyclic strength of individual structural components or local micro-volumes of the metal. Microvolumes of the metal exposed to the leaching flow easily, as they damaged the cyclic loading and cavitation effects weakened with neighbouring micro-volumes due to loss of connection, forming a low relief surface cavitation, activating the cavitation processes to an even greater extent [11].

The surface composite layer cBN-10%Co has high hardness and strength, which ensures its resistance to shock cyclic influences of collapsing caverns. The increased roughness of the surface of a solid surface composite in the form of strong columnar structures ensures the crushing of vacuum voids, redistributing their effect on the entire surface, and not concentrating them in certain zones. In addition, the gradient structure of the multilayer composite coating $Ti_{49}Ni_{51}$–$Ti_{33}Ni_{49}Zr_{18}$–cBN-10%Co Co makes it possible to create conditions for the relaxation of stresses created by the variable impact load of cavitation caverns and the manifestation of compensating internal forces due to thermo-elastic martensitic transformations of SME materials.

The increase in the performance characteristics of samples with a composite surface layer (Table 1) "steel—adhesive layer 1—relaxing SME layer 2—a functional reinforcing SME layer 3—functional wear layer 4" under conditions of cavitation erosion is explained by the special functional properties of SME materials, which are manifested in the ability to adapt to the loading conditions, which is provided by the formation technology and processing of the surface layer. The increase in the wear-and-fatigue characteristics of the surface composition is explained by the pseudoelasticity of the layer made of SME materials with the $Ti_{33}Ni_{49}Zr_{18}$ and the decrease in the stress concentration due to its relaxation capacity. The field of use of the above surface composition can be expanded by controlling the functional properties of the layers composing the

Table 2 Change in surface roughness (R_z/R_{max}) (µm)

	Process time (min)				
	0	30	60	90	120
Steel 1045	0,27/1,56	5,23/18,54	9,56/28,14	13,88/44,48	16,25/47,2
Steel 20Cr13	0,35/1,72	1,05/4,18	1,43/12,65	4,87/21,42	8,48/32,18
Ni–$Ti_{49}Ni_{51}$–$Ti_{33}Ni_{49}Zr_{18}$ –cBN-10%Co	61,76/79,36	63,14/80,12	65,86/80,67	67,24/81,04	69,21/85,60

Fig. 5 Structural components of the criterion of the cavitation resistance of parts with composite multilayer coatings

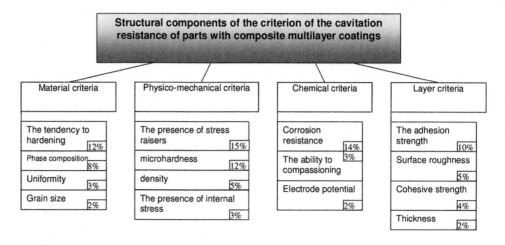

composition by varying the chemical composition and controlling the phase composition by thermal and thermo-mechanical processing.

The Analysis of Criterion of Cavitation Resistance

Performed research and analysis of the works of other authors [9, 11, 12, 13, 14] allowed us to analyze the structural components of the criterion of the cavitation resistance of parts with composite multilayer coatings, which is shown in Fig. 5. The main factors that affect cavitation resistance of parts with composite multilayer coatings, are defined by the following set: the presence of stress raisers (15%), corrosion resistance (14%), the tendency to hardening (12%), micro-hardness (12%) and the adhesion strength (10%).

Conclusion

The phase analysis, mechanical properties, cavitation erosion resistances, surface degradation studies of these coatings were investigated. The following conclusions can be drawn from the present study:

- The cavitation erosion damage in all the coatings occurrence is mainly due to generation and propagation of cracks induced by the cyclic micro-impact loads, which led to a larger number of hard carbide phase particles being continuously pulled off the surface of the coating and progressive damage occurs subsequently by fatigue.
- The cavitation resistance of the $Ti_{49}Ni_{51}-Ti_{33}Ni_{49}Zr_{18}$ coating according to the criterion of mass wear is 1,5–2 times higher than that of the base material without coating (steel 1045). The cavitation resistance of the

coating $Ti_{49}Ni_{51}-Ti_{33}Ni_{49}Zr_{18}-cBN-10\%Co$ according to the criterion of mass wear is 15–20 times higher than that of the base material.

- The proposed architecture of the multifunctional gradient composition, "steel–$Ti_{49}Ni_{51}$–$Ti_{33}Ni_{49}Zr_{18}$–cBN-10% Co", each layer of which has its functional purpose, allows to increase the service life of parts operating under conditions of cavitation-fatigue loading in corrosive environments.
- The roughness of the cavitated surface progressively increase with time and follows a similar trend with that of metal loss.

Acknowledgements The work was supported by the Russian Science Foundation (Agreement No. 15-19-00202).

References

1. Polovinkin VN Promising construction materials for special marine equipment, shipbuilding and military shipbuilding. http://www.proatom.ru/modules.php?name=News&file=article&sid=5778
2. Mikhailov AN, Mikhaylov DA, Grubka RM, Petrov MG (2015) Increase of longevity of machine parts on the basis of functionally oriented coatings. High Technol Eng 7(49):20–39
3. Suslov BM, Bazrov VF (2012) High technology in engineering. In: Suslova AG (ed) Mechanical engineering, Without language, pp 528
4. Blednova ZhM, Rusinov PO (2015) Intellectualization of surface layers, working under cyclic loading and reversing friction. In: Applied mechanics and materials, vol 798. Trans Tech Publications, Switzerland, pp 440–446
5. Rusinov PO, Blednova ZhM (2016) Structural and technological patterns of formation of surface nanostructured layers TiNiZr by high-speed flame spraying. In: Advanced materials and structural engineering, vol 978, pp 21–25
6. Rusinov PO, Blednova ZhM, Balaev EY, Dmitrenko DV (2016) Formation of composite layers TiNiZr-cBN-Co, working in conditions of cyclic loading and reverse friction. Procedia Struct Integr 2:1506–1513

7. Dmitrenko DV, Blednova ZhM, Balaev EYu (2016) Cavitation resistance of products with a composite surface layer of SME materials. In: International Conference "Vitality and structural materials science" "ZhivKom-2016". M.: IMASH RAS, pp 232–235

8. Blednova ZhM, Makhutov NA, Rusinov PO, Stepanenko MA (2015) Mechanical and tribological properties of the multifunctional composition "base-material with shape memory effect, formed under the conditions of high-energy effects, Factory laboratory. Diagn Mat 81(3):41–49

9. Wu SK, Lin HC, Yeh CH (2000) A comparison of the cavitation erosion resistance of TiNi alloys, SUS304 stainless steel and Ni-based self-fluxing alloy. Wear 244:85–93

10. Razorenov SV, Garkushin GV, Kannel GI, Kashin OA, Ratochka IV (2011) Behavior of nickel-titanium alloys with shape memory effect under conditions of shock wave loading. Phys Solid State 4:768–773

11. Vakulenko KV, Biblak IV, Cossack IB (2016) Computer simulation of cavitation erosion of steel X20Cr13 with coatings. In: Open information and computer integrated technologies, vol 72, pp 242–248

12. Petrov AI, Skobelev MM, Khanychev AG (2015) Investigation of comparative resistance to cavitation erosion of materials samples and coatings of the flowing part of hydraulic machines—Vestnik MSTU. Ser. Mech Eng 2:128–137

13. Pramod T, Kumar RK, Seetharamu S, Sampath Kumaran P (2016) Mechanism of material removal during cavitation erosion of HVOF coatings. Certif J 6(3):259–267

14. Yang L (2010) Cavitation erosion resistance of NiTi thin films produced by filtered arc deposition, University of Wollongong. http://ro.uow.edu.au/theses/3220

Microstructure and Mechanical Behaviors of Electron Beam Welded Ti$_{44}$Ni$_{47}$Nb$_9$ Shape Memory Alloys

Dan Yang, Haichang Jiang, Mingjiu Zhao, and Lijian Rong

Abstract

The microstructure, martensitic transformation behavior and mechanical properties of the Ti$_{44}$Ni$_{47}$Nb$_9$ electron beam welding joints were systemically studied in this research. The results showed that the microstructure of the fusion zone was composed of equiaxed grains with a small quantity of eutectic structure around the grain boundaries, and the grain size in the fusion zone was smaller than that in the as-cast microstructure. The ultimate tensile strength was a little lower than that of the base. The dissolution of niobium in the matrix obstructed the martensite transformation and decreased the martensite start temperature (M$_s$). After being deformed at the low temperature, the welding joints exhibited good wide transformation hysteresis, which was similar to that of the base metal.

Keywords

Ti$_{44}$Ni$_{47}$Nb$_9$ shape memory alloys • Electron beam welding • Martensitic transformation

Introduction

Ti–Ni–Nb shape memory alloys (SMAs) have attracted considerable attention due to the wide transformation hysteresis and excellent shape memory effect. These alloys have been successfully applied as couplings and fasteners [1–4]. The transformation hysteresis of Ti$_{44}$Ni$_{47}$Nb$_9$ SMAs will reach 150 K after a proper deformation [5]. As a result, processed connectors can be kept at room temperature rather than in the liquid nitrogen or any other low-temperature

medium, which makes them very convenient to be used in engineering applications, like the pipe-joints in fighters, missiles and armored vehicles [6–8].

With an enlarged set of application for Ti–Ni–Nb SMAs, the weldability of Ti–Ni–Nb SMAs has been widely studied. Wu et al. [9] studied the microstructure and mechanical behaviors of Ti–Ni–Nb SMAs joint welded by wire argon arc welding. The welding joints showed characteristics evident of brittle rupture and the ultimate tensile strength was far below that of the base metal. Moreover, the argon arc welding produced a set of other problems, including the broad heat affected zone and big welding deformation. Han et al. [10] analyzed the effect of welding parameters of precise pulse resistance butt welding on Ti–Ni–Nb SMAs. The results indicated that the correct parameters could make welding joints with higher strength and fine grains, but the welding was difficult to control. Chen et al. [11] made a conclusion that laser welding was appropriate for joining Ti–Ni–Nb SMAs and the strength of the welding joint was higher. However, the behavior of martensitic transformation was not studied. The maximum difficulty of welding Ti–Ni-based SMAs is to ensure not only the strength of the welding joint but also the shape memory effect [12]. For the traditional fusion welding methods, the heat treatment zone is broad, which will decrease the tensile strength and affect the shape memory effect of the welding joints [13].

Characteristics of electron beam welding (EBW) are a vacuum environment and high depth-to-width ratio, which can ultimately ensure the properties of SMAs. Under a vacuum environment in particular, harmful elements such as oxygen and nitrogen are not introduced into the weld zone during welding, which avoids the generation of gas pores and the decrease of mechanical properties. Moreover, we have studied the microstructure and mechanical behaviors of electron beam welded Ni$_{50.8}$Ti$_{49.2}$ shape memory alloys and the results proved that EBW is suitable for joining Ni$_{50.8}$Ti$_{49.2}$ SMAs [14]. As a result, we studied the microstructure and properties of Ti$_{44}$Ni$_{47}$Nb$_9$ SMAs joints welded by EBW in this research.

D. Yang · H. Jiang (✉) · M. Zhao · L. Rong
Division of Materials for Special Environments, Institute of Metal Research, Chinese Academy of Sciences, 72 Wenhua Road, Shenyang, 110016, China
e-mail: hcjiang@imr.ac.cn

© The Minerals, Metals & Materials Society 2018
A. P. Stebner and G. B. Olson (eds.), *Proceedings of the International Conference on Martensitic Transformations: Chicago*,
The Minerals, Metals & Materials Series, https://doi.org/10.1007/978-3-319-76968-4_35

Experimental

Materials

For this study, $Ti_{44}Ni_{47}Nb_9$ (at.%) ingots melted by vacuum arc furnace were processed into sheets with a thickness of 4 mm, including the process of homogenization, hot forging, hot rolling and a solid solution treatment. Then the sheets were cut into small sizes (100 × 60 × 4 mm) by a wire-cut electric discharge machine. Prior to electron beam welding, the oxide layer of the sheets to be welded should be mechanically removed from the sheet surface.

Electron Beam Welding Machines

For the electron beam welding experiments, a workable electron beam welding machine from the French TECHMETA company was employed. The maximum electron gun power and acceleration voltage were 15 kW and 60 kV, respectively.

Testing Methods

A scanning electron microscope (SEM) was employed to analyze the microstructure of the weld metal. The phase transformation characteristics were tested by the differential scanning calorimetry (DSC) method, which was operated using a PerkinElmer Diamond DSC in a helium atmosphere ranging from 120 to 375 K under a controlled cooling/heating rate of 10 K/min. Samples for the DSC tests were cut from the positions of DSC sample 1, 2 and 3, as shown in Fig. 1. They represented the base metal, the weld seam and the welding joint, respectively. Tensile tests were carried out to examine the ultimate tensile strength.

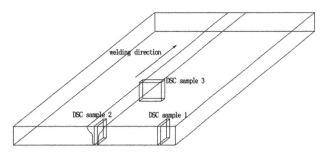

Fig. 1 The pattern for obtaining DSC samples: DSC sample 1: the base metal, DSC sample 2: the fusion zone and DSC sample 3: the welding joint

Results

Microstructure

Figure 2a shows the surface morphology of the welding joint. It can be seen that a surface freed from any defects can be obtained by the EBW method. Figure 2b exhibits the cross-sectional view of the welding joint. The welding joint is a typical T-type weld and its upper surface is about 3 mm in width. The red squares represent the corresponding position of Fig. 2c and d, respectively. It can be seen from Fig. 2c that the microstructure in the fusion zone is typical equiaxed dendrites. Because of the heat effect, the Nb element of the heat affected zone (HAZ) is evenly distributed, which is different from the base metal, as shown in Fig. 2d.

Martensitic Transformation Behavior

Figure 3 shows the DSC results of the base metal, the fusion zone and the welding joint of $Ti_{44}Ni_{47}Nb_9$ shape memory alloys. The DSC specimens were obtained from the position of samples 1–3, as shown in Fig. 1. During the cooling process, the peak represents the transformation of the parent austenite into the martensite, which is an exothermic process. On the other hand, it represents the reversible martensitic transformation during the heating process, called endothermic peak. It can be seen from the figure that there is only one step of reverse martensitic transformation in both the base metal and the fusion zone, while the welding joint owns the two-step transformation. That is because the DSC sample of the welding joint is made of the base metal and the fusion

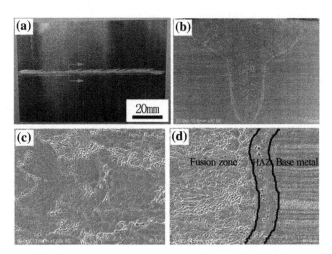

Fig. 2 a Surface pattern, **b** cross section microstructure, **c** the SE image of the fusion zone, **d** the SE image near the fusion line

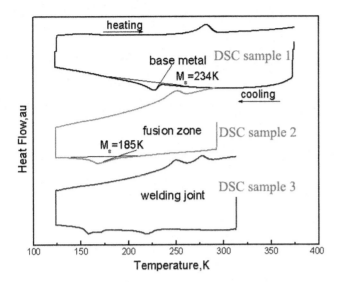

Fig. 3 The DSC results of the base metal, the fusion zone and the welding joint

zone, as shown in Fig. 1. The two-stage transformation peak represents the martensitic transformation of a different position. Moreover, it can be clearly seen that the M_s of the fusion zone is 49 K lower than the M_s value of the base metal.

Tensile Properties and Fracture Morphology

Figure 4 shows the tensile strain-stress curves of the base metal and the welding joint in austenite (a) or martensitic

states (b). It can be seen that the four curves have the same tendency. The material will deform plastically at first, as shown in stage I. Then the curve has a yield platform, as shown in stage II. However, the yield platform is different at room or low temperature. At room temperature, both the base metal and the welding joint are in the austenitic state, and the platform represents the macroscopic plastic deformation caused by stress-induced martensite. At low temperature, the experimental temperature is lower than the martensite finish temperature (M_f), and materials are all in the martensitic state. So, the platform represents the reorientation of thermal-induced martensite variants. The critical stress of the reorientation is much smaller than the stress-induced martensite, so yield strength at low temperature is lower than that at room temperature for both the base metal and the welding joint. Besides that, the yield strength of the welding joint is higher than that of the base metal at the same temperature due to the effect of welding thermal stress and the decrease of M_s. It is worth noting that the tensile strength of the welding joint can reach more than 95% whether tests are operated at room or low temperature, as shown in Table 1. The fracture surface of the base metal and the welding joint at RT or LT are exhibited in Fig. 5. It can be seen from the pictures that both the base metal and the welding joints show a typical ductile dimple fracture. The result indicates that the welding joints have excellent tensile strength and ductility. In a word, the welding joint obtained by EBW has good comprehensive mechanical properties.

Fig. 4 The tensile strain-stress curve of the base metal and the welding joint in austenite (**a**) or martensite states (**b**)

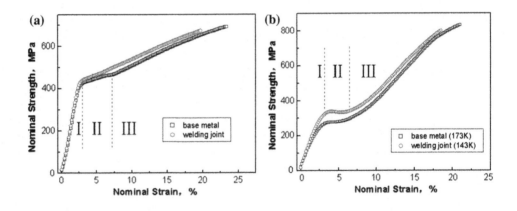

Table 1 The ultimate tensile strength of the base metal and welding joint at room and low temperature

Testing temperature		σ_b (MPa)
Room temperature	Base metal	692
	Welding joint	673
Low temperature	Base metal	838
	Welding joint	803

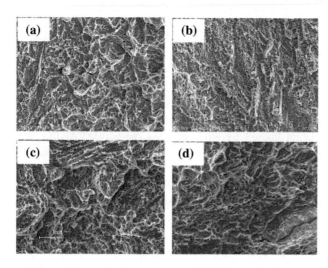

Fig. 5 The morphology: **a** base metal at RT **b** welding joint at RT **c** base metal at 173 K **d** welding joint at 143 K

The Transformation Hysteresis of the Welding Joints

As the A_s' (the converse martensite start temperature after deformation at low temperature) of Ti–Ni–Nb SMAs is above room temperature when the Ti–Ni–Nb SMA is deformed at a specific temperature, it is possible to keep the union joints made of Ti–Ni–Nb SMAs at room temperature rather than in liquid nitrogen. Therefore, studying the reverse martensitic transformation temperatures after deformation at low temperature (A_s' and A_f') provide an important basis for the applications of Ti–Ni–Nb SMAs. For convenience, a lamellar sample was cut from the low-temperature tensile fractures of the base metal and the welding joint. Then the samples were heated to 423 K by DSC to test the phase transformation temperature. The DSC results are shown in Fig. 6 and the (reverse) martensitic transformation temperatures before or after deformation are displayed in Table 2. Before deformation, the hysteresis width of the weld seam was 48 K, which is slightly higher than that of the base metal. After deformation at low temperature, the phase transformation hysteresis of both the base metal and the weld seam were broadened. The A_s' difference between the base metal and the weld seam could

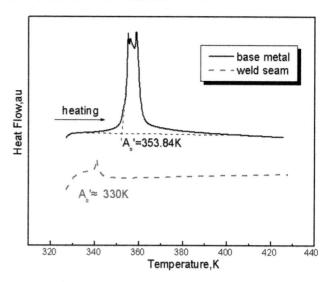

Fig. 6 The reverse stress-induced martensitic transformation DSC curves of the base metal (173 K, 22% deformation) and the weld seam (143 K, 17% deformation)

be caused by the difference of the deformed temperature and the amount of deformation. More than anything, it can be seen that the welding joints have the wide hysteresis no less than the base metal.

Conclusions

The $Ti_{44}Ni_{47}Nb_9$ welding joints welded by the electron beam welding were studied systematically, including the microstructure, the martensitic transformation, the mechanical properties and the transformation hysteresis. The fusion zone was made of equiaxed dendrites, and there is a small quantity of niobium-enriched phase in the grain boundary. A quantity of niobium is dissolved in the matrix and the microhardness of the weld seam increases. However, the special growth pattern makes the centerline the weakest area and decreases the ultimate tensile strength. The DSC results indicated that M_s of the fusion zone is 49 K lower than that of the base metal because the dissolved niobium is obstructing the martensite transformation. Fortunately, the welding joints showed wide hysteresis.

Table 2 The martensitic transformation temperatures of $Ti_{44}Ni_{47}Nb_9$ SMAs

Specimen conditions	$A_s'(K)$	$A_f'(K)$	$M_s(K)$	$M_f(K)$	$A_s(K)$	$A_f(K)$	$A_s–M_s(K)$	$A_s'–M_s(K)$
Base metal before deformation			228.56	209.01	266.65	292.01	**38.09**	
Weld seam before deformation			185.92	158.96	234.66	260.49	**48.74**	
Base metal with 22% deformation (173 K)	353.84	361.19	228.56					**125.28**
Weld seam with 17% deformation (143 K)	330	342.28	185.92					**144.08**

Acknowledgements The authors gratefully acknowledge the financial support for this research provided by the National Natural Science Foundation of China (Grant No. 51001100).

References

1. Xu ZY, Jiang BH (2000) Shape memory materials, Shanghai Jiao Tong Press
2. Wang L, Yan DS, Jiang ZM, Rong LJ (2004) Research and development of Ti–Ni–Nb shape memory alloy pipe-joint with wide hysteresis. J Mat Eng 7:60–63
3. Zhao LC, Duerig TW, Justi S et al (1990) The study of Niobium-rich precipitates in a Ni–Ti–Nb shape memory alloy. Scr Metall Mater 24:221–226
4. Jiang HC, Chen Y et al (2014) Effect of Mo addition on the microstructure and properties of TiNiNb alloy. Acta Metall Sin 27 (2):217–222
5. He XM, Rong LJ, Yan DS, Li YY (2006) Deformation dependence of transformation behavior of Ti–Ni–Nb wide hysteresis shape memory alloy. Rare Metal Mat Eng 35(9):227–231
6. Cao YH (2000) The development of shape memory alloys and applications in the fields of the missile and aerospace. Winged Missiles J 10:60–63
7. Moochul S (2012) Seismic retrofit and repair of reinforced concrete bridge columns using shape memory alloy spirals, Urbana, Illinois
8. Chen W, Li YL, Cai W et al (2005) Applications of Ti–Ni–Nb shape memory alloy tube coupling in the armored vehicles. Mech Eng 5:18–21
9. Wu Y, Meng XL et al (2005) Microstructure and mechanical behaviors of TiNiNb wide hysteresis shape memory alloy wire argon arc welding joint. Mater Sci Technol 13(3):312–319
10. Han LJ, Zhao XH, Zhao L, Zhang F (1999) Analysis of welding parameters of precise pulse resistance butt welding on TiNiNb shape memory alloys. Mater Sci Technol 7:76–79
11. Chen G, Zhao XK et al (2008) Microstructure and properties of laser welded joint of TiNiNb wide hysteresis shape memory alloy. Hot Work Technol 37(17):99–101
12. Liu M, Li YY et al (1997) Influence of precipitates on the two way shape memory effect. Acta Metall Sin 10(3):166–173
13. Lu WW, Chen YH et al (2014) Research progress of engineering application and welding technology of new type shape memory alloy of TiNiNb. Aeronaut Manuf Technol (1/2): 94–97
14. Yang D, Jiang HC, Zhao MJ, Rong LJ (2014) Microstructure and mechanical behaviors of electron beam welded NiTi shape memory alloys. Mater Des 57:21–25

Relationship Between Applied Stress and Hydrogen-Related Fracture Behavior in Martensitic Steel

Yasunari Takeda, Takashi Yonemura, Yuji Momotani, Akinobu Shibata, and Nobuhiro Tsuji

Abstract

In the present study, a relationship between the applied stress and the hydrogen-related fracture behavior in an 8Ni-0.1C martensitic steel was investigated by means of constant loading tensile tests under various applied stresses ranging from 400 to 1000 MPa. The time to fracture in hydrogen charged specimens increased from 0.5 to 30.5 s with decreasing the applied stress from 1000 to 400 MPa. Area fractions of brittle fracture surfaces, especially that of intergranular fracture surface, increased as the applied stress decreased. Orientation analysis using EBSD showed that micro-cracks observed in the vicinity of the main crack initiated along prior austenite grain boundaries regardless of the applied stress. The hydrogen-related fracture processes under various applied stresses were reconstructed by fracture surface topography analysis and the results indicated that intergranular fracture was dominant under the lower applied stress, while the fracture mode changed from intergranular fracture to quasi-cleavage fracture with progress of crack propagation under the higher applied stress.

Keywords

Martensitic steel • Hydrogen embrittlement
Fracture surface • EBSD
Fracture surface topography analysis

Introduction

Hydrogen embrittlement is the phenomenon that materials become very brittle due to the existence of hydrogen. Since high strength steels, especially martensitic steels, show high susceptibility to hydrogen embrittlement, this is one of the serious issues for practical applications of high strength steels. There are mainly two fracture modes in hydrogen embrittlement of martensitic steels: intergranular fracture and quasi-cleavage fracture. Previous studies [1–5] reported that hydrogen-related intergranular fracture occurred along prior austenite grain boundaries. Momotani et al. proposed that hydrogen accumulated at prior austenite grain boundaries during deformation, resulting in the intergranular fracture [6]. Quasi-cleavage fracture occurs inside prior austenite grains but not along typical cleavage planes. Nagao et al. observed the microstructure beneath the hydrogen-induced quasi-cleavage fracture surface in martensitic steels [7]. Their results suggested that the quasi-cleavage fracture path was along lath boundaries. Shibata et al. found through a crystallographic orientation analysis that the quasi-cleavage fracture occurred on {011} plane [8–10]. To reveal the distinct condition that controls the fracture mode in hydrogen embrittlement is one of the important keys for understanding the mechanism of hydrogen embrittlement. It has been reported that the fraction of intergranular fracture increased with increasing hydrogen content [3], impurities content [11, 12] or decreasing the tempering temperature [2]. Deformation conditions, such as strain rate, stress concentration factor and applied stress, are also the important factors which affect hydrogen-related fracture behavior. Momotani et al. reported that the fraction of brittle fracture surfaces significantly

Y. Takeda (✉) · T. Yonemura · Y. Momotani · A. Shibata
N. Tsuji
Department of Materials Science and Engineering, Kyoto
University, Kyoto, Japan
e-mail: takeda.yasunari.62v@st.kyoto-u.ac.jp

T. Yonemura
e-mail: yonemura.takashi@tsujilab.mtl.kyoto-u.ac.jp

Y. Momotani
e-mail: momotani.yuji@tsujilab.mtl.kyoto-u.ac.jp

A. Shibata
e-mail: shibata.akinobu.5x@kyoto-u.ac.jp

N. Tsuji
e-mail: nobuhiro-tsuji@mtl.kyoto-u.ac.jp

A. Shibata · N. Tsuji
Elements Strategy Initiative for Structural Materials, Kyoto
University, Kyoto, Japan

© The Minerals, Metals & Materials Society 2018
A. P. Stebner and G. B. Olson (eds.), *Proceedings of the International Conference on Martensitic Transformations: Chicago*,
The Minerals, Metals & Materials Series, https://doi.org/10.1007/978-3-319-76968-4_36

increased with decreasing the strain rate [6]. Wang et al. carried out slow strain rate tensile tests using the specimens with various stress concentration factors, and found that the intergranular fracture region increased as the stress concentration factor increased [3]. However, the effect of the applied stress on hydrogen-related fracture modes in martensitic steel has not been quantitatively clarified in detail. In the present study, the relationship between the applied stress and the hydrogen-related fracture mode in martensitic steels was investigated by means of constant loading tensile tests under various applied stresses.

Experimental Procedure

An 8Ni-0.1C steel was used in the present study. The chemical composition of the steel is shown in Table 1. A cast ingot of the steel was homogenized at 1100 °C for 9 h followed by air cooling. The homogenized ingot was cold-rolled to a thickness of 1.7 mm. Then the sheet was austenitized at 1000 °C for 30 min followed by ice brine quenching and sub-zero cooling in liquid nitrogen to obtain a fully martensitic structure. Specimens for constant loading tensile tests were cut from the heat-treated sheets and mechanically polished on both surfaces until the final thickness of 1 mm was reached to remove the decarburized layers formed during the heat treatments. The gauge length, width, and thickness of the specimen were 10 mm, 5 mm, and 1 mm, respectively. After polishing the surfaces of the specimen with a 3 μm diamond suspension, the specimen was cathodically pre-charged with hydrogen in a 3% NaCl aqueous solution including 3 g L^{-1} NH_4SCN at a current density of 3 A m^{-2} for 24 h. The diffusible hydrogen content after hydrogen pre-charging measured by a thermal desorption analysis was 1.72 wt. ppm. After the hydrogen pre-charging, constant loading tensile tests were carried out under the various applied stresses ranging from 400 to 1000 MPa at ambient temperature. To prevent a desorption of hydrogen, hydrogen charging at a current density of 3 A m^{-2} was continued during the constant loading tensile tests. Fracture surfaces were observed using scanning electron microscopy (SEM, JEOL: JSM-7800F). To reveal microstructural features of hydrogen-related cracks, electron backscattering diffraction (EBSD) analysis was carried out using SEM (JEOL: JSM-7100F). Hydrogen-related fracture process during the constant loading tensile test was simulated by fracture surface topography analysis (FRASTA). A FRASTA technique is a procedure for reconstructing a fracture process in microscopic level [13]. There are three

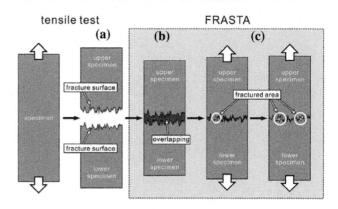

Fig. 1 Schematic illustration showing the method of fracture surface topography analysis (FRASTA) [14]

steps in the FRASTA method: (1) obtain topographic maps of conjugate area in opposing fracture surfaces, (2) overlap the two topographic maps computationally until there is no gap between them and (3) increase the relative distance between the two topographic maps. A schematic illustration of FRASTA method is shown in Fig. 1 [14].

Results and Discussion

Figure 2 shows an optical microscopy image of the heat-treated specimen. The microstructure exhibits a typical lath martensite structure consisting of laths, blocks, packets and prior austenite grains. The stress-strain curve of the uncharged specimen at ambient temperature at a strain rate

Fig. 2 Optical microscopy image of the heat-treated specimen

	C	Ni	Mn	Al	Si	P	S	Fe
Table 1 Chemical composition of the steel used in the present study (wt%)	0.116	7.94	0.01	0.033	0.005	0.001	0.0015	Bal.

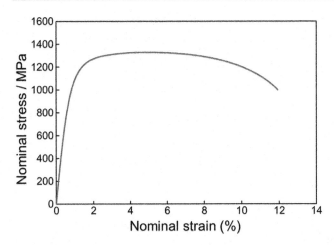

Fig. 3 Nominal stress–nominal strain curve of the uncharged specimen at ambient temperature at a strain rate of 8.3×10^{-6} s^{-1}

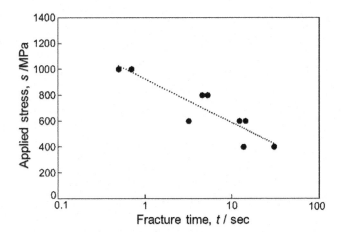

Fig. 4 Relationship between the applied stress and the fracture time in the constant loading tensile tests

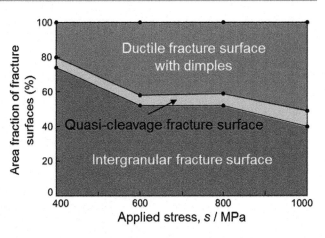

Fig. 6 Area fractions of different fracture surfaces after the constant loading tensile tests under various applied stresses

of 8.3×10^{-6} s^{-1} is presented in Fig. 3. The 0.2% proof stress and ultimate tensile strength measured from the stress-strain curve are 928 MPa and 1287 MPa, respectively.

Figure 4 shows the relationship between the applied stress and the fracture time in the constant loading tensile tests. While the applied stress of 1000 MPa is higher than the 0.2% proof stress, the other applied stresses ranging from 400 to 800 MPa are within elastic deformation region of the uncharged specimen. It can be found from Fig. 4 that the fracture time increased from 0.5 to 30.5 s with decreasing the applied stress from 1000 to 400 MPa. Figure 5 are SEM images showing typical fracture surfaces after the constant loading tensile test (applied stress = 1000 MPa). Intergranular fracture surface, quasi-cleavage fracture surface and ductile fracture surface with dimple patterns were observed in Fig. 5a–c, respectively. Figure 6 summarizes the relationship between the area fractions of various types of fracture surfaces and the applied stresses. As the applied stress decreased, the area fraction of brittle fracture surfaces,

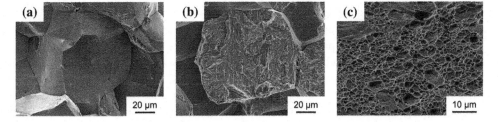

Fig. 5 SEM images showing typical fracture surfaces observed after the constant loading tensile tests (applied stress = 1000 MPa); **a** intergranular fracture surface, **b** quasi-cleavage fracture surface and **c** ductile fracture surface with dimples

Fig. 7 **a**, **b** SEM images and **c**, **d** corresponding EBSD orientation maps showing the micro-cracks observed in the vicinity of the main cracks in the specimens tested under the applied stresses of **a**, **c** 400 MPa and **b**, **d** 1000 MPa, respectively. Packet boundaries and prior austenite grain boundaries are drawn by white lines and white broken lines, respectively

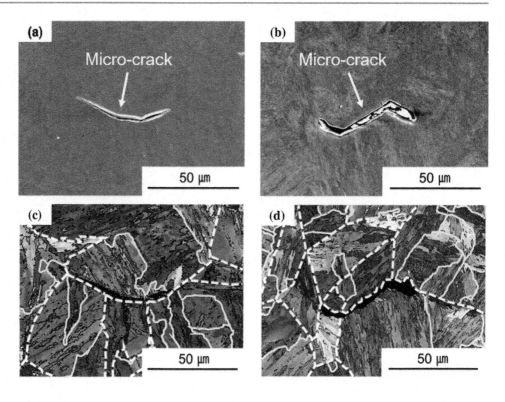

particularly that of intergranular fracture surfaces, significantly increased. The results shown in Figs. 4 and 6 indicate that the higher applied stress accelerated hydrogen-related fracture, but decreased the area fraction of brittle fracture surfaces.

Figure 7 are SEM images showing micro-cracks observed in the vicinity of the main cracks in the specimens tested under the applied stresses of (a) 400 and (b) 1000 MPa, respectively. Figure 7c, d shows EBSD orientation maps of the same regions as Fig. 7a, b. In these figures, packet boundaries and prior austenite grain boundaries identified by crystallographic orientation analysis are drawn in white lines and white broken lines, respectively. It is obvious from these figures that the micro-cracks formed along prior austenite grain boundaries in both cases. The results indicate that the hydrogen-related fracture initiated along prior austenite grain boundaries regardless of the applied stress.

Figure 8 summarizes changes in the fraction of each fracture surface under the applied stresses of (a) 400 MPa and (b) 1000 MPa, respectively, evaluated by FRASTA. The horizontal axis shows the distance (L) between the conjugate topography maps of fracture surfaces. L is defined as 0 at which the first cracking occurs in the specimen. Under the applied stress of 400 MPa (Fig. 8a), intergranular fracture mainly occurred at the early stages of fracture and the fraction of ductile fracture surface with dimples increased as the fracture proceeded. It was reported that hydrogen accumulated at prior austenite grain boundaries during deformation [6]. Therefore, the results suggest that under the lower applied stress, such as a stress in elastic region, hydrogen accumulation at prior austenite grain boundaries around the crack tip is necessary for the crack propagation. When the applied stress was 1000 MPa (Fig. 8b), on the other hand, not only intergranular fracture but also quasi-cleavage fracture and ductile fracture occurred at the early stages of fracture. It is considered that under the higher applied stress, such as a stress higher than the 0.2% proof stress, cracks can propagate easily even in transgranular manner once micro-cracks form. Therefore, the fracture time decreased and the fraction of transgranular fracture surfaces, such as quasi-cleavage fracture surface and ductile fracture surface, increased as the applied stress increased.

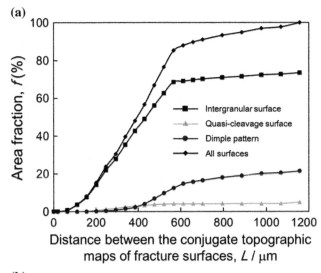

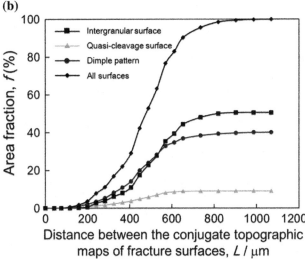

Fig. 8 Changes in fraction of each fracture mode under the applied stress of **a** 400 MPa and **b** 1000 MPa summarized as a function of the distance between the conjugate topography maps of fracture surfaces

Summary

In the present study, constant loading tensile tests under various applied stresses were carried out to reveal the relationship between the applied stress and the hydrogen-related fracture behavior. The main results obtained are as follows:

- The fracture time increased with decreasing the applied stress. As the applied stress decreased, the fractions of brittle fracture surfaces, especially that of intergranular fracture surfaces, increased. The results indicated that the hydrogen-related fracture was accelerated with increasing the applied stress, while the fracture mode became more brittle as the applied stress decreased.
- According to the EBSD orientation analysis, it was found that hydrogen-related cracking initiated along prior grain boundaries regardless of the applied stress.
- Hydrogen-related fracture processes during the constant loading tensile tests were reconstructed by FRASTA. Under the lower applied stress, the crack propagation proceeded mainly by intergranular fracture. Under the higher applied stress, on the other hand, the crack propagation occurred not only by intergranular fracture but also quasi-cleavage fracture and ductile fracture.

Acknowledgements This study was financially supported by a Grant-in-Aid for Scientific Research (B) (No. 15H04158) and the Elements Strategy Initiative for Structural Materials (ESISM) through the Ministry of Education, Culture, Sports, Science and Technology (MEXT), Japan, and Industry-Academia Collaborative R&D Program 'Heterogeneous Structure Control: Towards Innovative Development of Metallic Structural Materials' (No. 20100113) through the Japan Science and Technology Agency.

References

1. Banerji SK, McMahon C, Feng HC (1978) Intergranular fracture in 4340-type steels: effects of impurities and hydrogen. Metall Mater Trans A 9(2):237–247
2. Craig BD, Krauss G (1980) The structure of tempered martensite and its susceptibility to hydrogen stress cracking. Metall Trans A 11(11):1799–1808
3. Wang M, Akiyama E, Tsuzaki K (2005) Effect of hydrogen and stress concentration on the notch tensile strength of AISI 4135 steel. Mater Sci Eng A 398(1):37–46
4. Wang M, Akiyama E, Tsuzaki K (2007) Effect of hydrogen on the fracture behavior of high strength steel during slow strain rate test. Corros Sci 49(11):4081–4097
5. Wang G et al (2013) Hydrogen embrittlement assessment of ultra-high strength steel 30CrMnSiNi2. Corros Sci 77:273–280
6. Momotani Y et al (2017) Effect of strain rate on hydrogen embrittlement in low-carbon martensitic steel. Int J Hydrog Energy 42(5):3371–3379
7. Nagao A et al (2012) The role of hydrogen in hydrogen embrittlement fracture of lath martensitic steel. Acta Mater 60 (13):5182–5189
8. Shibata A, Takahashi H, Tsuji N (2012) Microstructural and crystallographic features of hydrogen-related crack propagation in low carbon martensitic steel. ISIJ Int 52(2):208–212
9. Shibata A et al (2015) Characterization of hydrogen-related fracture behavior in As-quenched low-carbon martensitic steel and tempered medium-carbon martensitic steel. Metall Mater Trans A 46(12):5685–5696
10. Shibata A et al (2017) Microstructural and crystallographic features of hydrogen-related fracture in lath martensitic steels. Mater Sci Technol 33(13):1524–1532
11. Takeda Y, McMahon CJ (1981) Strain controlled vs stress controlled hydrogen induced fracture in a quenched and tempered steel. Metall Trans A 12(7):1255–1266
12. Nagumo M, Matsuda H (2002) Function of hydrogen in intergranular fracture of martensitic steels. Philos Mag A 82(17–18):3415–3425
13. Kobayashi T, Shockey DA (1987) A fractographic investigation of thermal embrittlement in cast duplex stainless steel. Metall Trans A 18(11):1941–1949
14. Shibata A et al (2017) Fracture surface topography analysis of the hydrogen-related fracture propagation process in martensitic steel. Int J Fract 205(1):73–82

Part VIII
Advanced Processing Techniques: Additive, Porous, and Others

Contact Pressure and Residual Strain of Resistance Spot Welding on Mild Steel Sheet Metal

Hua Zhong, Xiaodong Wan, Yuanxun Wang, and Yiping Chen

Abstract

Coupled electrical–thermal and thermo-elastic–plastic analyses were performed to analyze the behavior of the mechanical features during the resistance spot welding (RSW) process including the squeeze, heating, and hold steps, and to prepare for further structural analysis for large, complex structures with a large quantity of resistance spot welds. A two-dimensional axisymmetric thermo-elastic-plastic FEM model was developed and analyzed in the commercial FEM program, ANSYS. The analysis was based on the transient temperature field obtained from a transient electrical–thermal simulation of the RSW process conducted by the authors. The distribution and change of the contact pressure at the electrode–workpiece interface and faying surface, the residual stress, and the residual plastic strain distribution of the weldment were obtained through the analysis.

Keywords

Resistance spot welding • Temperature field
Contact pressure • Residual strain

H. Zhong · X. Wan
School of Mechanical Science and Engineering, Huazhong University of Science and Technology, Wuhan, 430074, China

Y. Wang (✉) · Y. Chen (✉)
School of Civil Engineering and Mechanics, Huazhong University of Science and Technology, Wuhan, 430074, China
e-mail: wangyuanxun@hust.edu.cn

Y. Chen
e-mail: ypchen88@hust.edu.cn

H. Zhong
HangYu Aerospace Life-Support Industries Co Ltd, Xiangyang, 441002, China

Introduction

The resistance spot welding (RSW) process is widely used in sheet metal joining due to its advantages in welding efficiency and suitability for automation. For example, a modern vehicle typically contains 2000–5000 spot welds. Thus, the behavior of the RSW process is extremely important to the quality of the entire welding production.

During the RSW process, deformation, stress, and strain will be generated and changed in the weldment due to the electrode force and Joule heating, and residual stress and strain will retain in the weldment after welding. It is a complex process in which coupled interactions exist with electrical, thermal, mechanical, and metallurgical phenomena, and even surface behaviors. Numerous studies of the mechanical features for such a complex process have been performed on all kinds of welding conditions and materials, using both theoretical and experimental methods [5, 7, 12]. It can be concluded from these studies that spot welding failure is likely related to many parameters of the RSW process, e.g., residual stress, welding parameters, welding schedule, sheet thickness, welding gap, welding nugget size, and material properties. These parameters also affect the fatigue life of the welded joint. Results of the relative research studies show that the fatigue strength is mainly controlled by the residual stress, welding gap, and stress concentration at the notch of the welding nugget [1, 10, 14].

Recently, numerical methods have proved to be powerful tools in studying these interactions, especially the finite element analysis (FEA) method, which can deal with nonlinear behaviors and complex boundary conditions. It has become the most important method for the analysis of the RSW process [8, 9, 15] developed a real-time control method in RSW and obtained direct correlations between welding nugget formation and expansion displacement of electrodes. Nied [13] developed the first FEA model for the RSW process, investigated the effect of the geometry of the electrode on the workpiece, and predicted the deformation

© The Minerals, Metals & Materials Society 2018
A. P. Stebner and G. B. Olson (eds.), *Proceedings of the International Conference on Martensitic Transformations: Chicago*, The Minerals, Metals & Materials Series, https://doi.org/10.1007/978-3-319-76968-4_37

and stresses. However, the developed model was restricted to elastic deformation and could not calculate the contact areas at the electrode–workpiece interface and faying surface. Therefore, many researchers developed more sophisticated FEA models that incorporate the contact status, phase changing, and coupled field effects into the simulation of RSW [3, 6, 17]. The iterative method was employed to simulate the interactions between coupled electrical, thermal, and structural fields [4, 11]. In this method, the stress field and contact status were initially obtained from the thermo-mechanical analysis, and then the temperature field was obtained from the fully coupled electrical–thermal analysis based on the contact area at the electrode–workpiece interface and faying surface. The calculated temperature field was then passed back to the thermo-structural analysis to update the stress field and contact status. The iterative method can provide the temperature field, the electric potential field, and the stress and strain distributions of the spot welding in one calculation, but the simulation of transient processes with such a methodology would probably require a large amount of computing time. On the other hand, although so many studies have been conducted, we still do not have adequate and accurate information of the RSW process due to its inherent complexity. More detailed analyses are needed of the mechanical features during the whole process of RSW.

The objective of this paper is to develop a multi-coupled method to analyze the welding deformation of the RSW process in order to reduce the computing time with the minimum loss of computing accuracy. The behavior of the mechanical features during the RSW process including the squeeze, heating, and hold steps are very important for further structural analysis for large, complex structures with a large quantity of resistance spot welds.

Model and Mesh

For the solution of the welding deformation of the RSW process in this research, an axisymmetric model was developed and solved using the FEA method based on ANSYS code. The two-dimensional axisymmetric model is illustrated schematically in Fig. 1, where x and y represent the faying surface and the axisymmetric axis, respectively. Its corresponding dimensions are $OE = HI = 1.5$ mm for two sheets of equal thickness, $OI = EH = 15$ mm for the spot welding area, $PA = FG = 5$ mm and $AG = 18$ mm for the cooling water area, $PB = 11$ mm for the radius of the electrode, $EF = 12.5$ mm, $ED = 3$ mm, $OP = 32$ mm and $\alpha = 30°$, while the lower half of the model is mirror symmetric about the faying surface of the two sheet metals.

The model was meshed using three types of elements, as shown in Fig. 2. The solid elements were employed to simulate the thermo-elastic–plastic behavior of the sheets

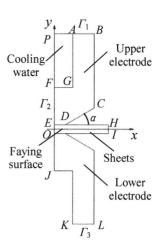

Fig. 1 Schematic diagram for the model of RSW

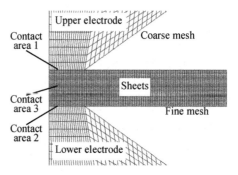

Fig. 2 Mesh generation of the developed model

and electrodes. The contact pair elements were employed to simulate the contact areas. There were three contact areas in the model. Contact areas 1 and 2 represented the electrode–workpiece interface, and contact area 3 represented the faying surface. They were all assumed to be in contact with two deformable surfaces, and these surfaces were allowed to undergo small sliding. In order to obtain reliable results, fine meshes were generated near these contact areas, while the meshes of other areas were relatively coarse.

Governing Equations and Boundary Conditions

The constitutive equations of the materials for axisymmetric transient thermal analysis based on thermo-elastic–plastic theory can be written as

$$d\{\sigma\} = [D]d\{\varepsilon\} - \{C\}dT \tag{1}$$

$$\{C\} = -[D]\left(\{\alpha\} + \frac{\partial[D]^{-1}}{\partial T}\{\sigma\}\right) \tag{2}$$

where $\{\sigma\}$ is the stress vector, $\{\varepsilon\}$ is the strain vector, $\{\alpha\}$ is the thermal expansion coefficient, and $[D]$ is the elastic–plastic matrix with $[D] = [D]_e$ in the elastic area,and $[D] = [D]_{ep} = [D]_e - [D]_p$ in the plastic area, in which $[D]_e$ is the elastic matrix and $[D]_p$ is the plastic matrix.

For the structural analysis, the stress equilibrium equation is given by

$$\nabla \sigma(r, t) + b(r, t) = 0 \qquad (3)$$

Where σ is the stress, b is the body force and r is the coordinate vector. For the transient thermal analysis, the following boundary conditions are specified on the surface of Γ_1, Γ_2, and Γ_3 (see Fig. 1).

- Γ_1 (AB): $\sigma_y = -q$, where q is the uniform pressure which can be determined according to the electrode force and the section area of the electrode.
- Γ_2 (FEOJ): $U_x = 0$, where U_x is the displacement in x direction.
- Γ_3 (KL): $U_y = 0$, where U_y is the displacement in y direction.

Welding Parameters and Material Properties

The welding parameters used in this analysis are welding current of 50 Hz, sine wave AC current of 12.2 kA, weld time of 13 cycles (0.26 s), electrode force of 3 kN, and holding time of 3 cycles (0.06 s).

Because the materials are subjected to a wide range of temperatures, most of the material properties are considered as temperature-dependent. The most important property in the analysis of the RSW process is the contact resistivity of the faying surface. The contact resistivity is a dependent function on contact pressure, temperature, and average yield strength of two contact materials. Vogler and Sheppard [16] pointed out that the contact resistance decreases as the contact pressure increases. Using a curve fitting procedure, Babu et al. [2] developed an empirical model for establishing the desired relationship of contact resistance against pressure and temperature. During the RSW process, the contact resistivity distribution influences the current density pattern, which affects the temperature field through Joule heating, while the temperature field then influences the mechanical pressure distribution through thermal expansion, related to the interface resistivity. Therefore, this is a highly non-linear problem involving the complex interaction of thermal, electrical, and mechanical phenomena. To simplify the problem, many researchers took the contact resistivity as a function of temperature [3, 6, 11, 17]. This simplification is reasonable since, firstly, the load is constant in a specified

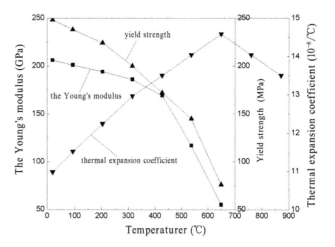

Fig. 3 Mechanical and thermal properties of the mild steel

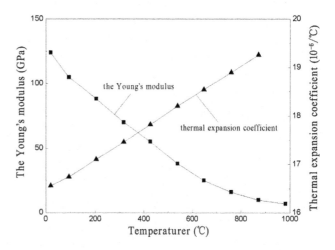

Fig. 4 Mechanical and thermal properties of the copper electrode

RSW process and, secondly, the yield strength of the materials, which determines the contact status in the contact area, is essentially influenced by temperature. With this simplification, the computing time can be reduced greatly. Therefore, in this research, the temperature-dependent contact resistance was imposed at the faying surface. The mechanical and thermal properties of both the copper electrode and mild steel sheet workpiece are shown in Figs. 3 and 4.

Thermo-Elastic–Plastic Analysis

The temperature field and its changing of the sheet metal RSW have been obtained and well discussed through the coupled electrical–thermal analysis by Hou et al. [8, 9]. Figure 5 shows the temperature-changing histories at the center of the weld nugget (point O in Fig. 1) and the center and the edge of the electrode–workpiece interface (points E

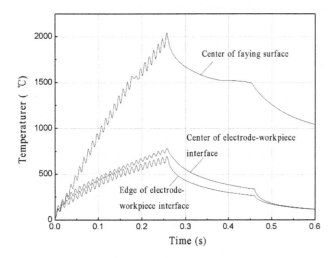

Fig. 5 Temperature-changing histories of the RSW process

and D in Fig. 1) from which we can see the changing of temperature and the heat affected zone (HAZ) during the RSW process.

For the thermo-elastic–plastic analysis, some hypotheses are cited. The mechanical properties stress and strain related with the welding temperature are linearly changed in a small time increment. Elastic stress, plastic stress, and temperature stress are separable. Strain stiffening occurs in the plastic field and obeys the theory of rheology. The Mises Yield Criterion is used for the material yield strength. The welding thermo-elastic–plastic analysis is constructed by the strain-displacement relationship or compatibility condition, stress-strain relationship or constitutive relationship, equilibrium condition, and boundary conditions. The constitutive equations of the material in the temperature field can be written as Eq. (1). In the elastic field, $[D] = [D]_e$ is the elastic matrix, and

$$\{C\} = \{C\}_e = [D]_e \left(\{\alpha\} + \frac{\partial [D]_e^{-1}}{\partial T} \{\sigma\} \right) \qquad (4)$$

$\{\alpha\}$ is the thermal expansion coefficient. In the plastic field, $[D] = [D]_{ep}$, and

$$[D]_{ep} = [D]_e - [D]_e \left\{ \frac{\partial f}{\partial \sigma} \right\} \left\{ \frac{\partial f}{\partial \sigma} \right\}^T [D]_e / S \qquad (5)$$

$$S = \left\{ \frac{\partial f}{\partial \sigma} \right\}^T [D]_e \left\{ \frac{\partial f}{\partial \sigma} \right\} + \left(\frac{\partial f_0}{\partial K} \right) \left\{ \frac{\partial K}{\partial \varepsilon} \right\}^T \left\{ \frac{\partial f}{\partial \sigma} \right\} \qquad (6)$$

$$\{C\} = \{C\}_{ep} = [D]_{ep}\{\alpha\} + [D]_{ep}\frac{\partial [D]_e^{-1}}{\partial T}\{\alpha\} - [D]_e\frac{\partial f}{\partial \sigma}\frac{\partial f_0}{\partial T} / S \qquad (7)$$

where $f(\sigma_x, \sigma_y, \ldots)$ is the yield function. The material would yield at the value $f_0(\sigma_s, T, K)$ at the temperature T and the strain hardening exponential K. At loading, the consistency condition must be content, so,

$$\left\{ \frac{\partial f}{\partial \sigma} \right\}^T \{d\sigma\} = \left(\frac{\partial f_0}{\partial K} \right) \left\{ \frac{\partial K}{\partial \varepsilon} \right\}^T \{d\varepsilon\} + \frac{\partial f_0}{\partial T} dT \qquad (8)$$

For any cell of the weldment, at time τ, temperature is T, force at the node is $\{F\}_e$, node displacement is $\{\delta\}$, strain is $\{\varepsilon\}$, stress is $\{\sigma\}$, and at time $\tau + d\tau$, there are $T + dT$, $\{F + dF\}_e$, $\{\delta + d\delta\}$, $\{\varepsilon + d\varepsilon\}$, $\{\sigma + d\sigma\}$. Using the virtual displacement principle, we can obtain

$$\{d\delta\}^T \{F + dF\}_e = \iint_{\Delta V} \{d\delta\}^T [B]^T (\{\sigma\} + [D]\{d\varepsilon\} - \{C\}dT)dV$$

$$= \{d\delta\}^T \iint_{\Delta V} [B]^T (\{\sigma\} + [D]\{d\varepsilon\} - \{C\}dT)dV \qquad (9)$$

It is equilibrium at time τ, so,

$$\{F\}_e = \iint_{\Delta V} [B]^T \{\sigma\} dV \qquad (10)$$

Then, Eq. (1) can be expressed as

$$\{dF\}_e = \iint_{\Delta V} [B]^T ([D]\{d\varepsilon\} - \{C\}dT)dV \qquad (11)$$

or written as

$$\{dF\}_e + \{dR\}_e = [K]_e\{d\delta\} \qquad (12)$$

where the original strain equivalent node force is

$$\{dR\}_e = \iint_{\Delta V} [B]^T \{C\}dTdV \qquad (13)$$

Cell stiffness matrix is

$$[K]_e = \iint_{\Delta V} [B]^T [D][B]dV \qquad (14)$$

Given the different $[D]$ and $\{C\}$ according to the elastic or plastic state, the equivalent node load and stiffness matrices can be obtained. Then, posting them to the general stiffness matrix and the general load vector, the algebra equations of the node displacement can be obtained by

$$[K]\{d\delta\} = \{dF\} \qquad (15)$$

where

$$[K] = \sum [K]_e \qquad (16)$$

$$\{dF\} = \sum (\{dF\}_e + \{dR\}_e) \qquad (17)$$

For welding problems, $\sum \{dF\}_e$ is often zero, then,

$$\{dF\} = \sum \{dR\}_e \qquad (18)$$

From the algebra equations (15), the node displacement can be obtained, and then the node stress can be obtained through the constitutive equations.

To correctly load the temperature field, the FEA model and mesh in the thermo-elastic–plastic analysis are identical with those in the temperature field analysis as shown in Fig. 1. But the boundary conditions and the property of the mesh cells must be changed correspondingly. For the plastic material of the mild steel and copper electrode, the double linear thermo-elastic strengthen material model is adopted, and the physical equations are

$$\begin{cases} \varepsilon = \frac{\sigma}{E} & (|\sigma| \leq \sigma_s) \\ \varepsilon = \frac{\sigma}{E} + (|\sigma| - \sigma_s)\left(\frac{1}{E'} - \frac{1}{E}\right) sign\, \sigma & (|\sigma| > \sigma_s) \end{cases} \qquad (19)$$

in which the material mechanical parameters, including the Young's modulus E, shear modulus E', and yield strength σ_s, which changed with temperature, are shown in Figs. 3 and 4. The shear modulus E' is taken as 1/10 of the Young's modulus E at corresponding temperature.

The electrode force was loaded at the contact surface (EH and OI in Fig. 1) of the electrode and the mild steel sheets as a uniform load. The load boundary condition was $\sigma_y = -q$, where q was the load intensity obtained by the contact pressure. Loading the temperature field at the corresponding time as the node body load, the APDL (analysis parametric design language) loop language of ANSYS was used in the loading process.

Results and Discussion

Contact Pressure

The contact pressures at the faying surface of the two sheet metals and the electrode–workpiece interface can be divided into three steps. Figure 6 shows the contact pressures on the faying surface of the two sheet metals at different heating cycles. During the squeeze step (marked as cycle 0), a maximum contact pressure of 83 MPa was attained near the edge of contact area, and the pressure at the faying surface was a little changed. As the heating cycles started (electrifying), the pressure near the edge of the contact area was decreased, while the pressure at the center of the contact area increased quickly. Cycle 1 to cycle 3 compose the second step. At cycle 3, the maximum contact pressure at the center of the faying surface of the two sheet metals reached 155 MPa, with a contact area radius of 3 mm. Then, the pressure near the edge of the contact area started to increase, became and kept larger than that of the center of the contact area during the consequent cycles at the third step from cycle 4 to cycle 8. The peak value of 168 MPa appeared at cycle 5, with the a contact area radius of 3.1 mm. At the same time, the location of the maximum pressure value moved outward

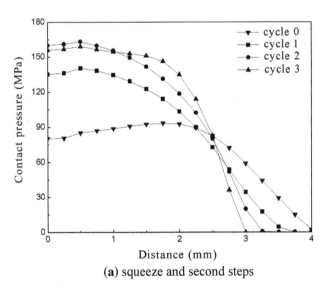

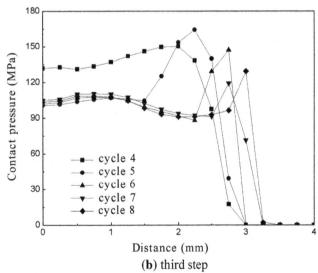

(a) squeeze and second steps   (b) third step

Fig. 6 Contact pressure at the faying surface

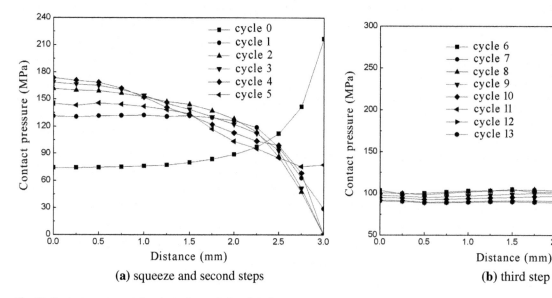

Fig. 7 Contact pressure at the electrode–workpiece interface

from the contact area. The high contact pressure near the edge of the contact area is a benefit to the RSW process because it can prevent liquid metal expulsion during welding. The contact area on the faying surface of the two sheet metals was nearly constant during the heating cycles. The radius of the contact area was varied in a narrow range from 3–3.5 mm.

The distributions and change of the contact pressure due to the thermal expansion of the RSW welding joint and the material properties were subjected to a wide range of temperatures. At the second step, the temperature at the center of the faying surface of the two sheet metals was increased rapidly and the temperature near the edge was quite low, so the thermal expansion of the material in the center of the faying surface was much bigger than that near the edge. The contact pressure in the center of the faying surface was larger. At the third step, the Young's modulus and yield strength of the material at the center of the faying surface decreased obviously with the increase of the temperature.

The contact pressure distributions on the electrode–workpiece interface are shown in Fig. 7. Obviously, it has a different pattern from that on the faying surface of the two sheet metals. During the squeeze step (cycle 0), the pressure was relatively uniform near the center of the contact area but very steep at the edge of the contact area. As the heating cycles started, the pressure profile changed. The pressure near the center of the contact area increased while the pressure at the edge of the contact area decreased quickly. Since the temperature at the center of the contact area increased quickly, the material along the axisymmetric axis

expanded, while the temperature at the edge of the contact area was still low and the material was unchanged. This status was kept for 4 cycles, and then another change occurred. The distribution pattern became similar to that of the squeeze step. This is probably because the temperature near the center of the contact area became so high that the Young's modulus of the material decreased, while the temperature at the edge of the contact area increased and led to thermal expansion. The contact pressure distribution was kept on this status up to the end of holding step, which means the electrode edge was kept working under large stress. In other words, there was a stress concentration at the edge of the electrode, which could be the vital cause for the abrasion of the electrode tip.

Residual Stress

The stress field in the weldment during the RSW process is very complex. The normal stress σ_y has an important influence on the form of the weld nugget. Figure 8 shows the distribution of normal stress σ_y at the squeeze and weld nugget forming steps. It can be seen that there was mainly compressive stress in the contact area, and the maximum stress was about 172 MPa at the edge of the electrode–workpiece interface. Figure 8b, c show the distribution of normal stress σ_y at cycle 9 and cycle 13, when the weld nugget started to form and completely formed (the end of holding step). The stress field became more complex due to the generation of thermal stress.

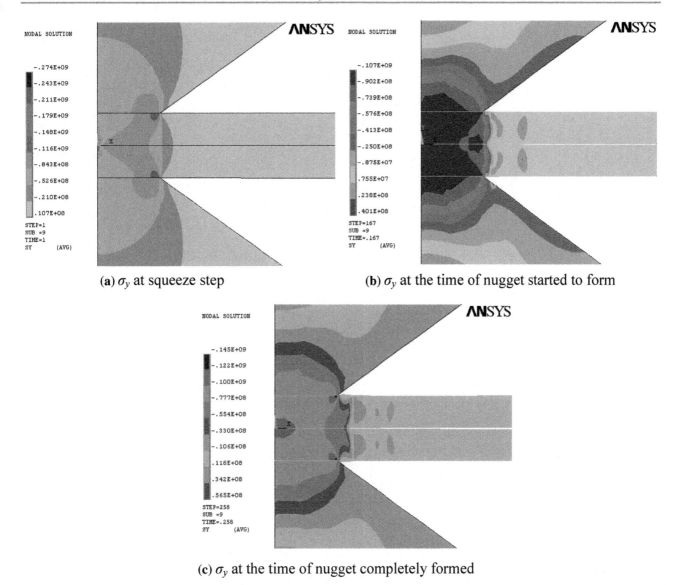

(a) σ_y at squeeze step

(b) σ_y at the time of nugget started to form

(c) σ_y at the time of nugget completely formed

Fig. 8 Normal stress σ_y in the weldment during the RSW process

Residual Plastic Strain and Welding Deformation

The distribution and change of the strain, especially the plastic strain, is very important to the residual stress and deformation of the weldment. The strain field in the weldment during the RSW process is also very complex. The welding residual stress is produced in the welded joint as a result of plastic deformation caused by non-uniform thermal expansion and contraction due to non-uniform temperature distribution in the welding process. Figure 9 shows the residual plastic strain of the weldment in radial and normal orientation after welding. It can be seen that the radial residual plastic strain is mostly compression strain and the normal strain is tension strain on the contrarily. The maximum of the residual plastic strain occurred near the edge of

the contact area through the thickness of the sheets. Comparing Fig. 10, the distribution of residual plastic strain of the weldment in radial and normal orientation at the highest temperature after RSW welding, we can give the following discussion. During the RSW process, the material in the center part of the welding joint expanded with electrifying and heating up, but embarrassed by the material around in the radial orientation. So, a larger compression plastic strain was produced as shown in Fig. 10a. The maximum radial compression plastic strain was about 0.063955 at the highest temperature after welding.

Embarrassed in the radial orientation, the deformation of the material was turned to the normal orientation during the RSW process. Compressed by the electrode pressure in the radial orientation (but the pressure was less than the

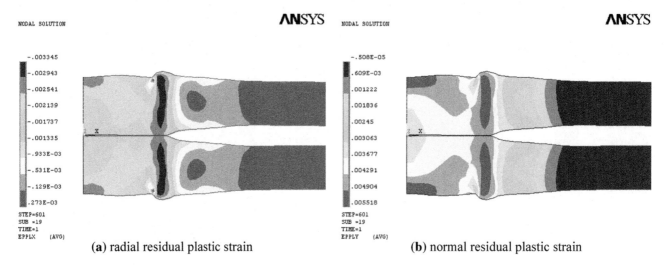

(a) radial residual plastic strain   **(b)** normal residual plastic strain

Fig. 9 Residual plastic strain of the weldment after welding

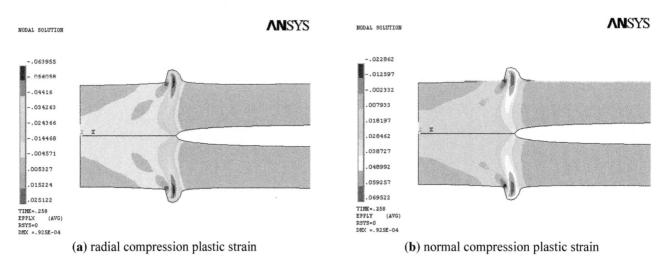

(a) radial compression plastic strain   **(b)** normal compression plastic strain

Fig. 10 Residual plastic strain of the weldment at highest temperature after welding

resistance of the thermal expansion of the material of the welding joint), the normal dimension increased and the normal tension plastic strain was produced. The material near the edge of the electrode was extruded and larger tension plastic strain was produced as shown in Fig. 10b. The maximum normal tension plastic strain was about 0.069522 at the highest temperature after welding.

Once electrifying after welding stopped, compression of the material started. Embarrassed by the material around in the radial orientation, tension plastic strain was produced and the radial compression plastic strain could be counteracted partly but not completely, so the residual compression plastic strain remained. Figure 9a shows that the residual compression plastic strain was only about 0.003345; most of the residual compression plastic strain was eliminated. In the normal orientation, compression plastic strain was produced by the electrode pressure, most of the tension plastic strain was eliminated, and residual tension plastic strain remained. Figure 9b shows that the residual tension plastic strain was about 0.005518.

Conclusions

The determinations of the contact pressure at the faying surface and electrode–workpiece interface are important aspects of the numerical analysis. The distributions of the contact pressure at the faying surface are very important during the RSW process, because they determine the distribution of the electric resistance. The contact resistance distribution influences the current density pattern, which affects the temperature field through Joule heating, while the temperature field then influences the mechanical behavior of the weldment. The distributions of the contact pressure at the electrode–workpiece interface have little influence on the temperature field, however, they have great influence on the useful life of the electrode.

The deformation of the weldment is produced due to the residual plastic strain. The residual plastic strain and the welding deformation are symmetrical if the two sheet workpieces have equal thickness. Research studies show that the symmetrical residual plastic strain has a very small influence on the deformation of the welding structure.

Acknowledgments Yuanxun Wang: Project supported by the financial support from the National Natural Science Foundation of China (11072083).

References

1. Adib H, Gilgert J, Pluvinage G (2004) Fatigue life duration prediction for welded spots by volumetric method. Int J Fatigue 26:81–94
2. Babu SS, Santella M, Feng LZ, Riemer BW, Cohron JW (2001) Empirical model of effects of pressure and temperature on electrical contact resistance of metals. Sci Technol Weld Join 6:126–132
3. Browne DJ, Chandler HW, Evans J, Wen TJ (1995) Computer simulation of resistance spot welding in aluminum: part I. Weld J 74:339s–344s
4. Chang BH, Li M, Zhou VY (2001) Comparative study of small scale and 'large scale' resistance spot welding. Sci Technol Weld Join 6:273–280
5. Chao YJ (2003) Ultimate strength and failure mechanism of resistance spot weld subjected to tensile, shear, or combined tensile/shear loads. J Eng Mater Technol Trans ASME 125:125–132
6. Cho HS, Cho YJ (1989) A study of the thermal behavior in resistance spot welds. Weld J 68:236s–244s
7. Deng X, Chen W, Shi G (2000) Three-dimensional finite element analysis of the mechanical behavior of spot welds. Finite Elem Anal Des 35:17–39
8. Hou ZG, Kim IS, Wang YX, Li CZ, Chen CY (2007) Finite element analysis for the mechanical features of resistance spot welding process. J Mater Process Tech 185:160–165
9. Hou ZG, Wang YX, Li CZ, Chen CY (2006) A multi-coupled finite element analysis of resistance spot welding process. Acta Mech Solida Sin 19:86–94
10. Kang H, Barkey ME, Lee Y (2000) Evaluation of multiaxial spot weld fatigue parameters for proportional loading. Int J Fatigue 22:691–702
11. Khan JA, Xu L, Chao Y (1999) Prediction of nugget development during resistance spot welding using coupled thermal–electrical–mechanical model. Sci Technol Weld Join 4:201–207
12. Lin SH, Pan J, Tyan T, Prasad P (2003) A general failure criterion for spot welds under combined loading conditions. Int J Solids Struct 40:5539–5564
13. Nied HA (1984) The finite element modeling of the resistance spot welding process. Weld J 63:123s–132s
14. Pan N, Sheppard S (2002) Spot welds fatigue life prediction with cyclic strain range. Int J Fatigue 24:519–528
15. Tsai CL, Dai WL, Dickinson DW, Papritan JC (1991) Analysis and development of a real-time control methodology in resistance spot welding. Weld J. 70:339s–351s
16. Vogler M, Sheppard S (1993) Electrical contact resistance under high loads and elevated temperatures. Weld J. 72:231s–238s
17. Yeung KS, Thornton PH (1999) Transient thermal analysis of spot welding electrodes. Weld J. 78:1s–6s

Strain-Induced Martensitic Transformation in a Co-Cr-W-Mo Alloy Probed by Nanoindentation

Irmgard Weißensteiner, Patrick Voigt, Helmut Clemens, and Verena Maier-Kiener

Abstract

Co-Cr-W-Mo alloys have been used for dental implants for many decades. Thermomechanical processing in combination with heat treatments leads to a significant improvement of the materials properties regarding application and manufacturing. During hot isostatic pressing (HIP) of the cast prematerial, a change in the chemical composition of the constituent phases occurs, which influences the susceptibility of the metastable fcc Co matrix to strain-induced martensitic transformation (SIMT). The purpose of this study was to identify the effect of the HIP process on SIMT. The local mechanical behavior of preselected grains in {111}, {110}, and {100} orientations was analyzed by nanoindentation, and the chemical composition of the primary dendrites was studied by atom probe tomography (APT). Due to HIP, the chemical composition of the matrix slightly changed, which in turn changes the fcc $\rightarrow$ hcp transformation temperature and, connected with that, the tendency to SIMT. Further, the present study revealed a slight orientation anisotropy of the Young's modulus in the as-cast condition and a general drop in hardness after HIP.

Keywords

Co-Cr implant alloy
Strain-induced martensitic transformation
Nanoindentation

Introduction

Co-Cr alloys are broadly developed and well established in the dental implantology due to the high resistance against corrosion in contact with human body fluids and a sufficiently high wear resistance [1, 2]. Due to economic reasons, the initial material for implants is usually cast and subsequently HIPed. Because the material shows a certain texture after the casting process, mechanical anisotropy must be considered in implant design. As state-of-the-art method, Co-Cr implants are manufactured by a combined CAD/CAM process, where a significant amount of strain is applied to the material. Due to the low or even negative stacking fault energy (SFE) of Co-Cr alloys [2–5], the deformation is accomplished by localized gliding on distinct gliding planes, twinning, and strain-induced martensitic transformation from the metastable high temperature fcc structure to the stable ε-hcp phase [6–9]. The SFE and in turn the tendency to SIMT of the Co-Cr solid solution is strongly influenced by the chemical composition, especially the addition of C and N is reported to be able to suppress the formation of ε-Co in Co-Cr-Mo alloys [2, 4, 10, 11].

Experimental Procedure

The starting material with the nominal composition of Co-24Cr-8W-3Mo (m.%) was cast in ingots of Ø 100 mm and subsequently HIPed at 1200 °C for 240 min at a pressure of 100 MPa in order to close residual casting porosity. For metallographic examination the samples were ground, polished, and finally macro-etched in a solution of saturated Fe(III)Cl in HCl (32%) for 6–8 min. The grain size was determined according to the linear intercept method by employing the software package Stream Motion 1.9.3 provided by Olympus. The macroscopic hardness was tested by Vickers hardness measurements. The mechanical properties were investigated by nanoindentation in the cast as well as

I. Weißensteiner (✉) · H. Clemens · V. Maier-Kiener
Department of Physical Metallurgy and Materials Testing,
Montanuniversität Leoben, Leoben, Austria
e-mail: irmgard.weissensteiner@unileoben.ac.at

P. Voigt
Titanium Solutions GmbH, Bremen, Germany

© The Minerals, Metals & Materials Society 2018
A. P. Stebner and G. B. Olson (eds.), *Proceedings of the International Conference on Martensitic Transformations: Chicago*,
The Minerals, Metals & Materials Series, https://doi.org/10.1007/978-3-319-76968-4_38

the HIPed condition. To this end, the samples were carefully cut in a laboratory cutting device, ground, and polished with a small polishing pressure. The final polishing step was performed with an oxide polishing suspension (OP-U from Struers), which was mixed with H_2O_2 in order to remove any deformation layers from the prior preparation.

Electron backscatter diffraction (EBSD) analyses were performed prior to and after the indentation process in order to identify grains in {111}, {110}, and {100} orientation parallel to the sample surface. For this procedure, a dual beam scanning electron microscope (SEM) of the type Versa 3D from FEI was employed, which is equipped with an EDAX Hikari XP camera. For the data acquisition, the software package OIM Data Collection 7 was used, and for the data evaluation, OIM Analysis 7.3 was employed. The datasets had a step-size of 80 nm. A grain confidence index standardization clean-up was performed, and all data points with a confidence index below 0.05 were discarded.

Nanoindentation testing was performed with a platform Nanoindenter G200 (Keysight Tec Inc, Santa Rosa, CA, USA) equipped with a three-sided Berkovich diamond indentation tip and a continuous stiffness measurement (CSM) unit. The latter allows the depth-dependent determination of the contact stiffness, and thus the evaluation of hardness and Young's modulus (Poisson ratio 0.31), by superimposing a sinusoidal displacement signal during indentation. Indentation experiments were conducted in constant strain rate mode with an applied indentation strain rate of 0.05 s^{-1} to a maximum indentation depth of 1000 nm, and the data analysis was performed according to the Oliver-Pharr analysis [12]. Finally, the mechanical properties were averaged between 800 and 950 nm indentation depth. For all tests, neighboring indentations were set with distances of more than 20 times indentation depth

(20 μm) to each other in order to exclude any overlapping of the individual plastically deformed zones [13].

The APT measurements were performed on a Cameca LEAP 3000X HR at 60 K, with an applied pulse rate of 200 kHz, a pulse fraction of 20%, and an evaporation rate of 1%.

Results and Discussion

A coarse microstructure of the material in as-cast condition and after HIP was revealed by a macro-etching method (Fig. 1 a, b). The examination of the grain size in this area by means of linear intercept method showed a size of 1.4 ± 0.9 mm and 1.7 ± 0.8 mm for as-cast and HIPed material, respectively. SEM investigations in combination with energy dispersive X-ray spectroscopy and EBSD analyses identified the microstructural constituents as coarse primary dendrites with small amounts (<5% in total) of interdendritic Laves and χ–phase as well as Nb-Mo nitrides [14]. Although the material appears to undergo no microstructural changes on the macroscopic level during the HIP process, the morphology of the interdendritic phases suggests that partial dissolution as well as re-precipitation took place. However, this change in phase fraction and microstructural morphology does not influence the macroscopic hardness; the cast material exhibits 306 ± 10 HV10 and after the HIP process, the hardness was determined to be 304 ± 11 HV10.

In order to correlate changes in microstructure with the mechanical properties of the alloy, nanoindentation experiments were performed in preselected grains with {111}, {110}, and {100} planes parallel to the sample surface. In all selected grains, at least nine indentations were set, and the

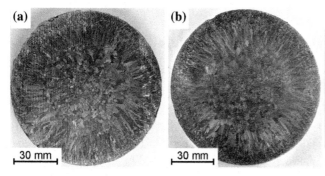
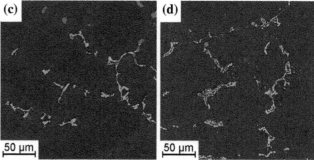

Fig. 1 Microstructure of the **a** as-cast and **b** HIPed condition (light-optical micrographs); SEM images taken in backscattered electron mode of the **c** cast and **d** HIPed material [14] (parts of this figure are reprinted from Practical Metallography 53, 450 by I. Weißensteiner, P. Voigt, V. Maier-Kiener, and H. Clemens, © Carl Hanser Verlag GmbH & Co.KG, Munich)

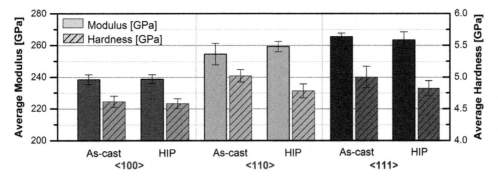

Fig. 2 Nanoindentation data of grains with varying crystal orientation parallel to the sample surface: {100}, {110}, and {111}. Hardness and modulus values are the average of 9 indents for both sample conditions, i.e., as-cast and HIP. The full bars represent the values of the elastic modulus whereas the shaded bars depict the average hardness values

average values for hardness and Young's modulus are presented in Fig. 2. A comparison of the orientation-dependent Young's moduli confirms the general trend observed for any fcc materials, where $E_{111} > E_{110} > E_{100}$ [15]. However, it is important to mention that measured elastic anisotropies are reduced during nanoindentation testing due to the multi-axial loading situation of the used pyramidal indentation tip [16]. Regarding the production route influences, the HIP process does not induce any significant change in the modulus values. In contrast to that, the hardness slightly decreases in all three analyzed crystallographic orientations.

After testing, the plastically deformed zone around the residual impressions were analyzed in detail by EBSD. For both materials conditions, the indentations in a grain with {100} orientation are shown exemplarily in Fig. 3. The grayscale background of the EBSD analyses around the marked indents correlates to the pattern quality of the individual data points, which is, for example, decreased due to surface roughness caused not only by the indentation pile-up, but also by a high dislocation density. Thus, the darker lines correspond to the gliding bands with accumulated deformation. Primarily in the as-cast condition, there are areas within the plastically deformed zone around the

impressions that exhibit hcp crystal structure; these data points are colored green in Fig. 3 c, f. Regions that fulfill the Shoji-Nishiyama orientation relationship, i.e., $(111)_{fcc} \| (11-20)_{hcp}$ and $<101>_{fcc} \| <0001>_{hcp}$ [17], are colored yellow. For an equal area around the indents ($25 \times 25~\mu m^2$), identical preparation procedure, and identical measurement as well as same analysis parameters, the area fraction of data points with a confidence index of >0.15 that are assigned to the hcp phase decreases from 1.7% (for the as-cast condition) to 0.7% (HIP-condition). The corresponding mechanical data, such as load-displacement or hardness and Young's modulus over indentation depth, however, show no obvious differences where drops might occur during phase transformation. Thus, up to now no local mechanical indications in hardness, modulus, or load can be directly connected to the present SIMT due to the implied plastic strain of 8% during indentation.

Due to the HIP process, the alloying elements are partially redistributed. This results in a change of the chemical composition of the primary dendritic areas, which exhibit an fcc structure. By means of APT, these changes could be analyzed more accurately when compared to SEM-EDS. Figure 4a shows the 3D reconstruction of the measured tips,

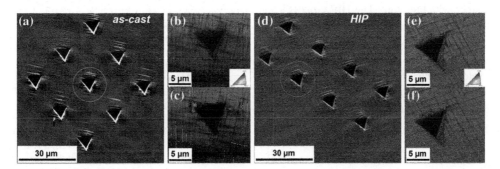

Fig. 3 Comparison of nanoindentations in as-cast **a–c** and HIPed condition **d–f**. **a** and **d** show the respective SEM images of the field of indents. The color code of the EBSD data in **b** and **e** shows the orientation and the overlaid greyscale corresponds to the pattern quality. In **c** and **f** the color code represents the crystal structure; fcc points are colored red, hcp points are colored green. Prior to the testing, the grains in {100} orientation were completely fcc; after testing the as-cast condition shows a higher fraction of data points of hcp structure. The yellow lines along the phase interfaces denote sites that fulfill the Shoji-Nishiyama orientation relationship [17]

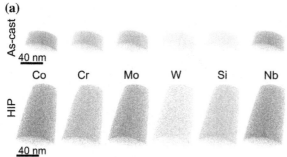

Fig. 4 **a** 3D-reconstruction of the APT tips of the primary dendritic matrix in as-cast and HIP condition. The distribution of the alloying elements is homogeneous in both tips; **b** change in chemical composition of the matrix due to the HIP process. Full bars show the content of the individual element in as-cast condition, whereas the shaded bars represent the chemical composition in the HIPed condition

where both exhibit a homogeneous distribution of the elements without any clustering. In Fig. 4b the corresponding change of the amount of the individual alloying elements is depicted. The color code represents the effect of the individual elements on the stacking fault energy of the solid solution according to [2]. The elements indicated in green color increase the fcc → hcp transformation temperature and, thereby, also the tendency to SIMT. In contrast, the elements marked in red stabilize the fcc structure. For a rough estimation of the effect on the stability of the fcc matrix, the compositional change can be combined with the individual shifts of the transformation temperature taken from [2], which finally results in a decrease of the fcc → hcp transformation temperature of about 70 K. This is in accordance with the lower amount of hexagonal phase after plastic deformation during indentation observed in the HIPed condition (Fig. 3).

Conclusions

In this study, the effect of the HIP process on the mechanical properties and SIMT of a Co-24Cr-8W-3Mo dental alloy were investigated by EBSD and nanoindentation. The alloy shows an anisotropy of the elastic modulus, which is qualitatively in accordance with literature. The HIP process leads to a decrease of the hardness values in all measured orientations, whereas no overall trend in the Young's modulus could be observed. The effective strain of 8% applied during nanoindentation leads to a change of the crystal structure in the plastically deformed zone around the indentations. It was proven that the transformed area fraction in the as-cast condition is significantly higher than that in the HIPed specimen. This is explained by the change of the chemical composition of the fcc matrix during the HIP process. By means of APT an increase of the fcc stabilizing elements in the matrix was detected, which corresponds to a drop of the

fcc → hcp transformation temperature of about 70 K, which, in turn, leads to a decrease of the tendency to SIMT as observed by EBSD.

Acknowledgements The authors thank Patrick Voigt and the Hanseatische Warenhandelsgesellschaft for providing the sample material. Parts of this work were also funded by the Styrian and the Tyrolean Provincial Government, represented by Steirische Wirtschaftsförderungsgesellschaft mbH and Standortagentur Tirol, within the framework of the COMET Funding Programme (837900, MPPE A7.19) is appreciated.

References

1. Yamanaka K, Mori M, Chiba A (2014) Mater Sci Eng C 40:127
2. Narushima T, Ueda K (2015) Alfirano. Springer, Berlin, Heidelberg, pp. 157–178
3. Koizumi Y, Suzuki S, Yamanaka K, Lee B-S, Sato K, Li Y, Kurosu S, Matsumoto H, Chiba A (2013) Acta Mater 61:1648
4. Yamanaka K, Mori M, Kuramoto K, Chiba A (2014) Mater Des 55:987
5. Yamanaka K, Mori M, Chiba A (2014) Mater Sci Eng, A 592:173
6. Remy L (1977) Metall Trans A 8:253
7. Stringfellow RG, Parks DM, Olson GB (1992) Acta Metall Mater 40:1703
8. Yamanaka K, Mori M, Koizumi Y, Chiba A (2014) J Mech Behav Biomed Mater 32:52
9. Rémy L, Pineau A (1976) Mater Sci Eng 26:123
10. Yamanaka K, Mori M, Chiba A (2014) J Mech Behav Biomed Mater 37:274
11. Li YP, Yu JS, Kurosu S, Koizumi Y, Matsumoto H, Chiba A (2012) Mater Chem Phys 133:29
12. Oliver WC, Pharr GM (1992) J Mater Res 7:1564
13. Hay JL, Pharr GM (2000) ASM Handb 8:232
14. Weißensteiner I, Voigt P, Maier-Kiener V, Clemens H (2016) Pract Metallogr 53:450
15. Rösler J, Bäker M, Harders H (n.d.) Mechanical Behaviour of Engineering Materials, Springer, Berlin, Heidelberg, pp. 31–62
16. Vlassak JJ, Nix WD (1993) Philos Mag A 67:1045
17. Li Y, Yamashita Y, Tang N, Liu B, Kurosu S, Matsumoto H, Koizumi Y, Chiba A (2012) Mater Chem Phys 135:849

Estimation of Solute Carbon Concentration by Electrical Resistivity Method in Low-Carbon Martensitic Steel

Toshihiro Tsuchiyama, Taiga Taniguchi, Daichi Akama, Setsuo Takaki, Kenji Kaneko, Masahide Yoshimura, Masaaki Fujioka, and Ryuji Uemori

Abstract

The concentration of solute carbon in as-quenched tempered low-carbon martensitic steels (Fe-2%Mn-0.3% C) were estimated from the electrical resistivity. It was found that the electrical resistivity decreased gradually with the increase of the tempering period, and its decreasing rate was enlarged by raising the tempering temperature. The decrement in electrical resistivity was mainly due to the decrease in the amount of solute carbon caused by carbide precipitation. An empirical equation was then applied to convert the electrical resistivity to the solute carbon concentration, where the densities of dislocation and that of grain boundary were also taken into account. Quantitative analysis for a specimen tempered at 373 K for 3.0 ks revealed that the concentration of solute carbon was decreased by 0.005 mass% during the tempering. This estimated value agreed well with the amount of precipitated carbide ($Fe_{2.5}C$) measured by TEM observation. As a result, it was concluded that the solute carbon concentration could be estimated quantitatively from the electrical resistivity measurement in as-quenched and tempered martensitic steel.

Keywords

Electrical resistivity • Low-carbon steel
Martensite • Tempering • Solute carbon
Age hardening

T. Tsuchiyama (✉) · D. Akama · S. Takaki · K. Kaneko
Department of Materials Science and Engineering, Kyushu University, 744, Moto-Oka, Nishi-Ku, Fukuoka, 819-0395, Japan
e-mail: toshi@zaiko.kyushu-u.ac.jp

D. Akama
e-mail: akama@zaiko.kyushu-u.ac.jp

S. Takaki
e-mail: takaki@zaiko.kyushu-u.ac.jp

K. Kaneko
e-mail: kaneko@zaiko.kyushu-u.ac.jp

T. Tsuchiyama · D. Akama · S. Takaki
International Institute for Carbon-Neutral Energy Research (WPI-I2CNER), Kyushu University, 744, Moto-Oka, Nishi-Ku, Fukuoka, 819-0395, Japan

T. Taniguchi
Kyushu University, 744, Moto-Oka, Nishi-Ku, Fukuoka, 819-0395, Japan

M. Yoshimura · M. Fujioka
Steel Research Laboratories, Nippon Steel & Sumitomo Metal Corporation, 20-1 Shintomi, Futtsu, Chiba Prefecture 293-8511, Japan

R. Uemori
Research Center for Steel, Kyushu University, 744, Moto-Oka, Nishi-Ku, Fukuoka, 819-0395, Japan
e-mail: uemori@zaiko.kyushu-u.ac.jp

Introduction

Recent development of analytical instruments such as high-resolution transmission electron microscopy (TEM) and 3D atom probe has revealed the atomic distribution of carbon in martensite. For example, Hutchinson [1] reported the inhomogeneous distribution of carbon in martensite even under as-quenched condition. In particular, it was found that most of carbon precipitated as carbides or segregated at lath boundaries and dislocations through auto-tempering in the case of low-carbon steels with a relatively high Ms temperature (0.12%C steel). This result suggests that the amount of carbon present in solid solution (solute carbon concentration) is dependent on the degree of auto-tempering, and thus, it is influenced by the Ms temperature or cooling rate on quenching. When such a quenched martensite is subjected to subsequent tempering process, the remaining solute carbon would gradually precipitate and segregate further, which affects the mechanical property of the steel. Therefore,

© The Minerals, Metals & Materials Society 2018
A. P. Stebner and G. B. Olson (eds.), *Proceedings of the International Conference on Martensitic Transformations: Chicago*,
The Minerals, Metals & Materials Series, https://doi.org/10.1007/978-3-319-76968-4_39

quantitative analysis methods measuring the amounts of solute carbon must be established for understanding the strengthening mechanism of martensitic steels.

Electrical resistivity measurement is one of the most promising methods which responds directly and sensitively to the concentration of elements present in solid solution [2–5]. Although the electrical resistivity is significantly affected by not only solute elements but also lattice defects such as grain boundaries and dislocations [4, 5], it would be possible to estimate the solute carbon concentration if the effects of lattice defects and other solute elements could be separated.

In this study, electrical resistivity measurements were applied to a low-carbon martensitic steel for estimating the concentration of solute carbon in as-quenched specimen and its change during low temperature tempering. The results obtained by the electrical resistivity method was then verified by comparing the estimated amount of carbide between this method and observation result by TEM observation.

Experimental Procedure

A low-carbon steel with chemical composition of Fe-0.31C-2.0Mn-0.5Si (mass%) was used in this study. The steel was produced by vacuum melting, and then annealed at 1423 K for 3.6 ks, followed by hot rolling to 15 mm thick plate. Specimens cut from the steel plate (10 × 10 × 60 mm) were solution-treated at 1373 K for 1.8 ks, followed by water quenching to obtain full martensite structure. The cooling rate on the water-quenching was measured at 450 K/s at a temperature around Ms temperature of 653 K. The solution-treated specimens were subjected to tempering at different temperatures, and then the changes in hardness and microstructure were investigated after the heat treatment. The microstructures were observed with TEM (JEM-3200 FSK, JEOL) at an operating voltage of 300 kV. Thin foil specimens for the TEM observations were obtained by electropolishing at room temperature using an automatic twinjet electropolisher at a voltage of 35 V with electrolyte consisted of 10% perchloric acid and 90% acetic acid. Electrical resistivity was measured by four-point probe method [2] at 77 K and treated with a unit of specific resistance [Ωm]. For obtaining the continuous resistivity change during tempering, the measurement and additional heat treatment were alternately conducted without disconnecting the lead lines for the measurements of electrical resistivity.

Results and Discussion

Change in Hardness and Microstructure During Low Temperature Tempering in 0.3C Martensitic Steel

Figure 1 shows changes in hardness of quenched and tempered 0.3C steel as a function of tempering time. The hardness of the specimens tempered at 523 K decreased monotonically during the tempering as usually reported, while that tempered at 373 K exhibited age hardening. The peak hardness was obtained when the aging time reached around 3.0 ks, and soon followed by over-aging softening by prolonged tempering. Speich et al. [6] reported a similar age hardening phenomenon from a specimen tempered at 373 K in 0.18%C martensitic steel, and inferred that the carbon segregation to dislocations was attributed to the increase of hardness. On the other hand, Danoix et al. [7] found that spinodal (ordered) regimes formed within the martensite matrix during the 373 K tempering. These facts indicate a certain amount of solute carbon should have remained in the as-quenched martensite matrix though some of the added carbon might have already segregated or precipitated before tempering due to auto-tempering as reported by Hutchinson [1].

Figure 2 shows TEM images of (a) as-quenched, (b) peak-aged, and (c) over-aged specimens. Although a small amount of tiny carbides was observed, coarse carbide particles were hardly seen in the as-quenched specimen, as shown in Fig. 2a. On the other hand, a dense distribution of fine carbide particles is clearly observed in the peak-aged

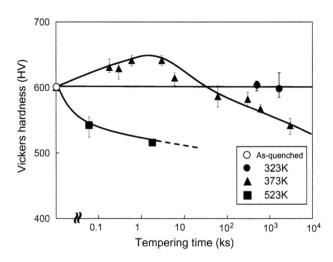

Fig. 1 Changes in hardness of quenched and tempered 0.3%C steel as a function of tempering time

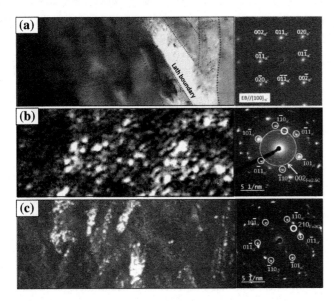

Fig. 2 TEM images of **a** as-quenched, **b** peak-aged (373 K-3.0 ks), and **c** over-aged (523 K-1.8 ks) 0.3%C steels

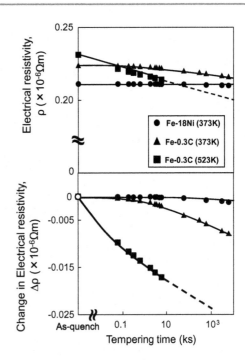

Fig. 3 Changes in electrical resistivity, ρ, and difference from as-quenched state, Δρ, as a function of tempering time in 0.3%C steel tempered at 373 K and 523 K, and Fe-18%Ni alloy tempered at 373 K

specimen, as shown in Fig. 2b. As a result of structure analysis with electron diffraction patterns, the fine carbides were found to be χ carbides (Fe$_{2.5}$C) which is one of the metastable carbides in carbon steel [8]. Its average size and number density were measured with this picture at 2 nm and 1.5 × 10^5/μm^3, respectively. The age hardening observed in Fig. 1 could have been caused by precipitation of this fine carbide particles. In the over-aged specimen (c), coarse stable carbides, namely cementite (Fe$_3$C), was frequently observed.

Results of Electrical Resistivity Measurements

Figure 3 shows the changes in electrical resistivity, ρ, and the difference from as-quenched state, Δρ, as a function of tempering time in the tempered specimens. In addition to the low temperature tempering of 0.3C steels at 373 and 523 K, results on carbon-free Fe-18%Ni alloy (18Ni steel) tempered at 373 K were also presented to differentiate the effect of the changes in dislocation and grain boundary densities from the effect of solute carbon. The values of ρ for the 0.3C steel are found to decrease with increasing tempering time and the decreasing rate in ρ is enhanced by raising the tempering temperature. On the other hand, the Δρ values for 18Ni steel is almost negligible, which suggests that the effects of the changes in dislocation and grain boundary densities during such a low temperature tempering is small enough, and the Δρ in 0.3C steels mainly corresponds to changes in the amount of solute carbon. Araki et al. [9] have demonstrated that the segregation of carbon at grain boundaries has insignificant effect on electrical resistivity in a previous

study. Assuming that the segregation at dislocations is also insignificant, similar to grain boundary segregation, the Δρ can only derive from the change in the amount of solute carbon in solid solution in the martensite.

It should be noted here that the values of ρ of as-quenched specimens shown in Fig. 3 is not consistent and shows considerable scatter. The reason for this behavior has not been strictly clarified; however, some internal strain or defects generated on water-quenching should vary depending on the size and location of the specimen, and this may sensitively influence the measured value of electrical resistivity. Such scattering never occurred when we used 5% manganese steels with a sufficient hardenability which were air-cooled to obtain full martensitic structure.

Estimation of Solute Carbon Concentration in As-Quenched and Tempered Specimens

In order to convert the measured electrical resistivity into carbon concentration, the following empirical relationship, originally proposed for ferritic steels [3], was applied to the martensitic steel.

$$\rho = 0.704 + 13.3[\%Si] + 0.04[\%Mn] + 30[\%C] \quad (\mu\Omega\,cm) \tag{1}$$

By assigning the chemical composition of each element and the measured electrical resistivity to Eq. 1, the solute

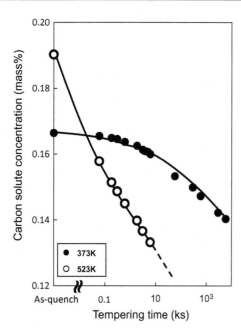

Fig. 4 Calculated solute carbon concentrations plotted as a function of tempering time for 373 and 523 K tempered 0.3%C steel

carbon concentration could be roughly estimated. The calculated solute carbon concentrations were plotted as a function of tempering time for 373 and 523 K tempered specimens in Fig. 4. The solute carbon concentration is gradually decreased with increasing time. As for the as-quenched specimens (the points on the vertical axis), the solute carbon concentration is estimated at around 0.18%, although this value contains the as-mentioned scatter due to internal strain or defects.

Now we focus on the peak-age condition, at 373 K for 3.0 ks. Under this condition, the change in solute carbon concentration from the as-quenched state is found to be approximately 0.005 mass%. If this carbon was all consumed to form χ carbides ($Fe_{2.5}C$), the volume fraction of the carbide would be calculated at 0.063 vol.% with consideration of its density and structure. Since the size of carbide particles is already known from Fig. 1 (2 nm), the number density can be simply calculated at $1.6 \times 10^5/\mu m^3$.

This calculated value agrees well with the microstructure observed with TEM, suggesting that the electrical resistivity measurement is an effective method for quantitative estimation of the solute carbon concentration in as-quenched and tempered martensitic steel.

Conclusions

(1) Age hardening behavior was confirmed in the Fe-0.3% C steel tempered at 373 K with its peak hardness after 3.0 ks tempering. This phenomenon suggests that solute carbon was present in solid solution in the as-quenched martensite, which then precipitates as carbides during the tempering.

(2) It was found that the change of electrical resistivity was directly correlated with the concentration of solute carbon during the tempering.

(3) The estimation of the number density of precipitated carbides by electrical resistivity measurement agreed well with TEM observation, which demonstrates the validity of this method for estimating solute carbon concentration in martensitic steels.

References

1. Hutchinson B, Hagström J, Karlsson O, Lindell D, Tornberg M, Lindberg F, Thuvander M (2011) Acta Mater 59:5845
2. Komatsu S, Fujikawa S (1997) J Jpn Inst Light Met 47:170
3. Joubouji K (2011) J Inst Nucl Saf Syst 18:228
4. Karolik AS, Luhvich AA (1994) J Phys Condens Matter 6:873
5. Miyajima Y, Komatsu S, Mitsuhara M, Hata S, Nakashima H, Tsuji N (2015) Phil Mag. https://doi.org/10.1080/14786435.2015.1021400
6. Speich GR, Leslie WC (1972) Metall Trans 3:1043
7. Danoix F, Zapolsky H, Allain S, Gouné M (2015) Proc Int Conf PTM, TMS, 537
8. Imai Y, Ogura T, Inoue A (1972) Tetsu-to-Hagané 58:726
9. Araki S, Fujii K, Akama D, Tsuchiyama T, Takaki S, Ohmura T, Takahashi J (2017) Tetsu-to-Hagané 103:491

Comparison of Highly Mobile Twin Boundaries in Cu–Ni–Al and Ni–Mn–Ga Shape Memory Single Crystals

M. Vronka, M. Karlík, Y. Ge, and O. Heczko

Abstract

We compared twinning systems in Cu–Ni–Al and Ni–Mn–Ga single crystals from macroscale down to atomic scale. Using a newly developed formalism, we studied the twinning stress or mobility and microstructure of the equivalent twin boundaries. In Cu–Ni–Al, compound twinning exhibits the twinning stress of 1–2 MPa, and Type II twinning stress is approximately 20 MPa, which is much higher than the twinning stress for Type II in Ni–Mn–Ga (0.1–0.3 MPa). No temperature dependence was found for twinning stress of Type II in either alloy. Transmission electron microscopy revealed that in contrast to Ni–Mn–Ga, there was no internal twinning in Cu–Ni–Al, only stacking faults. The highest density of stacking faults was observed in the presence of Type I twin boundaries. The extremely low twinning stress may be associated with the deep hierarchy of twinning in Ni–Mn–Ga.

Keywords

SMA • Single crystals • Martensite • Twinning stress TEM

M. Vronka (✉) · O. Heczko
Institute of Physics of the Czech Academy of Sciences, Na Slovance 2, 182 21 Prague 8, Czech Republic
e-mail: vronka@fzu.cz

O. Heczko
e-mail: heczko@fzu.cz

M. Vronka · M. Karlík
Faculty of Nuclear Sciences and Physical Engineering, Department of Materials, Czech Technical University in Prague, Trojanova 13, 120 00 Prague, Czech Republic
e-mail: miroslav.karlik@fjfi.cvut.cz

M. Karlík
Department of Physics of Materials, Charles University, Ke Karlovu 5, 121 16 Prague 2, Czech Republic

Y. Ge
Department of Materials Science and Engineering, School of Chemical Technology, Aalto University, 16200 00076 Aalto, Finland
e-mail: yanling.ge@aalto.fi

Introduction

The group of effects that provide giant deformations induced by magnetic fields are known under the common name magnetic shape memory (MSM) [1]. If this deformation is caused by phase transformation, we speak about magnetic field–induced phase transformation. If this deformation takes place in one phase by lattice reorientation, the effect is called magnetically induced reorientation (MIR) [2]. The existential condition for MIR is extremely low twinning stress or, inversely, very highly mobile twin boundaries [2], which we will focus further on. This extremely high mobility of twin boundaries is rather exceptional, and the off-stoichiometric Ni–Mn–Ga single crystal is the main example and the most promising material for MIR [1, 2]. Similar highly mobile twin boundaries were observed in a few non-magnetic shape memory alloys [3, 4] such as Cu–Ni–Al, which belongs among the most studied systems of shape memory alloys (SMAs) in the past four decades (for review see [5]). We suggested that the Cu–Ni–Al single crystal could be the best non-magnetic analog to Ni–Mn–Ga [6]. Using this analogy may help to clarify the origin of the high mobility of twin boundaries [6].

The parent cubic austenite phase has $L2_1$ long-range ordered structure for Ni–Mn–Ga and DO_3 structure for Cu–Ni–Al, which can be considered as a degenerate $L2_1$ structure [3, 7]. With a decrease in temperature, one of the several martensitic phases occurs: 10M, 14M, or NM martensites for Ni–Mn–Ga, and 2H or 18R modulated martensites for Cu–Ni–Al [1, 3]. In this study we will further focus on 10M and 2H structures. The lattice parameters of these martensite types and the parent austenite phase are listed in Table 1.

© The Minerals, Metals & Materials Society 2018
A. P. Stebner and G. B. Olson (eds.), *Proceedings of the International Conference on Martensitic Transformations: Chicago*, The Minerals, Metals & Materials Series, https://doi.org/10.1007/978-3-319-76968-4_40

Table 1 Lattice parameters of cubic austenite phase and monoclinic martensite phase in Ni–Mn–Ga and Cu–Ni–Al [6]

Alloy	Austenite lattice parameters (Å)	Martensite lattice parameters (Å) monoclinic description			
	a_0	a	b	c	γ (°)
$Ni_{50.7}Mn_{27.7}Ga_{21.6}$	5.840	5.960	5.943	5.607	90.29
$Cu_{69.4}Ni_{3.44}Al_{27.17}$	5.841	6.089	6.089	5.368	92.07

In 10M modulated martensite of Ni–Mn–Ga, five different types of twins can be observed: compound a-b twin laminate, monoclinic twins, Type I and Type II twins, and so-called "non-conventional type of twinning" [1]. The lowest twinning stress of approximately 0.2 MPa, which means the highest mobility of twinning boundaries, was measured for Type II twinning boundaries [8–10]. The twinning stress for Type I boundaries is about 1–2 MPa [8–10]. The 2H martensite of Cu–Ni–Al consists of a mixture of compound, and Type I and Type II twin boundaries [11, 12]. Here, the lowest twinning stress occurs for the compound twinning at about 1–2 MPa [3, 6]. The twinning stress for Type II twinning is approximately ten times higher, and Type I twins have not been found operational during reorientation [3, 6].

If we describe the 2H martensite of Cu–Ni–Al using the monoclinic system [6], the full comparison to the twinning system of Ni–Mn–Ga can be made. All lattice parameters for the study were taken from [6], and they are shown in Table 1.

For comparison, the twinning planes for all possible twinning types in Cu–Ni–Al 2H and Ni–Mn–Ga 10M martensite are shown in Table 2. The Type I and compound twinning planes are the same for both Cu–Ni–Al 2H and Ni–Mn–Ga 10M martensites [6]. The difference is in the nomenclature, as the monoclinic twinning is directly known as compound twinning in Cu–Ni–Al. The only exception is that the twinning planes for Type II twinning are inclined from Type I twinning planes by 12.89° for Cu–Ni–Al [6] and approximately by 6° for Ni–Mn–Ga [8].

The motivation of the presented research of twinning boundaries in Cu–Ni–Al as a non-magnetic analog to Ni–Mn–Ga is to help highlight the origin of the high mobility of twin boundaries. Above that, this work further develops the presented analogy between these martensites [6], including the comparison of twinning boundaries on the nanoscale level by transmission electron microscopy. Based on the presented results and previous published works about Ni–Mn–Ga, we discuss the twinning behavior of martensite phases in these two different single crystals exhibiting similar high mobility of twinning boundaries.

Experiment

Single crystals of $Ni_{50.7}Mn_{27.7}Ga_{21.6}$ and $Cu_{69.4}Ni_{3.4}Al_{27.2}$ (at.%) were grown by the Bridgman technique and used for this study. Compression tests were performed in an INSTRON 1362 electromechanical testing machine, and tensile tests were carried out by a Deben Microtest 2kN tensile stage inside a Tescan Fera3 scanning electron microscope. The conventional TEM and HRTEM observations were carried out by an FEI Tecnai F20 field emission gun transmission electron microscope operated at 200 kV with a double tilt specimen holder. The samples for the TEM studies were thinned to the final thickness by double jet electropolishing at −20° in a solution of 30% HNO_3 in methyl alcohol.

Table 2 Twinning planes for all possible twinning types in 2H martensite of Cu–Ni–Al and 10M martensite of Ni–Mn–Ga in conformity with nomenclature, which describes Cu–Ni–Al 2H martensite by monoclinic lattice [6, 8]

	Twinning planes
Cu–Ni–Al	2H martensite
Type I twin	(1 0 1), (0 1 −1), (0 1 1) or (0 1 −1)
Type II twin[a]	(0.69 0.22 0.69), (0.69 0.22 −0.69), (0.22 0.69 0.69) or (0.22 0.69 −0.69)
Compound twin	(1 0 0) or (0 1 0)
Ni–Mn–Ga	10M martensite
Type I twin	(1 0 1), (0 1 −1), (0 1 1) or (0 1 −1)
Type II twin[b]	(0.07 0.71 0.71), (0.07 0.71 −0.71), (0.71 0.06 0.71) or (0.71 0.06 −0.71)
Modulation twin	(1 0 0) or (0 1 0)
Compound twin	(1 −1 0) or (1 1 0)

[a]These planes inclined by 12.89° from corresponding Type I planes
[b]These planes inclined approximately 6° from corresponding Type I planes

Results and Discussion

The stress-strain curves were measured during the compression and tension tests of several Cu–Ni–Al samples. Few plateaus corresponding to the pseudoplastical twinning mechanism were exhibited on the curves. These plateaus could be assigned to different types of twinning. All curves were repeatable for many cycles. The plateau at approximately 1–2 MPa belonged to compound twinning, and the plateau at about 20 MPa belonged to Type II twinning. This identification and magnitude of twinning stress are in agreement with [3]. The full interpretation was done and the stress-strain curves are shown in [6]. By comparison, in the Ni–Mn–Ga single crystal, the twinning stress of the Type II twin boundary was in the range 0.1–0.3 MPa, in agreement with literature [8, 9]. The twinning stress of compound twins has not been measured yet; however, from the reorientation experiment it is clear that the stress must be smaller than 4 MPa [13].

Although the monoclinic structures of martensite are comparable for both alloys, in Cu–Ni–Al 2H martensite, Type II twinning exhibits a hundred times larger twinning stress compared to that in Ni–Mn–Ga 10 M martensite. Despite the higher twinning stress, the geometry of the twin boundary in both alloys is similar. The main difference is the existence of a deeper twinning hierarchy in Ni–Mn–Ga, such as compound a-b twin laminate and nanolamination [14], which may influence the twinning stress.

Type I twins have never been observed during reorientation in Cu–Ni–Al. Laminates of Type I twins at the habit planes were observed only during stress-induced martensitic transformation in tension [15]. As we could not find this type of twinning on any of the samples during compression, we also prepared an elongated sample for tensile test. Before the test, the sample was in the as-transformed state with a spontaneously formed twinned microstructure. This microstructure, as will be shown later, contained a mix of Type I and II twins.

Figure 1 shows the stress-strain curve during tensile test on this sample. At approximately 120 MPa the jagged plateau occurred. From SEM observation, this was assigned to the transformation from the 2H martensite to the 18R martensite structure and confirmed by X-ray analysis. The transformation interface spread from the edges to the central part of the sample. At the end of the plateau, the whole sample was in 18R phase. No movement of twinning boundaries was observed during the tensile tests. After releasing the stress, the sample partially transformed back to 2H martensite structure, as apparent from the return branch of the stress-strain curve (Fig. 1).

This experiment suggests that the twinning stress for Type I twins must be higher than at least 120 MPa, which is five times higher than the twinning stress for Type II twins. This is again consistent with the presented analogy with Ni–Mn–Ga.

Furthermore, no temperature dependence was observed in the measurement range for both compound and Type II twinning in Cu–Ni–Al. The same results for Type II twins were also observed in the Ni–Mn–Ga system [16]. Thus, despite the different magnitude of the twinning stress, the temperature behavior of Type II twinning is the same for both alloys. Unfortunately, the temperature dependence experiment for Type I twins in Cu–Ni–Al could not be done due to the inability to observe the movement of these twins. In the Ni–Mn–Ga system, the twinning stress of Type I twins strongly increases with decreasing temperature [16, 17]. However, there is no report about twinning stress and temperature dependence of compound twin boundaries [6]. According to the analogy, the similar temperature independence of compound twinning stress in the Ni–Mn–Ga system can be expected, and the similar strong temperature dependence of Type I twinning stress in the Cu–Ni–Al system is assumed.

In contrast to the results from mechanical testing, all three possible types of twinning in Cu–Ni–Al were observed on

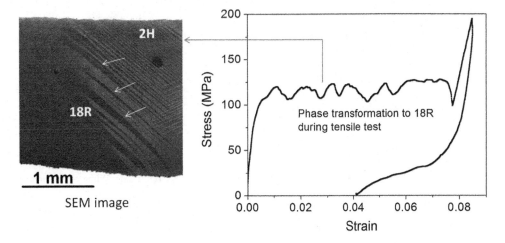

Fig. 1 Measured stress-strain curve during tensile test of Cu–Ni–Al elongated sample. The plateau at approximately 120 MPa corresponds to phase transformation to 18R. The phase interface is shown in the electron micrograph with white arrows, which mark the interface. In 2H phase the twinned structure is observed as nearly diagonal bands

TEM foils. Figure 2 shows the typical example of the compound, Type I and Type II twins visualized by bright field image of diffraction contrast and by high-resolution transmission electron microscopy (HRTEM) for a sample in the as-transformed state. Figure 2a, b show compound twins, which are the common feature of 2H martensite Cu–Ni–Al structure. These twin boundaries are apparently very flexible, as they may be found curved on TEM foils. No internal twinning was observed between them, only stacking faults with low density. Figure 2c, d show Type I twins. The twinning boundaries are always straight with dense stacking faults. The stacking faults cause steps in twinning boundaries (Fig. 2d). The Type I twinning plane between steps formed by stacking faults is a well-defined plane along {101}. Figure 2e, f show Type II twins. Unfortunately, we were able to orient only one side of the twins. Despite this, it is clear that the twinning boundaries are straight with much less density of stacking faults compared to those of Type I.

In general, it seems that the density of irregular stacking faults is much less if no Type I twinning is present. No steps were observed at compound twinning boundaries compared to Type I twinning boundaries. These steps originating from stacking faults may be present at lower density also at Type II twinning boundaries, but the inability to orient both twins prevented any observation and confirmation. Only compound twinning boundaries are conjugate planes and were found curved similarly to the modulation twinning boundaries in Ni–Mn–Ga [18]. Comparing Type I with Type II twinning, it was observed that the widths of Type I twin bands are much narrower than those of Type II twins. Furthermore, Type I twins were found inside Type II twin bands. This kind of hierarchy was not observed in modulated Ni–Mn–Ga.

However, the main difference between Ni–Mn–Ga and Cu–Ni–Al twinned microstructure is that in Ni–Mn–Ga, the internal twinning is much more common than stacking faults [18, 19]. Considering the mobility, it suggests that internal twins do not impede the mobility of the twinning boundaries as much as stacking faults or steps do at the boundaries. That is also consistent with the observation that Type I twins contain the most stacking faults and perhaps therefore they are the least mobile or even totally blocked in Cu–Ni–Al.

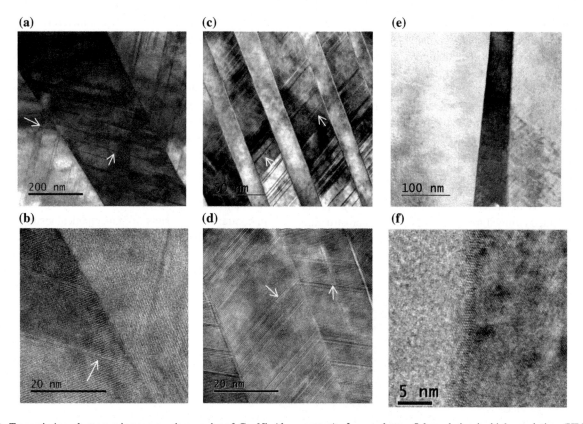

Fig. 2 Transmission electron microscopy micrographs of Cu–Ni–Al twinning boundaries: **a** compound twinning (bright field contrast), **b** single twin boundary (HRTEM), **c** type I twinning (bright field contrast), **d** several type I boundaries in high resolution (HRTEM), **e** type II twinning (bright field contrast), and **f** single type II boundary (HRTEM). White arrows mark stacking faults

Conclusions

The developed formalism allows full comparison of the twinning behavior of single crystals of Cu–Ni–Al and Ni–Mn–Ga [6]. Owing to the twin geometry (Table 2), the behavior is expected to be very similar. The twinning stress of Type II and compound twinning in Cu–Ni–Al martensite were found to be independent of temperature, as was Type II twinning in Ni–Mn–Ga. On the other hand, the magnitude of twinning stress of Type II twinning is approximately 100 times lower in Ni–Mn–Ga.

According to TEM observations, very flexible and not always straight compound twins are the common feature of 2H Cu–Ni–Al and 10M Ni–Mn–Ga martensites. In contrast to Ni–Mn–Ga, no internal twinning was observed in compound twins; only stacking faults were observed. These faults are, however, of much lower density compared to other types of twinning in Cu–Ni–Al. The dense stacking faults and the impossibility of forming a deeper twinning hierarchy compared to Ni–Mn–Ga (such as compound a-b twin laminate) may affect the mobility of twinning boundaries in Cu–Ni–Al. Further studies on the effect of twinning hierarchy on twin boundary mobility are needed.

Acknowledgements This work has been financially supported by the Czech Science Foundation grant No. 14-36566G (AdMat), and furthermore by project LM2015087 of the Czech Ministry of Education, Youth and Sports. M. V. would like to acknowledge financial support by the grant SGS16/249/OHK4/3T/14. M. K. would like to acknowledge financial support of the ERDF in the frame of the project No. CZ.02.1.01/0.0/0.0/15_003/0000485.

References

1. Heczko O, Scheerbaum N, Gutfleisch O (2009) Magnetic shape memory phenomena. In: Liu JP et al (eds) Nanoscale magnetic materials and applications. Springer Science + Business Media, New York, pp 399–439
2. Heczko O (2014) Magnetic shape memory effect and highly mobile twin boundaries. Mater Sci Technol 30:1559–1578
3. Novak V, Sittner P, Ignacova S, Cernoch T (1999) Transformation behavior of prism shaped shape memory alloy single crystals. Mater Sci Eng A-Struct Mater Prop Microstruct Process 438:755–762
4. Basinski ZS, Christian JW (1954) Crystallography of deformation by twin boundary movements in indium-thallium alloys. Acta Metall 2:101–116
5. Otsuka K, Wayman CM (1998) Shape Memory Materials. Cambridge University Press, Cambridge
6. Vronka M, Seiner H, Heczko O (2017) Temperature dependence of twinning stress—analogy between Cu–Ni–Al and Ni–Mn–Ga shape memory single crystals. Philos Mag 97:1479–1497
7. Overholser RW, Wuttig M, Neumann DA (1999) Chemical ordering in Ni–Mn–Ga Heusler alloys. Scr Mater 40:1095–1102
8. Straka L, Heczko O, Sozinov A (2011) Highly mobile twinned interface in 10M modulated Ni–Mn–Ga martensite: analysis beyond the tetragonal approximation of lattice. Acta Mater 59:7450–7464
9. Kellis D, Smith A, Ullakko K, Müllner P (2012) Oriented single crystals of Ni–Mn–Ga with very low switching field. J Cryst Growth 359:64–68
10. Straka L, Hanninen H, Heczko O (2011) Temperature dependence of single twin boundary motion in Ni–Mn–Ga martensite. Appl Phys Lett 98:141902
11. Bhattacharya K (2004) Microstructure of martensite: how it forms and how it gives rise to the shape memory effect. Oxford University Press, Oxford
12. Sari U, Aksoy I (2006) Electron microscopy of 2H and 18R martensites in Cu-11.92 wt% Al-3.78 wt% Ni shape memory alloy. J Alloy Compd 417:138–142
13. Chulist R, Straka L, Sozinov A, Lippmann T, Skrotzki W (2013) Modulation reorientation in 10M Ni–Mn–Ga martensite. Scr Mater 68:671–674
14. Seiner H, Straka L, Heczko O (2014) A microstructural model of motion of macro-twin interfaces in Ni–Mn–Ga 10M martensite. J Mech Phys Solids 64:198–211
15. Shield TW (1995) Orientation dependence of the pseudoelastic behaviour of single crystals of Cu–Al–Ni in tension. J Mech Phys Solids 43:869–895
16. Straka L, Soroka A, Seiner H, Hanninen H, Sozinov A (2012) Temperature dependence of twinning stress of Type I and Type II twins in 10M modulated Ni–Mn–Ga martensite. Scr Mater 67:25–28
17. Heczko O, Kopecky V, Sozinov A, Straka L (2013) Magnetic shape memory effect at 1.7 K. Appl Phys Lett 103:072405
18. Ge Y, Zarubova N, Heczko O, Hannula SP (2015) Stress-induced transition from modulated 14M to non-modulated martensite in Ni–Mn–Ga alloy. Acta Mater 90:151–160
19. Zarubova N, Ge Y, Heczko O, Hannula SP (2013) In situ TEM study of deformation twinning in Ni–Mn–Ga non-modulated martensite. Acta Mater 61:5290–5299

Multiphase Microstructure and Extended Martensitic Phase Transformation in Directionally Solidified and Heat Treated Ni$_{44}$Co$_6$Mn$_{39}$Sn$_{11}$ Metamagnetic Shape Memory Alloy

P. Czaja, R. Chulist, M. Szlezynger, M. Fitta, and W. Maziarz

Abstract

Directionally solidified Ni–Co–Mn–Sn alloy shows a multiphase solidification microstructure relatable primarily to the varying Co–Mn/Sn ratio. Thermal treatment at 1220 K lasting for 72 h encourages chemical homogeneity with average stoichiometry of Ni$_{45.1}$Co$_{6.2}$Mn$_{37.2}$Sn$_{11.5}$. At room temperature, despite the chemical uniformity, the as-homogenized alloy shows a multiphase microstructure with coexisting L2$_1$ austenite and 6M and 4O martensite phases. The martensite phase preferentially locates at grain boundaries. The onset of the martensitic transition temperature is estimated at 402 K, which overlaps with the Curie transition of austenite. The martensitic transition appears to initially take place at the grain boundaries and then it extends to low temperature as the volume of the grains transforms to martensite.

Keywords

Martensite • Microstructure • SEM • TEM

Introduction

Ni–Co–Mn–Sn metamagnetic shape memory alloys are exceptional smart materials from both applications and a fundamental point of view given the combination of their thermoelastic and magnetic properties closely linked to the reversible martensitic phase transformation (MPT). Initial research into Ni–Co–Mn–Sn was triggered by reports on magnetic field induced strain recovery ($\sim$1%) in polycrystalline Ni$_{43}$Co$_7$Mn$_{39}$Sn$_{11}$ [1], and it has been further stimulated by reports on considerable cooling power afforded by Ni$_{47}$Co$_{3.1}$Mn$_{36.6}$Sn$_{13.3}$ (196 J/kg) [2] and Ni$_{43}$Mn$_{42}$Co$_4$Sn$_{11}$ (242 J/kg) [3]. Complex superparamagnetic and superspin glass states found in the martensite phase in Ni$_{43.5}$Co$_{6.5}$Mn$_{39}$Sn$_{11}$ [4] have further fed the curiosity in this system as has the kinetic arrest phenomenon reported for the Ni$_{37}$Co$_{11}$Mn$_{42.5}$Sn$_{9.5}$ and Ni$_{37}$Co$_{11}$Mn$_{43}$Sn$_9$ [5] as well as a low ($\sim$6K) thermal hysteresis in Ni$_{45}$Co$_5$Mn$_{40}$Sn$_{10}$ allowed by the favorable lattice compatibility between austenite and martensite with the middle eigenvalue of the transformation stretch matrix $\lambda_2 = 1.0032$ [6]. Generally the Ni–Co–Mn–Sn system's magnetic properties are largely inherited from the influence of Co substitution into the ternary Ni–Mn–Sn [7]. Depending on the replaced element, the introduction of Co can lead to the increase in saturation magnetization of austenite while simultaneously bringing down the MPT temperature, and thus overall it can encourage magnetization discrepancy (ΔM) across the MPT, rendering the inverse MPT more susceptible to the magnetic field ($\Delta M \cdot H$) [7, 8]. Nonetheless, incorporation of an excess amount of Co into Ni–Mn–Sn may compromise the L2$_1$ phase stability, which frequently leads to the precipitation of a γ phase and suppression of MPT [9, 10]. Most of the literature so far has been concerned with polycrystalline Ni–Co–Mn–Sn produced either by arc- or induction-melting and it has often omitted the finer details of phase microstructure following the solidification process and homogenization heat treatment. Most recently, directional solidification has also been proven effective for production of metamagnetic Ni–Mn–Sn based alloys with enhanced magnetocaloric and mechanical properties [11, 12]. This contribution is therefore centered on examination of the microstructure and MPT in directionally solidified and heat treated Ni$_{44}$Co$_6$Mn$_{39}$Sn$_{11}$ alloy. The alloy shows a multiphase solidification microstructure abatable by 72 h anneal at 1220 K. Some peculiar features of MPT are also unveiled.

P. Czaja ($\boxtimes$) · R. Chulist · M. Szlezynger · W. Maziarz
Institute of Metallurgy and Materials Science Polish Academy of Sciences, 25 Reymonta St., 30-059 Kraków, Poland
e-mail: p.czaja@imim.pl

M. Fitta
The Henryk Niewodniczanski Institute of Nuclear Physics, Polish Academy of Sciences, 152 Radzikowskiego Str., 31-342 Kraków, Poland

© The Minerals, Metals & Materials Society 2018
A. P. Stebner and G. B. Olson (eds.), *Proceedings of the International Conference on Martensitic Transformations: Chicago*, The Minerals, Metals & Materials Series, https://doi.org/10.1007/978-3-319-76968-4_41

Experimental

A polycrystalline master alloy with $Ni_{44}Co_6Mn_{39}Sn_{11}$ nominal composition was produced from pure ($\geq 99.99\%$) elements by conventional induction casting under protective argon gas. Subsequently, the master ingot was placed in an Al_2O_3 crucible and remelted at 1570 K. Then it underwent directional solidification (Bridgman) with a 10 mm/h crucible pull-out speed. A slice of an ingot was removed for examination of an as-cast microstructure. The remaining part was placed in an evacuated quartz ampoule and heat treated for 72 h at 1220 K in order to encourage chemical homogeneity. Following heat treatment, the ingot was slow cooled with the furnace to 670 K and then cooled in air. The as-cast and homogenized microstructure and chemical composition were evaluated with a FEI E-SEM XL30 scanning electron microscope (SEM) furnished with an X-ray energy dispersive spectrometer (EDX). The crystal structure was confirmed with synchrotron high-energy X-ray radiation (87.1 keV, $\lambda = 0.142342$ Å) in a transmission geometry at the beamline Petra P07B at DESY, Germany. It was also investigated with a FEI TECNAI G2 F20 (200 kV) transmission electron microscope (TEM) coupled with an energy dispersive X-ray (EDX) microanalyser and a High Angle Annular Dark Field Detector (HAADF). Thin foils for TEM examination were prepared with TenuPol-5 double jet electropolisher using an electrolyte of nitric acid (20%) and methanol (80%) at 240 K. Thermal effects were elaborated with a Mettler DSC 823 differential scanning calorimeter (DSC) with a 10 K/min cooling/heating ramp. The dc mass magnetic susceptibility and magnetization were measured in the temperature range from 2 K up to 380 K and in magnetic fields up to 90 kOe using the Vibrating Sample Magnetometer (VSM) option of the Quantum Design Physical Property Measurement System (PPMS-9).

Results and Discussion

Upon directional solidification, the Ni–Co–Mn–Sn alloy shows a dendritic, multiphase microstructure as evidenced by Fig. 1a, b. The backscatter electron diffraction (BSE) image (Fig. 1b) reveals distinct contrast differences with lighter grey areas corresponding to the body of the dendrites (Fig. 1a) and with a darker grey shade accentuating the interdendritic phase. According to the EDS analysis, the average composition of the BSE brighter phase is $Ni_{40.7\pm0.8}Co_{5.2\pm0.2}Mn_{36.2\pm0.7}Sn_{17.9\pm0.4}$, whereas the in-between dendrite phase has the average composition of $Ni_{40.5\pm0.8}Co_{7.4\pm0.3}Mn_{41.1\pm0.8}Sn_{11.1\pm0.2}$. This phase inhomogeneity is thus primarily related to the variation in the Co–Mn/Sn ratio, while Ni is less prone to segregation. The brighter phase thus contains more Sn and less Co and Mn relative to the darker phase and its Co–Mn/Sn ratio is 2.3, whereas the same ratio for the darker phase amounts to 4.4. This behavior reflects the solidification behavior observed in Ni–Mn–Sn alloys [13]. Composition variation leads to occurrence at room temperature of austenite and martensite with varying electron-to-atom ratio (e/a), see Fig. 2 showing the bright field (BF) images with the corresponding selected area electron diffraction patterns (SADP) and scanning transmission electron microscopy (STEM) images with detailed averaged composition in micro-areas. In the area dominated by the austenite phase, $e/a = 8.176$, whereas in that with prevailing martensite phase, $e/a = 8.268$. The appearance of the phases in respective areas can be explained in relation to the well-established dependence of the MPT temperature on e/a. The martensite has been indexed according to the 6M modulated structure with the [210] zone axis. The austenite has been indexed with the L2$_1$ Heusler structure along the [111] zone axis. Thermal treatment at 1220 K lasting 72 h and completed with slow cooling

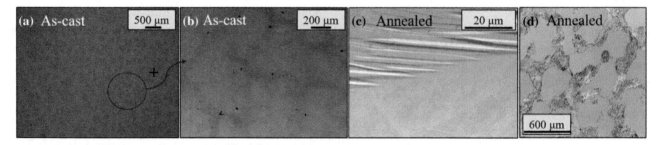

Fig. 1 Room temperature light microscopy, **a** backscattered electron (BSE), **b**, **c** electron backscattered diffraction (EBSD) **d** images of as-directionally solidified (**a, b**) and homogenized (**c, d**) Ni–Co–Mn–Sn alloy

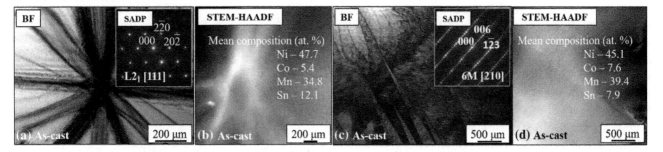

Fig. 2 Bright field (BF) images (**a, c**) together with corresponding selected area electron diffraction patterns (SADP), insets (**a, c**), and scanning transmission electron microscopy (STEM) images (**b, d**) taken from the as-solidified Ni–Co–Mn–Sn alloy

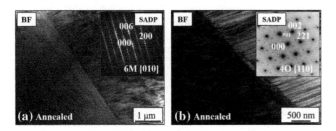

Fig. 3 BF and corresponding SADP (inset) images taken from the homogenized Ni$_{45.1}$Co$_{6.2}$Mn$_{37.2}$Sn$_{11.5}$ alloy

restored equilibrium phase homogeneity what manifests itself by uniform BSE contrast (Fig. 1c). The average composition is determined as Ni$_{45.1\pm0.9}$Co$_{6.2\pm0.2}$Mn$_{37.2\pm0.7}$Sn$_{11.5\pm0.2}$, which, except for Mn, is near the nominal composition. Depletion in Mn content by approximately 2 at.% may be ascribed to Mn volatilization during solidification and post-casting thermal treatment. Figure 1c also reveals that at room temperature the homogenized Ni–Co–Mn–Sn alloy has a dual-phase microstructure composed of the austenite and martensite phases, the latter given away by the plate-like martensitic features. The heterogeneous microstructure is also confirmed with EBSD (Fig. 1d), demonstrating that the martensitic plates are preferentially sited at the grain boundaries. TEM analysis of the heat treated alloy (Fig. 3) confirmed the presence of two types of martensite with 6M (Fig. 3a) and with 4O (Fig. 3b) modulated structures. A trace amount of a Mn-rich phase was also detected (not shown). The Mn-rich phase is frequently found in Ni–Mn-based alloys and is associated with hardly dissolvable MnS [14].

The alloy's structure is further corroborated with high-energy synchrotron radiation experiment. The results in the form of an integrated 2theta scan are shown in Fig. 4. The peaks presented in the figure can be well indexed according to the L2$_1$ austenite phase and the 4O martensite with small contribution from the NiMn phase [11].

DSC measurements (Fig. 5) allowed for the establishing of the critical martensite start (M_s), martensite finish (M_f), austenite start (A_s), and austenite finish (A_f) temperatures taken by linear extrapolation to the curves corresponding to

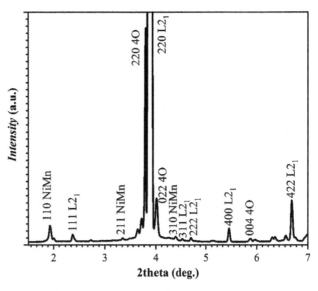

Fig. 4 Room temperature integrated 2theta scan for the annealed Ni$_{45.1}$Co$_{6.2}$Mn$_{37.2}$Sn$_{11.5}$ alloy

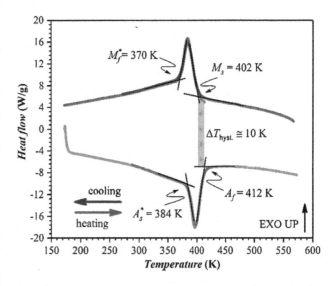

Fig. 5 DSC curves showing the forward and reverse martensitic transformation with the characteristic martensite start (M_s), martensite finish (M_f), austenite start (A_s) and austenite finish (A_f) temperatures for the Ni$_{45.1}$Co$_{6.2}$Mn$_{37.2}$Sn$_{11.5}$ heat treated alloy

exo- and endothermic peaks related to the forward and reverse MPT. The temperatures are as follows: $M_s = 402$ K, $M_f^* = 370$ K, $A_s^* = 384$ K and $A_f = 412$ K. The transformation hysteresis given as $\Delta T_{hyst.} = A_f - M_s$ is small and yields 10 K, consistent with Ref. [6]. The forward MPT peak temperature is equal to $T_{pA \to M} = 384.4$ K and the reverse MPT peaks at $T_{pM \to A} = 395.5$ K. The transformational entropy change (ΔS_{MPT}) determined as $\Delta S_{MPT} = L/T_{pA \leftrightarrow M}$, where L is the latent heat of the transformation and $T_{pA \leftrightarrow M}$ denotes the forward or reverse peak temperature, brings the following values $\Delta S_{MPT}^{A \to M} = 35.6$ J/kg K for the forward MPT and $\Delta S_{MPT}^{M \to A} = 30.9$ J/kg K for the reverse transition. The difference in the ΔS_{MPT} between the forward and reverse MPT can be understood in connection to the first-order nature of the martensitic transformation and the relative proximity of the M_s and A_f temperatures to the Curie transition of austenite (T_C^A), which then impacts the magnetic contribution ($\Delta S_{mag} > 0$) to the overall ΔS_{MPT} (<0) [15]. In fact, the T_C^A is not explicit on the DSC (Fig. 3) suggesting that most likely it takes place within the structural transformation interval and effectively is obscured by the exo- and endothermic peaks. According to [1] the T_C^A of $Ni_{43}Co_7Mn_{39}Sn_{11}$ alloy ($e/a = 8.1$) is found at 400 K while its $M_s \sim 335$ K, which permits good resolution between both transitions. In the present case the structural and magnetic transformations overlap due to the increase in the MPT temperature brought about by the increase in electron-to-atom ratio, $e/a = 8.13$, and the larger grain size stemming from the solidification conditions relative to [1], which then overall promotes the martensitic transformation while the T_C^A is less sensitive to both factors and henceforth is assumed not to have deviated profoundly from 400 K.

Interestingly, the determined characteristic A_s temperature contrasts with the room-temperature alloy's microstructure (Fig. 1c, d). The latter supplies stark evidence for the presence of austenite at temperatures well below $\ll A_s$. Careful reexamination of the DSC curves (Fig. 5), however, indicates that both forward and reverse MPT peaks tail off below M_f and A_s, suggesting that the transition is not complete at these temperatures and continues to about 200 K, hence, asterisk near A_s and M_f in Fig. 5. This supposition is supported by the magnetic susceptibility vs. temperature measurements (Fig. 6). The restricted experimental measurement range offered by the VSM instrument failed to capture the onset of the MPT, and hence Fig. 6 shows the zero field cooled (ZFC), field cooled (FC), and field heated (FH) curves tracing the progress of the transformation (up to 370 K) once it had already commenced or had not yet terminated, depending on the cooling or heating programme. The divergence between the FC and FH curves within the temperature range 370–200 K indicates the typical fingerprint of hysteresis for MPT in agreement with tailing in

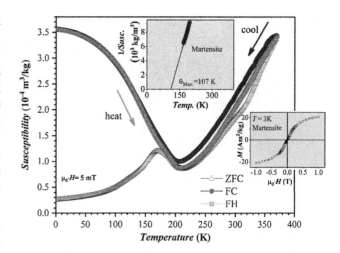

Fig. 6 Magnetic susceptibility versus temperature measured while zero field cooling (ZFC), field heating (FH), and field cooling (FC) the $Ni_{45.1}Co_{6.2}Mn_{37.2}Sn_{11.5}$ alloy. The top inset (grey) shows the inverse magnetic susceptibility versus temperature in the 196–168 K temperature interval. The bottom right inset (yellow) shows the magnetization versus magnetic field dependence ($-1 \leq \mu_0 \cdot H \geq 1$) measured at 3 K

Fig. 5. It is seen that with decreasing temperature, the magnetization decreases yet seemingly it does not go to zero, pointing to a non-paramagnetic state of martensite. Below 200 K the bifurcation between the ZFC and FC curves announced more complex magnetic interactions, often of antiferromagnetic and ferromagnetic components. The Curie temperature of the martensite phase (T_C^M), determined as the inflection point from the FC curve, is equal to $T_C^M = 147$ K. The top inset in Fig. 6 provides the inverse magnetic susceptibility vs. temperature dependence within the linear 168–196 K temperature range, which is extrapolated to zero in order to determine the paramagnetic Curie temperature of martensite ($\theta_{Mart.}$); it yielded $\theta_{Mart.} = 107$ K. The effective magnetic moment (μ_{eff}) has been obtained employing the $\mu_{eff}^2 = \frac{3Mk_B}{\mu_0 N_A b_{mas}}$, where M is the molecular mass, k_B is Boltzmann's constant, μ_0 is the permeability of vacuum, N_A is Avogadro's number and b_{mas} is the slope of the linear fit. The results produced $\mu_{eff} = 22 \pm 2.9$ μ_B f.u., which indicates complex magnetic interaction (bottom right inset in Fig. 6) discussed elsewhere [4].

Summary

Following directional solidification, the Ni–Co–Mn–Sn alloy shows a complex microstructure. The chemical homogeneity is promoted by the thermal annealing at 1220 K for 72 h. The average composition of the alloy is established as $Ni_{45.1}Co_{6.2}Mn_{37.2}Sn_{11.5}$ and at room temperature its microstructure is a mixture of austenite (L2$_1$) and martensite phases (6M, 4O). The martensite phase preferentially appears at grain boundaries. The martensitic transformation extends to low temperatures as the remaining inner grain areas transform to martensite. The martensite phase displays a complex magnetic microstructure.

Acknowledgements The authors would like to gratefully acknowledge financial support from the Polish National Science Centre for Research and Development (Project number: PBS/A5/36/2013).

References

1. Kainuma R, Imano Y, Ito W, Morito H, Sutou Y, Oikawa K, Fujita A, Ishida K, Okamoto S, Kitakami O, Kanomata T (2006) Metamagnetic shape memory effect in a Heusler-type $Ni_{43}Co_7Mn_{39}Sn_{11}$ polycrystalline alloy. Appl Phys Lett 88:192513-1-3

2. Krenke T, Duman E, Acet M, Moya X, Manosa L, Planes A (2007) Effect of Co and Fe on the inverse magnetocaloric properties of Ni–Mn–Sn. J Appl Phys 102:033903

3. Bruno NM, Yegin C, Karaman I, Chen J-H, Ross JH Jr, Liu J, Li J (2014) The effect of heat treatments on Ni43Mn42Co4Sn11 meta-magnetic shape memory alloys for magnetic refrigeration. Acta Mater 74:66–84

4. Cong DY, Roth S, Liu J, Luo Q, Potschke M, Hurrich C, Schultz L (2010) Superparamagnetic and superspin glass behaviors in the martensitic state of $Ni_{43.5}Co_{6.5}Mn_{39}Sn_{11}$ magnetic shape memory alloy. Appl Phys Lett 96:112504-1-3

5. Umetsu RY, Ito K, Ito W, Koyama K, Kanomata T, Ishida K, Kainuma R (2011) Kinetic arrest behavior in martensitic transformation of NiCoMnSn metamagnetic shape memory alloy. J Alloys Compd 509:1389–1393

6. Srivastava V, Chen X, James RD (2010) Hysteresis and unusual magnetic properties in the singular Heusler alloy $Ni_{45}Co_5Mn_{40}Sn_{10}$. Appl Phys Lett 97:014101-1-3

7. Khovalyo V, Koledov V, Shavrov V, Ohtsuka M, Miki H, Takagi T, Novosad V (2008) Influence of cobalt on phase transitions in $Ni_{50}Mn_{37}Sn_{13}$. Mater Sci Eng A 481–482:322–325

8. Ito W, Xu X, Umetsu RY, Kanomata T, Ishida K, Kainuma R (2010) Concentration dependance of magnetic moment in $Ni_{50-x}Co_xMn_{50-y}Z_y$ (Z = In, Sn) Heusler alloys. Appl Phys Lett 97:242512-1-3

9. Perez-Sierra AM, Pons J, Santamarta R, Vermaut P, Ochin P (2015) Solidification process and effect of thermal treatments on Ni–Co–Mn–Sn metamagnetic shape memory alloys. Acta Mater 93:164–174

10. Wójcik A, Maziarz W, Szczerba MJ, Sikora M, Hawełek Ł, Czaja P (2016) Influence of Fe addition on the martensitic transformation, structure and magnetic properties of metamagnetic Ni–Co–Mn–Sn alloys. Acta Phys Pol A 130:1026–1028

11. Czaja P, Chulist R, Szczerba MJ, Przewoźnik J, Olejnik E, Chrobak A, Maziarz W, Cesari E (2016) Magnetostructural transition and magnetocaloric effect in highly textured Ni–Mn–Sn alloy. J Appl Phys 119:165102-1-6

12. Czaja P, Szczerba MJ, Chulist R, Bałanda M, Przewoźnik J, Chumlyakov YI, Schell N, Kapusta CZ, Maziarz W (2016) Acta Mater 118:213–220

13. Schlagel DL, McCallum RW, Lograsso TA (2008) Influence of solidification microstructure on the magnetic properties of Ni–Mn–Sn Heusler alloys. J Alloys Compd 463:38–46

14. Passamani EC, Xavier F, Favre-Nicolin E, Larica C, Takeuchi AY, Castro IL, Proveti JR (2009) Magnetic properties of NiMn-based Heusler alloys influenced by Fe atoms replacing Mn. J Appl Phys 105:033919-1-8

15. Segui C, Cesari E (2012) Composition and atomic order effects on the structural and magnetic transformations in ferromagnetic Ni–Co–Mn–Ga shape memory alloys. J Appl Phys 111:043914-1-7

Effect of Heat-treatment Conditions on Microstructure and Shape Memory Properties of Ti–4.5Al–3V–2Fe–2Mo Alloy

Yuichi Matsuki, Hirobumi Tobe, and Eiichi Sato

Abstract

The effect of heat treatment conditions on microstructure and superelasticity of Ti–4.5Al–3V–2Fe–2Mo (mass%) alloy was investigated by scanning electron microscopy with electron backscattered diffraction (SEM/EBSD) and loading-unloading tensile tests. The presence or absence of solution treatment (ST) at 1223 K before heat treatment at 1073 K was considered. The SEM analysis showed that the specimen without solution treatment had globular α and β grains while the solution-treated specimen had needle-shaped α in coarsened β grains. The loading-unloading tensile tests revealed that the stress for inducing martensitic transformation and shape recovery strain were almost the same in both the specimens, although the grain sizes of β austenite were significantly different.

Keywords

Superelasticity • Alpha + beta titanium alloy
SP-700 • Texture • Heat-treatment

Y. Matsuki · H. Tobe (✉) · E. Sato
Institute of Space and Astronautical Science, Japan Aerospace Exploration Agency, 3-1-1 Yoshinodai, Chuo, Sagamihara, 252-5210, Japan
e-mail: tobe@isas.jaxa.jp

Y. Matsuki
e-mail: matsuki.yuichi@ac.jaxa.jp

E. Sato
e-mail: sato@isas.jaxa.jp

Y. Matsuki
Department of Materials Engineering, Graduate School of Engineering, The University of Tokyo, 7-3-1 Hongo, Bunkyo, Tokyo, 113-8656, Japan

Introduction

Ti–4.5Al–3V–2Fe–2Mo (mass%) alloy, commercially called SP-700, has been widely utilized in aerospace engineering as a structural material. SP-700 has superior cold workability and superplasticity to Ti–6Al–4V [1, 2], and has been adopted for spherical gas tank liners fabricated by superplastic blow forming and subsequent welding for rockets at ISAS/JAXA [3]. Recently, our research group revealed that superelasticity (SE) and shape memory effect (SME) can be induced in SP-700 after heat treatment at 1073 K and 1098 K for 10.8 ks, respectively, followed by quenching [4, 5]. The mechanism is as follows: Heat treatment at high temperature over 1023 K on the commercial SP-700 reduces the volume fraction of α phase. Since β stabilizers such as V, Fe, and Mo are rich in β phase, α phase reduction dilutes β stabilizers in β phase and consequently β phase is less stabilized. Then stress-induced martensitic transformation from β (bcc) to α″ (orthorhombic) occurs before slip deformation under loading. SE is attained at room temperature (RT) when the austenite transformation start temperature (A_s) is below RT. A much higher heat treatment temperature further increases A_s above RT and retains α″ martensite at RT, resulting in SME.

Unlike NiTi shape memory alloys, SP-700 is available as commercial, large standard-sized plates (∼1.5 m × 2.5 m) because of the good workability. Therefore, coexistence of cold workability, superplasticity, and SE or SME in large SP-700 plates is expected to enable novel SE or SME applications, which cannot be attained in NiTi. At ISAS/JAXA, foldable rocket/satellite parts designed by utilizing SE is under consideration to make an effective use of the limited space in rockets. Such design requires welding processes for manufacturing, though welding creates heterogeneous microstructure that consists of coarsened grains in the welded zone and fine grains in heat-affected zones [6]. Therefore, a solution treatment at above β transus to the whole part, including welded zones, should be considered for microstructure homogenization along with its effect on the SE property. In this study, the effect of heat treatment conditions, that is, solution treatment and subsequent intermediate-temperature heat treatment, on the microstructure and SE properties of SP-700 was investigated.

Experimental

A commercial SP-700 plate 1 mm in thickness supplied by JFE Steel Corporation by hot-rolling and subsequent cold-rolling was used in this study. The chemical composition of the as-received plate is shown in Table 1. Specimens for microstructural observation and tensile tests were cut using an electro-discharge machine and heat-treated under Ar atmosphere in quartz tubes followed by quenching into water by breaking the tubes. The heat treatment consists of two steps: First, presence or absence of ST at 1223 K for 10.8 ks and, second, subsequent heat treatment at 1073 K for 10.8 ks for adjusting α/β volume fraction ratio to induce SE. The microstructure, including texture of β, was analyzed by SEM/EBSD. Loading-unloading tensile tests were carried out at RT along rolling direction (RD) at a strain rate of 2.5×10^{-4} s^{-1} with tensile strain up to 3%. The gage length of the specimens was 60 mm.

Results and Discussion

Effect of Heat-Treatment Condition on Microstructure

Figure 1 shows SEM images of the as-received specimen and the specimens heat-treated under the conditions of 1073 K-10.8 ks and 1223 K-10.8 ks (ST) + 1073 K-10.8 ks. The two heat-treated specimens have almost the same α/β volume fraction, which is smaller than that of the as-received specimen and consistent with the reported α/β volume fraction of SP-700 heat-treated at 1073 K followed by quenching [2, 5]. Therefore, the duration of 10.8 ks at 1073 K is enough for the specimens to reach the phase equilibrium.

The 1073 K-10.8 ks specimen had globular α and β grains of about 3 μm in diameter. On the other hand, the ST + 1073 K-10.8 ks specimen had fine, needle-shaped α with 1 μm thickness, which precipitated in coarsened β

grains of about 500 μm diameter with the Burgers orientation relationship $(\{110\}_\beta//(0001)_\alpha, \langle 111 \rangle_\beta//[11\overline{2}0]_\alpha)$. After ST, the α phase in the as-received material totally disappeared during solution treatment and β phase transformed to twinned α'' martensite of 100% volume fraction during subsequent quenching [4]. In the ST + 1073 K-10.8 ks specimen, heat treatment at 1073 K decomposed α'' phase into $\alpha + \beta$ phase, and needle-shaped α was considered to be formed on the twin boundaries of α'' martensite.

Figure 2 shows the inverse pole figures (IPFs) of the β austenite for the RD and the normal direction (ND) of the rolling plane in the specimens before and after heat treatments. The IPFs in the as-received specimen show high axis densities at around $\langle 035 \rangle_\beta$ along the RD and $\langle 001 \rangle_\beta$ along the ND, respectively, indicating that the main component of the texture is $\{001\}_\beta \langle 530 \rangle_\beta$. A strong $\{001\}_\beta \langle 110 \rangle_\beta$ texture was formed by heat treatment at 1073 K for 10.8 ks, while a strong $\{113\}_\beta \langle 110 \rangle_\beta$ recrystallization texture was observed in the ST + 1073 K-10.8 ks specimen. Although textures are different for the heat-treated specimens, the crystal orientation along the RD is the same ($\langle 110 \rangle_\beta$). Similar $\{001\}_\beta \langle 110 \rangle_\beta$ deformation texture and $\{112\}_\beta \langle 110 \rangle_\beta$ recrystallization texture have been reported in metastable β-type titanium alloys [7].

Effect of Heat Treatment Condition on Superelasticity

Figure 3 shows the stress-strain curves of the specimens obtained by loading-unloading tensile tests with the tensile strain up to 3% at RT. The yield stress corresponds to stress-induced martensitic transformation. The stress for inducing martensitic transformation (σ_{SIM}: ~670 MPa) and shape recovery strains, including elastic and SE strains (~2%), were almost the same in both the heat-treated specimens. The σ_{SIM} and recovery strain are known to depend on the chemical composition related to β stability and lattice parameters of α'' martensite, crystal orientation, and grain size of β austenite. It is inferred that the chemical composition of β austenite is the same in both the heat-treated specimens because the α/β volume fraction ratios are almost the same. The heat-treated specimens have the same β crystal orientation of $\langle 110 \rangle_\beta$ in the RD with nearly the same texture intensity. Therefore, the effects of the chemical composition and texture on the σ_{SIM} and recovery strain are not significant.

Table 1 Chemical composition of the as-received SP-700 (mass%)

Ti	Al	V	Fe	Mo	O	C
Bal.	4.4	3.0	2.0	2.0	0.09	0.01

Fig. 1 SEM images of the as-received and heat-treated specimens (Bright: β Dark: α)

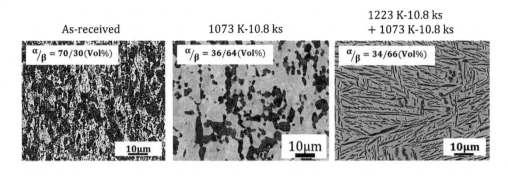

Fig. 2 IPFs of β austenite obtained before and after heat treatments

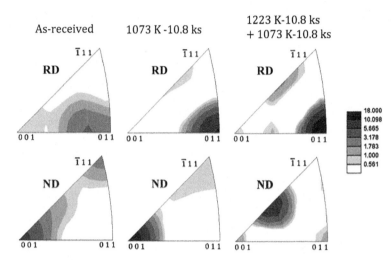

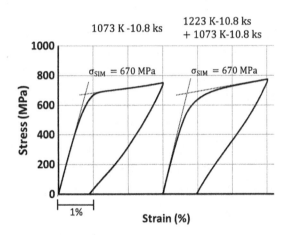

Fig. 3 Stress-strain curves of the heat-treated specimens loaded along RD

In cold-rolled metastable β-type titanium alloys, it has been reported that specimens solution-treated at around 1173 K tend to have inferior SE to ones heat-treated at an intermediate temperature of around 873 K due to larger grain sizes and lower dislocation densities [8]. Increasing grain sizes reduces the restriction of grain boundaries to martensitic transformation and increases transformation temperatures, resulting in a decrease in SE recovery strain. In addition, the strength of β austenite decreases with

increasing grain size, and dislocations are easily induced during stress-induced martensitic transformation, which also decreases SE recovery strain. However, shape recovery strain was retained in SP-700 subjected to solution treatment and subsequent heat treatment, although β austenite grains became large (~ 500 μm), as shown in Fig. 1. This is because needle-shaped α phase that precipitated in β grains contributed to suppress the increase in transformation temperatures and the introduction of dislocations by precipitation hardening.

Conclusions

In this study, the effect of heat treatment conditions on microstructure and superelasticity of Ti–4.5Al–3V–2Fe–2Mo (mass%) alloy was investigated. It was revealed that the specimen heat-treated at 1073 K for 10.8 ks had globular α and β grains and $\{001\}_{\beta}\langle110\rangle_{\beta}$ deformation texture, while the specimen solution-treated at 1223 K for 10.8 ks followed by heat treatment at 1073 K for 10.8 ks had needle-shaped α in coarsened β grains and $\{113\}_{\beta}\langle110\rangle_{\beta}$ recrystallization texture. The needle-shaped α restricted the size of the transformation regions in β austenite and also increased β strength by precipitation hardening, which suppressed the decrease in the stress for inducing martensitic transformation and shape recovery strain. Therefore, almost the same superelastic properties were obtained in both the heat-treated specimens along the rolling direction.

References

1. Ogawa A, Fukai H, Minakawa K, Ouchi C (1993) Beta titanium alloys in the 1990's. The Minerals, Metals & Materials Society, pp 513–524

2. Ishikawa M, Kuboyama O, Niikura M, Ouchi C (1993) Titanium'92 science and technology. The Minerals & materials Society, pp 141–148

3. Sato E, Sawai S, Uesugi K et al (2007) Mater Sci Forum 551–552:43–48

4. Tobe H, Sato E (2016) Proceedings of the 13th world conference on titanium. The Minerals, Metals & Materials Society, pp 1577–1580

5. Tobe H, Sato E (2016) J Jpn Inst Light Met 66:174–179

6. Chung W-C, Tsay L-W, Chen C (2009) Mater Trans 50:544–550

7. Kim HY, Sasaki T, Okutsu K, Kim JI, Inamura T, Hosoda H, Miyazaki S (2006) Acta Mater 54:423–433

8. Kim HY, Ikehara Y, Kim JI, Hosoda H, Miyazaki S (2006) Acta Mater 54:2419–2429

Porous Ni–Ti–Nb Shape Memory Alloys with Tunable Damping Performance Controlled by Martensitic Transformation

Shanshan Cao, Yuan-Yuan Li, Cai-You Zeng, and Xin-Ping Zhang

Abstract

Porous Ni–Ti–Nb shape memory alloys (SMAs) with designed porosities and compositions were prepared by the powder metallurgy technique. The phase-transformation-controlled damping behavior of the porous alloys was investigated by dynamic mechanical analysis (DMA). Systematic microstructural study indicates that both the pore configuration and Nb distribution in the matrix of porous Ni–Ti–Nb alloys have significant influence on the damping performance of the alloys. Increase in both pore size and porosity in micro-scale leads to obvious decrease of the internal friction, while addition of Nb brings dramatic increase in the damping capacity of the porous alloys. The damping capacity can be optimized by adjusting the Nb/NiTi ratio, which balances the competitive contribution of Nb/matrix and B2/B19′ interfaces to the internal friction during martensitic transformation. Moreover, β-Nb phase of lamellar structure in the matrix plays a greater role in the internal friction than that of granular shape by offering a large amount of interfaces.

Keywords

Porous Ni–Ti–Nb alloy • Martensitic transformation
Damping capacity

Introduction

Ni–Ti based SMAs have been widely studied and applied in multifunctional fields due to their outstanding properties of shape memory effect (SME) and superelasticity originating from the displacive martensitic transformation [1], which also yields unique damping performance [2–5]. The high damping capacity together with excellent mechanical properties and high corrosion resistance of these alloys enable them to be employed in energy absorption devices for civil constructions requiring high strength and durability [6, 7], which cannot be offered by traditional damping materials such as polymers and rubbers. Ternary Ni–Ti–Nb alloys, as typical Ni–Ti based SMAs well known for their wide martensitic transformation hysteresis, have been reported to have high damping capacity controlled by the martensitic transformation over a wide temperature range due to the large transformation hysteresis [8–11]. Actually, Ni–Ti–Nb alloys can be taken as the composites of the NiTi (Nb) matrix and β-Nb secondary phase [12]. The interfaces between the β-Nb particles and the matrix not only offer preferential sites for martensite nucleation, but also introduce a large number of boundaries during martensite propagating, and thus improve the damping capacity of the alloys [9]. On the other hand, in the past decades porous SMAs have drawn increasing interest for their low density, relatively low modulus and high permeability. These unique features associated with the porous structure can better fulfill the requirements of light-weight and mass transportation in the applications of aerospace and biomedicine [13]. Moreover, the martensitic-transformation-controlled damping capacity of the porous SMAs can also be tailored by their pore structure, and thus further broaden the application of the alloys as smart dampeners [14]. As for porous Ni–Ti–Nb SMAs, so far only limited attempts have been made to investigate the martensitic transformation behavior and mechanical properties of the alloys fabricated by various powder metallurgy methods [15, 16], and damping performance of submicron porous Ni–Ti–Nb alloy with restricted dimension prepared by dealloying [17]. The martensitic-transformation-controlled damping performance tuned by composition and pore structure of these alloys is still in need of systematic studies.

In the present work, porous Ni–Ti–Nb alloys of different nominal compositions and porosities were fabricated by the

S. Cao (✉) · Y. Y. Li · C. Y. Zeng · X. P. Zhang (✉)
School of Materials Science and Engineering, South China University of Technology, 510640 Guangzhou, China
e-mail: msscao@scut.edu.cn

X. P. Zhang
e-mail: mexzhang@scut.edu.cn

© The Minerals, Metals & Materials Society 2018
A. P. Stebner and G. B. Olson (eds.), *Proceedings of the International Conference on Martensitic Transformations: Chicago*,
The Minerals, Metals & Materials Series, https://doi.org/10.1007/978-3-319-76968-4_43

space-holder assisted conventional sintering technique. The pore features and microstructural characteristics of the as-sintered alloys were investigated together with evaluation of their damping capacity, so as to reveal the effect of Nb content and pore structure on the martensitic-transformation-controlled damping performance in the porous Ni–Ti–Nb alloys.

Experiments

Nickel powder (61 μm, 99.9% purity) and titanium powder (50 μm, 99.9% purity) with an equiatomic ratio were blended for 24 h, followed by mixing with Nb powder (50 μm, 99.7% purity) of different contents of 5 at.%, 7 at.%, 9 at.% and 15 at. %, respectively, for another 24 h to yield mixtures with nominal compositions of $(NiTi)_{(100-x)/2}Nb_x$ ($x = 0, 5, 7, 9, 15$). Afterwards, NH_4HCO_3 powder (150–200 μm, 99.99% purity), as the space-holder, of different amounts of 5 wt.%, 10 wt. %, 15 wt.% and 20 wt.%, respectively, was added to the above mixed powder with a nominal composition of $(NiTi)_{45.5}Nb_9$ and blended for 8 h. Then, the as-blended powders were cold compacted into green samples with a cuboid geometry of $5 \times 7 \times 25$ (thickness × width × length, mm^3) under a compressive stress of 100 MPa. The green samples were subjected to a gradient sintering in a quartz tube furnace under the protection of flowing argon (99.99% purity), with temperature elevated gradually from room temperature till 1050 °C and held for 3 h, followed by water quenching.

The pore structure and microstructural characteristics of the as-prepared porous Ni–Ti–Nb alloys were characterized by scanning electron microscope (SEM, Nano430, FEI) equipped with an energy-dispersive X-ray spectrometer (EDX, INCAX-act, Oxford). The damping performance of the alloys with specimen geometry of $1.2 \times 4 \times 25$ (thickness × width × length, mm^3) was analyzed by a dynamic mechanical analyzer (DMA, Q800, TA) using the single-cantilever mode at 10 μm amplitude and 0.2 Hz frequency in the temperature range of −120 °C to 120 °C with a heating/cooling rate of 5 °C/min.

Results and Discussion

Typical backscattered electron (BSE) images of the as-sintered $(NiTi)_{(100-x)/2}Nb_x$ alloys are shown in Fig. 1, in which their pore features and microstructure are clearly seen, and the phases are presented in different contrast levels, analyzed with the aid of EDX.

It can be found that dispersively distributed dark grey $NiTi_2$ particles exist in all as-sintered porous alloys, as typically shown in Fig. 1a. The Ti depletion induced by these particles finally yields a Ni-rich B2 matrix showing

light grey contrast. Since the solubility of Nb in NiTi is quite limited [10], Nb exists mainly in the form of β-Nb precipitates in the matrix exhibiting the brightest contrast as shown in Fig. 1b–d. For the porous Ni–Ti–Nb alloys with relatively low Nb content, most β-Nb precipitates form a eutectic structure with the NiTi B2 matrix and thus show a fine lamellar shape, such as the porous $(NiTi)_{46.5}Nb_7$ alloy presented in Fig. 1b. Figure 2 shows the details of β-Nb phase distribution in such a eutectic structure with quantitative morphological features. The lamella-shaped β-Nb precipitates with an average thickness of 190 ± 163 nm and an average interspace of 270 ± 238 nm separate the B2 matrix into small channels and thus introduce a large number of boundaries in the matrix due to the large specific area. However, as shown in Fig. 1c, with the increase of Nb content, more granula-shaped β-Nb precipitates with a diameter of around 20–30 μm appear in the matrix of the porous $(NiTi)_{45.5}Nb_9$ rather than the fine lamella-shaped ones, which also introduce interfaces in the matrix but certainly with a much smaller amount. On the other hand, according to Fig. 1a–c, the porous alloys without using space-holder contain irregular pores with a diameter of several tens of microns, which is close to the size of the raw powders, while the one shown in Fig. 1d with 20 wt.% space-holder contains not only micro-sized pores, but also much larger ones with average pore size of around 200 μm. Moreover, according to the data in Table 1, the porosity of the porous $(NiTi)_{45.5}Nb_9$ alloys measured by the relative density method obviously increases with the fraction of the space-holder.

Figure 3 shows the curves of damping capacity *versus* temperature for porous Ni–Ti–Nb alloys with different nominal compositions and space-holder fractions. Damping peaks upon cooling and heating can be observed in the damping curves of all porous alloys, which can be attributed to the martensitic transformation of the alloys. It is worth noting that the martensitic transformation of the porous alloys is not complete due to the limited cooling capacity of the instrument (DMA). Therefore, the peak temperatures upon cooling (M_p) and heating (A_p) are selected as the characteristic temperatures to describe the phase transformation behavior and damping performance of the alloys. According to Fig. 3a, porous Ni–Ti–Nb alloys with a nominal Nb content lower than 9 at.% shows a M_p close to −100 °C, while a higher Nb content, i.e., 15 at.%, lowers M_p to −112 °C. This can be explained by the fact that the preference of solute Nb for the Ti sublattice increases the Ni/Ti ratio and thus suppresses the martensitic transformation [8]. Besides, the porous $Ni_{50}Ti_{50}$ alloy has an A_p of −75 °C and thus yields a hysteresis (A_p–M_p) of 27 °C, while the addition of Nb leads to a shift of A_p of porous Ni–Ti–Nb alloys to the temperature range of 0 °C to −20 °C, and thus causes a much wider hysteresis of around 100 °C together with broader damping peaks. This can be

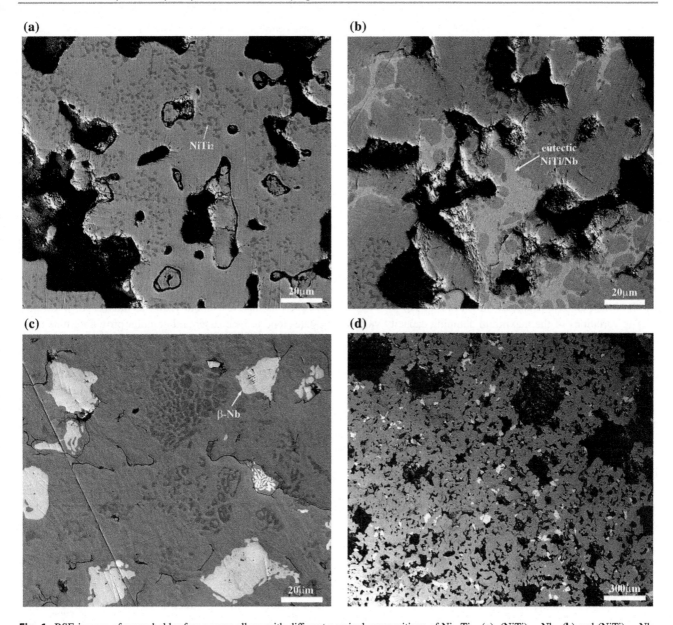

Fig. 1 BSE images of space-holder-free porous alloys with different nominal compositions of $Ni_{50}Ti_{50}$ (**a**), $(NiTi)_{46.5}Nb_7$ (**b**) and $(NiTi)_{45.5}Nb_9$ (**c**), together with a porous $(NiTi)_{45.5}Nb_9$ alloy with 20 wt.% space-holder (**d**)

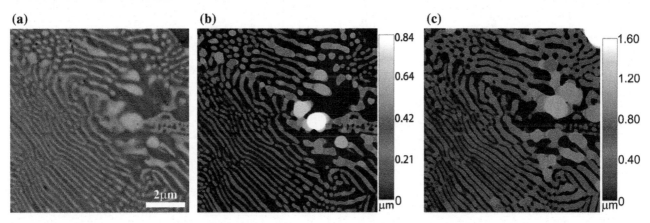

Fig. 2 The detailed view (**a**), thickness (**b**) and interspace (**c**) maps of the lamella-shaped Nb-rich phase in the matrix of the porous $Ni_{46.5}Ti_{46.5}Nb_7$ alloy

Table 1 Apparent density and porosity of porous $(NiTi)_{45.5}Nb_9$ alloys with different fractions of space-holder

NH_4HCO_3 (wt.%)	Apparent density (g/cm^3)	Porosity (%)
0	4.32	36.8
5	3.67	45.1
10	3.48	47.9
15	3.43	48.6
20	2.99	55.2

attributed to the β-Nb precipitates which impede the martensitic transformation and stabilize the martensite during the reverse transformation, leading to large transformation hysteresis of the porous Ni–Ti–Nb alloys [18]. On the other hand, use of the space-holder in the fabrication of the porous Ni–Ti–Nb alloys has little influence on the temperature and width of the damping peaks, as shown in Fig. 3b.

The variations of the maximum damping capacity values upon cooling and heating with Nb content and porosity are shown in Fig. 4. Clearly, the addition of Nb to porous NiTi alloys obviously improves the damping capacity with enhanced damping peaks during phase transformation, as shown in Fig. 4a. The maximum of the optimal damping capacity brought by a Nb content of 7 at.% is 5 times that of the binary $Ni_{50}Ti_{50}$ alloy upon cooling, and 8 times upon heating. Moreover, the maximum damping capacity values firstly increase significantly, and then tend to decrease with the Nb content of the porous Ni–Ti–Nb alloys. This can be attributed to the competitive contribution of the Nb/matrix and B2/B19′ interfaces to the internal friction during martensitic transformation, in addition to the morphological

characteristics of β-Nb related to Nb content. When the Nb content is relatively low, increasing Nb content leads to increased amount of fine lamella-shaped β-Nb precipitates, which introduce a large number of interfaces yielding strong internal friction with B2 and B19′ matrix during martensitic transformation, and thus increase the damping capacity. However, as the Nb content is further increased, β-Nb precipitates tend to be coarser and granular providing much less interfaces in the matrix. The interfaces introduced by these particles can no longer compensate the loss of B2/B19′ interfaces in the matrix, and thus lead to decreased damping capacity during martensitic transformation. Moreover, as shown in Fig. 4b, the maximum damping capacity values of the porous Ni–Ti–Nb alloys upon cooling and heating tend to decrease with the increase of the porosity, meaning that the pores of irregular shape in micro- or even submillimeter scale have negative effect on the damping capacity of the alloys. This can be understood from the fact that the limited interfaces provided by the relatively large pores cannot cover the loss of the B2/B19′ interfaces in the matrix during martensitic transformation.

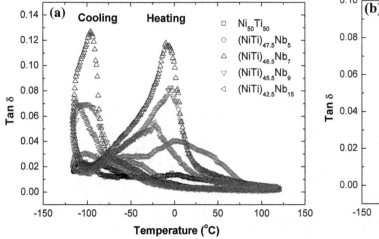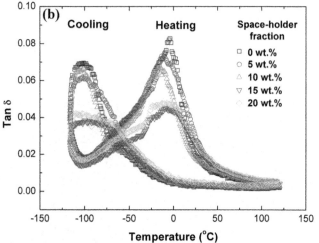

Fig. 3 Damping capacity of porous Ni–Ti–Nb alloys of different compositions (**a**) and $(NiTi)_{45.5}Nb_9$ alloys with different fractions of space-holder (**b**) as a function of temperature

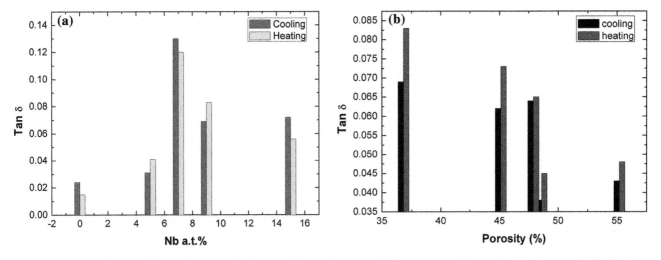

Fig. 4 Variations of the maximum damping capacity upon cooling and heating with Nb content (**a**) and porosity (**b**) of porous Ni–Ti–Nb alloys

Conclusion

The damping capacity of porous Ni–Ti–Nb alloys fabricated by powder metallurgy technique can be tuned by the composition and pore structure of the alloys. Porous Ni–Ti–Nb alloys exhibit outstanding damping performance controlled by martensitic transformation with much higher damping peaks and broader temperature range than that of the binary Ni–Ti ones. The competitive contribution of Nb/matrix and B2/B19′ interfaces to the internal friction leads to optimal damping capacity brought by the porous alloy with a nominal composition of $Ni_{46.5}Ti_{46.5}Nb_7$. Fine lamella-shaped β-Nb precipitates in porous alloys with relatively low Nb content provide a much larger number of interfaces than the coarser and granular ones induced by high Nb content, and thus make greater contribution to the damping capacity. Increasing pore size and porosity by using micro-sized space-holder in fabrication of the porous alloys leads to obvious decrease of damping capacity due to the loss of the B2/B19′ interfaces in the matrix during martensitic transformation.

Acknowledgements This research was supported by the National Natural Science Foundation of China under Grant Nos. 51401081 and 51571092, and Key Project Program of Guangdong Provincial Natural Science Foundation under Grant No. S2013020012805.

References

1. Otsuka K, Ren X (2005) Physical metallurgy of Ti–Ni-based shape memory alloys. Prog Mater Sci 50:511–678
2. Van Humbeeck J, Stoiber J, Delaey L, Gotthardt R (1995) The high damping capacity of shape memory alloys. Z Metallkd 86:176-183
3. Yoshida I, Monma D, Iino K, Otsuka K, Asai M, Tsuzuki H (2003) Damping properties of $Ti_{50}Ni_{50-x}Cu_x$ alloys utilizing martensitic transformation. J Alloys Compd 355(1):79–84
4. Wu SK, Lin HC (2003) Damping characteristics of TiNi binary and ternary shape memory alloys. J Alloy Compd 355(1):72–78
5. Van Humbeeck J (2003) Damping capacity of thermoelastic martensite in shape memory alloys. J Alloys Compd 355(1):58–64
6. Dolce M, Cardone D (2001) Mechanical behavior of shape memory alloys for seismic applications: 1. Martensite and austenite NiTi bars subjected to torsion. Int J Mech Sci 43(11):2631–2656
7. Golovin IS, Sinning HR (2004) Internal friction in metallic foams and some related cellular structures. Mater Sci Eng A 370(1):504–511
8. Piao M, Miyazaki S, Otsuka K (1992) Characteristics of deformation and transformation in $Ti_{44}Ni_{47}Nb_9$ shape memory alloy. Mater Trans 33:346–353
9. Cai W, Lu XL, Zhao LC (2005) Damping behavior of TiNi-based shape memory alloys. Mater Sci Eng A 394(1):78–82
10. Xiao F, Zhao XQ, Xu HB, Jiang HC, Rong LJ (2009) Damping capacity and mechanical property of NiTiNb shape memory alloys 45(1):18–24
11. Bao Z, Guo S, Xiao F, Zhao XQ (2011) Development of NiTiNb in-situ composite with high damping capacity and high yield strength. Prog Nat Sci Mater Inter 21:293–300
12. Zhang CS, Zhao LC, Duerig TW, Wayman CM (1990) Effects of deformation on the transformation hysteresis and shape memory effect in a $Ni_{47}Ti_{44}Nb_9$ alloy. Scripta Mater 24:1807–1812
13. Bansiddhi A, Sargeant TD, Stupp SI, Dunand DC (2008) Porous NiTi for bone implants: a review. Acta Biomater 4:773–782
14. Zhang YP, Zhang XP (2009) Internal friction behaviors of porous NiTi alloys with variable porosities. Chin J Nonferrous Met 19 (10):1872–1879
15. Bansiddhi A, Dunand DC (2009) Shape-memory NiTi-Nb foams. J Mater Res 24(6):2107–2117
16. Bansiddhi A, Dunand DC (2011) Niobium wires as space holder and sintering aid for porous NiTi. Adv Eng Mater 13(4):301–305
17. Guo W, Kato H (2015) Submicron-porous NiTi and NiTiNb shape memory alloys with high damping capacity fabricated by a new top-down process. Mater Des 78:74–79
18. Shi H, Pourbabak S, Van Humbeeck J, Schryvers D (2012) Electron microscopy study of Nb-rich nanoprecipitates in Ni-Ti-Nb and their influence on the martensitic transformation. Scripta Mater 67(12):939–942

Cryogenic Immersion Time Influence on Thermal and Mechanical Properties of a Ni$_{48}$-Ti$_{52}$ Shape Memory Alloy

Bartholomeu Ferreira da Cruz Filho

Abstract

Smart Memory Alloys (SMA) have great potential for application in various engineering and medicine fields. In applications in which these materials are used as actuators, there is a need for improvement of their properties such as resistance to thermomechanical fatigue and structural fatigue. This work investigates the potential of cryogenic treatment to change properties of SMA. Deep cryogenic treatment has been used for decades to enhance material properties in steels, such as the wear resistance increase in tool steels and fatigue life in ferrous materials. The objective of this study is to investigate the influence of deep cryogenic treatment on mechanical properties (elastic modulus, damping, and stiffness) and thermal properties (phase transformation temperatures and latent heat processing) of Ni$_{48}$-Ti$_{52}$ alloy. In this study, an experimental comparative analysis of these properties was carried out before and after the cryogenic treatment at −196 °C using different immersion times. The test samples were prepared and referred to as NiTi_CR (CPs as received), NiTi_TC12, NiTi_TC18, and NiTi_TC24 (CPs cryogenically treated by immersion −196 °C for 12, 18, and 24 h, respectively). The heating and cooling rate used was 18 °C/h. These thermal properties were measured by Differential Scanning Calorimetry and the mechanical properties by Impulse Excitation. Microstructural analysis was based on optical and electronic microscopy scanning and X-ray diffraction. The results showed that cryogenic treatment affects all the properties investigated, with emphasis on reducing the latent heat of transformation and increasing the damping factor. Microstructural analysis indicates that these changes may be associated with changes in grain size and precipitates.

B. F. da Cruz Filho (✉)
Brasilia, Brazil
e-mail: cruzfilhobf@gmail.com

Keywords

Criogeonic treatment • NiTi alloys • Ni$_{48}$-Ti$_{52}$ alloy Martensitic transformations

Introduction

The strong evidence that the effect of Deep Cryogenic Treatment (DCT) reduces wear on tool steels and increases the life of high-cycle fatigue in ferrous materials [1–3], and the fact that its effect on tool steels and other matter is a source of several papers on the subject [4–6] led the research of the effect of DCT in the thermal and mechanical properties of shape memory alloys (SMA).

Shape memory alloys exhibit the shape memory effect (SME) and/or pseudoelasticity (PE) with resilience of the deformation of about 10% when subjected to a heating, after having been deformed under full martensite transformation temperature (M_f)—(SME); A charging and discharging cycle over the complete austenite transformation temperature (A_f)—(PE).

The multifunctional characteristics of these alloys guarantee a high potential for application in various fields from medicine (with products such as "stents" and orthodontic appliances and guide wires) to space engineering [7].

The technological interest in the dynamic properties of SMA has focused primarily on the NiTi alloy. NiTi alloys demonstrate superior cushioning properties of SMA due to copper [8, 9] and its ability to absorb vibration, which is particularly important with regard to the integrity of mechanical systems. The high-capacity damping presented by SMA is related to the movement of interphase martensitic variants (depending on atomic mobility) and the control of twinning (disagreement on microstructure), which is directly proportional to the dislocation density in the structure [8, 9]. Damping is also dependent on external variables such as the frequency and amplitude of oscillation and some internal variables such as grain size, density of martensitic variants,

© The Minerals, Metals & Materials Society 2018
A. P. Stebner and G. B. Olson (eds.), *Proceedings of the International Conference on Martensitic Transformations: Chicago*, The Minerals, Metals & Materials Series, https://doi.org/10.1007/978-3-319-76968-4_44

and structural defects [9, 10]. SMA is expected in a high damping capacity and low modulus of elasticity in its martensitic state [9]. During the phase transformation occurs the presence of a peak of damping capacity and increased modulus or stiffness [8, 9].

The objective of this study was to evaluate how the time to soak in DCT influences the mechanical properties of Ni-Ti. Watching hardness, elastic modulus and damping of that alloy subjected to cyclic cryogenic treatment of 24 to −196 °C with soaks of 12, 18, and 24 h at a temperature of −196 °C and heating and cooling rates of 18 °C/h.

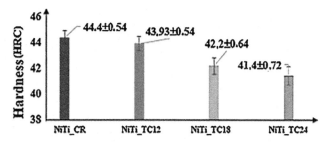

Fig. 1 HRC hardness of samples of Ni_{48}-Ti_{52} with and without DCT

Materials and Experimental Methods

The material investigated was a Ni_{48}-Ti_{52} alloy, manufactured by Minesis. The test specimens for the different tests were drawn from a cylindrical alloy bar, 9.55 mm in diameter and 1500 mm in length, martensitic at room temperature (24 °C). Its chemical composition (Ni 47.72%, Ti 51.86%, Al 0.11%, K 0.08%, S 0.06%, Ca 0.14% and Si 0.03%) was determined by Flurescence X-ray in the Multiuser Laboratory of the Chemistry Institute of UnB.

The methodology used was to submit the Ni_{48}-Ti_{52} test bodies (TC12, TC18 and TC24) to a DCT, starting at ambient temperature up to −196 °C, with a heating and cooling rate of 18 °C/h and soaking by 12, 18 and 24 h, respectively. These Ni_{48}-Ti_{52} test bodies (with and without DCT) were characterized by differential scanning calorimetry (latent heat and phase transformation temperature), mechanically by impulse excitation (modulus of elasticity and damping factor) and by the hardness test (HRC). These obtained properties were compared microstructurally, by scanning electron microscopy (SEM) and X-ray diffraction (XDR).

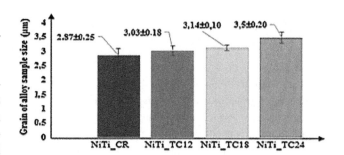

Fig. 2 Grain of alloy sample size Ni_{48}-Ti_{52} with and without DCT

NiTi_TC24) immersed for 12, 18 and 24 h showed percentage reduction (Δ%) of 0.93%, 4.74%, and 9.09%, respectively, in relation to the sample (NiTi_CR) without cryogenic treatment. Considering the margin of error of the equipment being around 1%, the small percentage variation of the NiTi_TC12 sample was considered practically negligible. The (Δ%) percentage reductions of the NiTi_TC18 and NiTi_TC24 samples of 4.74 and 9.09% can be explained by the evolution of the elastic modulus as a function of the temperature of the Ni-Ti alloy [12]. By observing the chemical composition of the material, the type of bond involved and considering the increase of grain size with the immersion time, the bond is weakened [13]. The damping factor of samples NiTi_TC12, NiTi_TC18 and NiTi_TC24 immersed for 12, 18 and 24 h, respectively, presented in Table 1 and Fig. 3, respectively, obtained a percentage increase of 58.94, 127.39 and 199.62%, relative to NiTi_CR, without DCT. It is assumed that internal variables such as grain size increase, variant density and structural defects may have been the cause of increases in DCT damping [10].

Results and Discussion

Influence of DCT on Mechanical Properties

Regarding hardness, NiTi_TC18 and NiTi_TC24 samples presented, respectively, 4.96 and 6.76% hardness reduction in relation to Niti_CR, while NiTi_TC12 CP hardness presented within the uncertainty error of hardness of the CP without DCT (Fig. 1). From this, it is also verified that the reductions of the hardness of the CPs with DCT relate to the grain increases corresponding to the respective immersion times of 12, 18 and 24 h at −196 °C (Fig. 2). Therefore, it can be considered that the Ni_{48}-Ti_{52} alloy hardness can be related to the Hall-Petch equation [11].

With respect to the modulus of elasticity (Table 1 and Fig. 3), DCT samples (NiTi_TC12, NiTi_TC18 and

Influence of DCT on Thermal Properties

Tables 2, 3 and 4 and Figs. 4, 5, 6, 7, 8 and 9 present the results of the DSC assays. From the results of the differential scanning calorimetry (DSC) tests applied on Ni_{48}-Ti52 test bodies, it was observed that the temperatures As, A_f, Ms, the phase transformation hysteresis (*Hist*) and the latent heat of transformation (ΔH_A and ΔH_M) of the cryogenically treated

Tabela 1 Mean mechanical properties of treated and untreated CPs

Corpo de provas	E (GPa)	Δ (%)	ζ_{TF} E-6	Δ (%)
NiTi_CR	56.10137	–	558.941397	–
NiTi_TC12	55.58139	0.93	883.612209	58.94
NiTi_TC18	53.44134	4.74	1270.953177	127.39
NiTi_TC24	51.00128	9.09	1674.684187	199.62

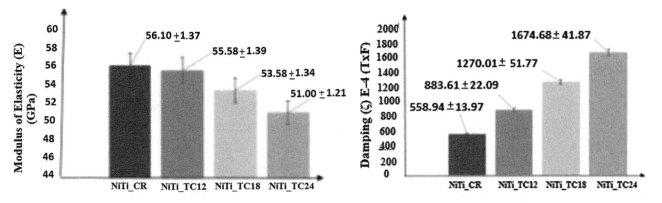

Fig. 3 Modulus of elasticity (E) and the damping Ni_{48}-Ti_{52} samples without and with DCT

Table 2 Phase transformation temperatures of treated and untreated CPs

CP'S	M_S (°C)	P_M (°C)	M_f (°C)	A_S (°C)	P_A (°C)	A_f (°C)
NiTi_CR	62.331	57.303	48.824	57.729	87.744	94.247
NiTi_TC12	64.032	45.023	20.010	30.015	48.024	84.042
NiTi_TC18	66.033	30.015	10.005	22.013	81.341	82.041
NiTi_TC24	70.035	43.422	6.003	20.010	58.329	70.035

Table 3 Hysteresis (Af-Ms) of treated and untreated CPs

CP'S	Hist (°C)	Δ%
NiTi_CR	31.9216	–
NiTi_TC12	20.0010	−37.3
NiTi_TC18	16.008	−49.9
NiTi_TC24	10.005	−68.7

Table 4 Latent head of phase transformation of Ni_{48}-$Ti52$ alloy witout and with DCT

CP'S	ΔH_M (J/g)	Δ%	ΔH_A (J/g)	Δ%
NiTi_CR	−22.57113	–	22.68113	–
NiTi_TC12	−9.77045	−56.71	9.62048	−57.84
NiTi_TC18	−5.37027	−76.21	5.04025	−77.78
NiTi_TC24	−4.96025	−78.02	4.40022	−80.60

bodies presented lower values than those of the untreated bodies, presented, respectively, in Figs. 4, 5, 7, 8 and 9. However, Ms temperatures of the DCT bodies (*NiTi_TC12, NiTi_TC18* and *NiTi_TC24*), shown in Fig. 6, increased respectively by 2.73, 5.94 and 12.36% relative to the body

without DCT. The temperatures As, A_f and M_f may have been reduced due to the presence of Ti_2Ni precipitates, because according to Ishida [14], the temperatures of the forward and reverse transformations in the transformation neighborhoods of the matrix as a whole are altered by the

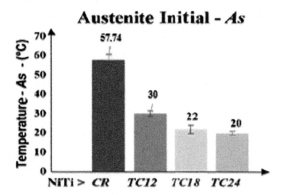

Fig. 4 Initial austenite temperature

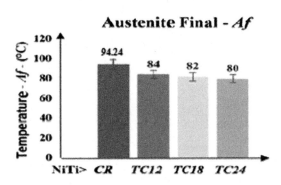

Fig. 5 Final austenite temperature

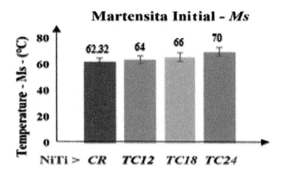

Fig. 6 Initial martensite temperature

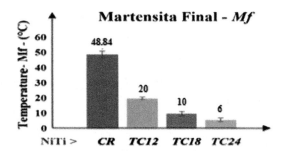

Fig. 7 Final martensite temperature

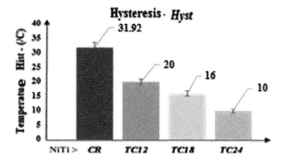

Fig. 8 Evolution of histeresis

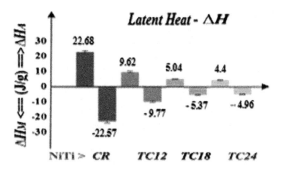

Fig. 9 Evolution of latent heat

presence of Ti_2Ni precipitates. Increases in Ms temperatures may be associated with the increase in grain size of cryogenically treated $CP's$. According to the literature the temperature Ms is affected by the size of the austenitic grain and, in general, the highest temperature Ms, corresponds to the largest austenitic grain size [15–17].

Analysis-Ray Diffraction (XRD)

From this data, diffraction X-ray Ni_{48}-Ti_{52} samples with and without DCT, were constructed with the result of the network parameters the diffractograms of Figs. 10, 11, 12 and 13. It is observed that both $NiTi\_CR$ sample and $NiTi\_TC12$, $NiTi\_TC18$ and $NiTi\_TC24$, showed monoclinic phases B19′, Phase R and Ti_2Ni precipitations at room temperature, a predominant characteristic of the Ni_{48}-Ti_{52} alloy.

Scanning Electron Microscopy (SEM) Analysis

Figures 14, 15, 16 and 17 present the images obtained by SEM. It is observed that all the samples presented martensitic needles, monoclinic phase B19′ and precipitates Ti_2Ni, at room temperature, characteristic of titanium-rich Ni_{48}-Ti_{52} alloy. The presence of precipitates may have caused changes in the properties presented after DCT

Fig. 10 XRD do CP NiTi_CR

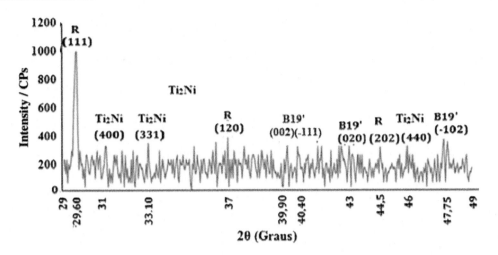

Fig. 11 XRD do CP NiTi_TC12

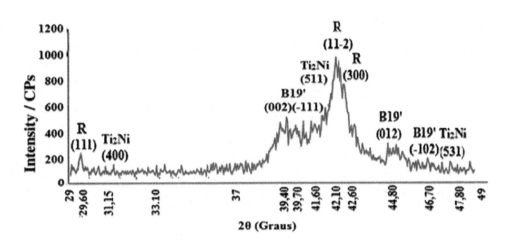

Fig. 12 XRD do CP NiTi_TC18

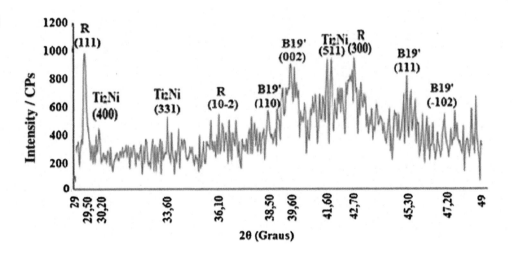

Fig. 13 XRD do CP NiTi_TC24

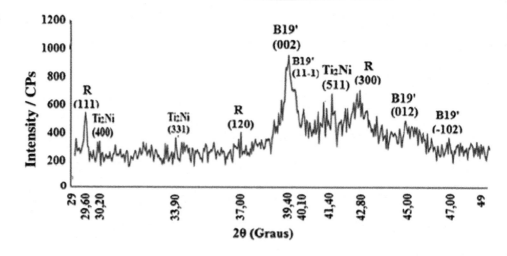

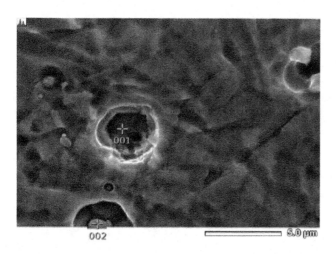

Fig. 14 SEM—NiTi_CR

Fig. 16 SEM—NiTi_TC18

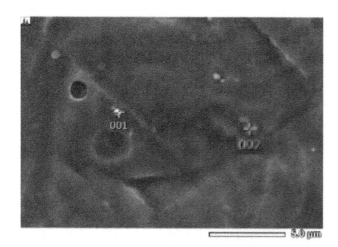

Fig. 15 SEM—NiTi_TC12

Fig. 17 SEM—NiTi_TC24

Conclusion

In this work, it is evaluated whether and how the cryogenic treatment at $-196\ °C$ and the cryogenic immersion time affect mechanical properties (modulus of elasticity, damping and hardness) and thermal properties (phase transformation temperatures and latent heat of transformation). By means of an experimental comparative analysis between these properties of Ni_{48}-Ti_{52} alloy as-received and cryogenically treated, it was observed that the cryogenic treatment applied in this alloy affects these properties. In lines, the hardness decreases, the damping increased, the modulus of elasticity decreases, the latent heat decreases and the thermal hysteresis decreases. A microstructural analysis and literature indicate that these changes may be associated with the appearance of the Ti_2Ni precipitate and the increase in grain size due to the cryogenic treatment applied.

References

1. Mariante GR (1999) Efeito do tratamento Criogênico nas propriedades mecânicas do aço rápido AISI M2, In: *Dissertação de mestrado—PPGEM-UFRJ*
2. Surberg CH, Stratton P, Lingenhöle K (2008) The effect of some heat treatment parameters on the dimensional stability of AISI D2. Cryogenics 48:42–47
3. Yen PL (1997). Formation of fine eta carbide in special cryogenic and tempering process key to improve properties of alloy steels. Ind Heat 14:40–44
4. Gobbi SJ (2009) Influência do tratamento criogênico na resistência ao desgaste do aço para Trabalho a frio AISI D2. Dissertação de Mestrado em Ciência Mecânicas, Publicação ENM.DM 132/09. Departamento de Engenharia Mecânica, Universidade de Brasília, Brasília, DF, 96p
5. Ashiuchi ES (2009) Influência do tratamento criogênico na fadiga sob condições de fretting no AL 7050-T7451. Dissertação de Mestrado em Ciências Mecânicas, Departamento de Engenharia Mecânica, Universidade de Brasília, Brasília, DF, 95p
6. Moreira et al (2009) Influência do tratamento criogênico na usinabilidade do aço rolamento ABNT 52100 temperado
7. Otsuka KE, Wayman CM (1998). *Shape memory materials.* Cambridge University Press, Cambridge, UK
8. Cai W, Lu XL, Zhao LC (2005) Damping behavior of TiNi-based shape memory alloys. Mater Sci Eng A 394:78–82
9. Silva NJ et al (2011) Comparative study of dynamic properties a NiTi alloy with shape memory and classical structural materials. Matérial (Rio J.), Rio de janeiro 16(4)
10. Van Humbeeck J (2003) Damping capacity of thermoelastic martensite in shape memory alloys. J Alloy Compd 355:58–64
11. Dieter GE (1988) Mechanical metallurgy. SI metric edn. McGraw-Hill Book Company, London, p 620
12. Cross WB, Kriotis AH, Ans Stimler FJ Nitinol characteization Study for NASA—langley Research Center under Contract NAS 1-7522 to define the structural and recovery proprieties of several Nitinol compositions possessing different transition temperature. The study was essentially a 19-month program conducted from July 1967 through January 1969
13. Saburi T (1998) Ti-Ni shape memory alloys. In: Otsuk K, Waymann CM Shape memory materials. Cambridge University Press, pp 49–46
14. Ishida A, Sato A, Miyazaki S (1997) Microstructure of Ti-Rich Ti-Ni thin films. In: Proceedings of the second international conference on shape memory and superelastic technologies, SMST-97. p 161
15. Nishiyama Z (1978) Martensitic transformation. In: Morris EF, Meshii M, Wayman CM (eds) Academic Press
16. Guimarães JRC, Rios PR (2010) Martensite start temperature and the austenite grain-size. J Mater Sci 45(4):1074–1077
17. Yang H-S, Bhadeshia HKDH (2009) Austenite grain size and the martensite-start temperature. Scripta Mater 60:493–495

The Martensitic Transformation in Indium-Thallium Alloys

T. R. Finlayson, G. J. McIntyre, and K. C. Rule

Abstract

The martensitic transformation in indium-thallium alloys is reviewed with a focus on the nature of the transformation as has been investigated using elastic and lattice vibrational properties. Recent measurements from thermal-neutron Laue and cold-neutron, triple-axis experiments are presented in an attempt to ratify the traditional explanation for the transformation as being driven by phonon softening for low-ζ $[\zeta\zeta0][\zeta\zeta0]$ (i.e., T1) phonons. No evidence for phonon softening has been found to as low as $\zeta = 0.02$ reciprocal lattice units on the T1 phonon branch. An alternative mechanism is advanced, for the transformation involving nucleation and growth on $\{111\}$ planes, which appears to be consistent with electron diffuse scattering in the system. Such a mechanism is also consistent with the transition being driven by the electronic free energy of the system, as has been demonstrated by other properties.

Keywords

Indium-thallium alloys · Phonon dispersion Transformation mechanism

T. R. Finlayson (✉)
Department of Chemical and Biomolecular Engineering, The University of Melbourne, Melbourne, VIC 3010, Australia
e-mail: trevorf@unimelb.edu.au

G. J. McIntyre · K. C. Rule
Australian Centre for Neutron Scattering, Australian Nuclear Science and Technology Organisation, Kirrawee DC, Locked Bag 2001, Sydney, NSW 2232, Australia
e-mail: gmi@ansto.gov.au

K. C. Rule
e-mail: kirrily@ansto.gov.au

Introduction

The face-centred cubic (fcc) to face-centred tetragonal (fct) martensitic transformation in indium-thallium alloys was first revealed via optical microscopy on the polished surface of an In-20 at.%Tl alloy by Bowles, Barrett, and Guttman and accompanying x-ray measurements [1]. Indeed, the transformation now features in text books to illustrate a transformation morphology displaying planar interfaces extending right across parent grains [2]. The transformation was suggested to proceed by homogeneous shear on one of a set of $\{110\}$ planes in $\langle110\rangle$ directions but in opposite sense in adjacent, and thus twin-related, regions. Within each planar region, finer-scale twinning on another set of $\{110\}$ planes at 60° to the first, creates an inhomogeneous shear. Subsequent reports [3, 4] also based on optical metallography and x-ray observations, supported this $\{110\}\langle1\bar{1}0\rangle$ double-shear mechanism.

Again, through the combination of optical metallography and x-ray diffraction, Pollack and King [5] showed that In-xat.%Tl alloys for $15.5 \leq x \leq 31$ at.% undergo this fcc to fct transformation with a martensite start temperature, M_s, varying from approximately 425 K for a 15.5 at.% alloy to close to 0 K for 31 at.%. In addition, Pollock and King published a phase diagram which showed that for alloys of less than 24 at.%Tl, M_s and the martensite finish temperature, M_f, were different by just a few degrees, but for higher Tl concentrations, the difference $M_s - M_f$ increased as the alloy concentration increased until, for alloys of greater than about 28 at.%Tl, they would remain two-phased (tetragonal and cubic) to the lowest temperatures.

In a series of papers [6–8] the elastic constants for various In-Tl alloys were measured using ultrasonic techniques. Particularly for the $1/2(c_{11} - c_{12})$ (or c') modulus, it was suggested that this shear modulus went to zero at the M_s temperature, although it must be borne in mind that often for c' measurements, the damping of the ultrasonic wave in the parent crystal as M_s was approached became so severe that

Fig. 1 The phonon-dispersion data at room temperature for #1067, In-24 at.%Tl, prepared by Bridgman growth in an evacuated quartz capsule. The data were measured on two crystal plates, spark machined from the cylindrical crystal, as explained in the text. The solid lines show the calculated frequencies for a 4NN Born-von Kármán force-constant model while the dash-dot curves are the theoretical calculation of Gunton and Saunders [9]. The thin lines show the initial slopes expected from the elastic constants for an alloy of this composition. The inset shows the low-q data along the T1 branch with a "guide-to-the-eye" line through the experimental points. (Figure reproduced from [20].)

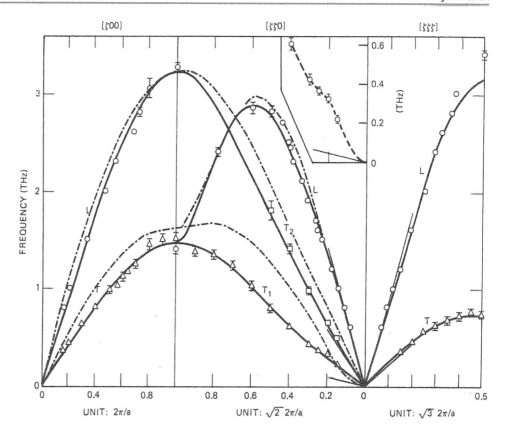

the value for $1/2(c_{11} - c_{12})$ had to be arrived at by independent measurements of c_{11} and c_{12}, followed by subtraction. Observing a number close to zero may have been an inevitable artifact.

However, using their measured elastic constants, Gunton and Saunders [9] performed a pseudopotential calculation for the phonon-dispersion relations. An important consequence of this calculation was the appearance of a strong positive curvature at low q along the $[\zeta\zeta 0][\zeta\bar{\zeta}0]$ branch (hereafter called the T1 branch).

Clearly, there was a need for experimental phonon-dispersion measurements to address this situation.

Phonon-Dispersion Measurements

Indium, in particular, and likewise an In-Tl single crystal, is a most unfavourable system on which to attempt neutron phonon-dispersion measurements on account of the high ratio of the absorption to coherent scattering cross sections. The only possibility as was already well known from measurements on pure indium (H. G. Smith, private communication, 1983), was to prepare large, flat-plate crystals so that the phonons could essentially be measured with the crystal in reflection. To this end, our first preparation was via a Bridgman growth for an In-24 at.%Tl alloy in an evacuated quartz capsule, 25 mm in diameter. The longitudinal axis for

the crystal growth proved to be [001], so that two crystal plates having (100)[001] and (110)[001] orientations could be spark machined from the single-crystal ingot. These two plates were sufficient to record all the high-symmetry phonons, although the (110)[001] plate had to be rotated to the (110)[1$\bar{1}$0] orientation (i.e., [1$\bar{1}$0] normal to the scattering plane), to enable the longitudinal and transverse [$\zeta\zeta\zeta$] phonons to be measured.

The resultant room-temperature data are shown in Fig. 1 together with a most satisfactory 4NN Born-von Kármán force-constant model [10]. Also shown on the Figure are the dispersion relations from the above pseudopotential calculation [9], where, particularly along the T1 branch, the pseudopotential does not provide a good fit to the measured T1 phonons, although it does show the low-q curvature, in order to fit the measured elastic constant.

Single-Crystal Limitations

The limitation to measuring the T1, low-q phonons was the quality of our two crystal plates, which restricted these measurements of the T1 branch to $\zeta = 0.15$ reciprocal lattice units (rlu). This lack of quality was believed to arise from solute segregation at the solidification interface during the Bridgman growth. Indeed, evidence for the lack of crystal quality could be seen from images of particular Bragg peaks

Fig. 2 Horizontal Bridgman furnace, designed to grow oriented, flat-plate crystals of indium-thallium alloys, approximately 3 mm thick, for inelastic neutron-scattering measurements. An oriented seed crystal would be located in the thin end of the crucible, which is water cooled. The base is maintained at a steady temperature slightly below the melting point of the alloy and a small resistive heater beneath the radiation shield is driven along the ingot. The whole furnace is contained beneath an evacuated glass bell jar

taken during the experiment. Thus, it was necessary to reduce the liquid solution and to grow a flat-plate directly from a small, oriented-seed crystal as part of the growth. The design of such a horizontal Bridgman furnace is shown in Fig. 2. Oriented, high-quality crystals, of approximate dimensions $60 \times 25 \times 3$ mm^3, were the result. Also, crystals with higher thallium contents, having a lower M_s temperature and a larger temperature range over which to search for phonon softening, were grown. The complete set of crystals for which data are presented in this paper, are listed in Table 1.

While most studies were continued on the HB3 spectrometer at the Oak Ridge National Laboratory, with a fixed final energy of 3.58 THz, we obtained some access to a cold-neutron, triple-axis spectrometer (H7) at the Brookhaven National Laboratory, where a T1 phonon at $\zeta = 0.03$ rlu could be resolved using fixed incident energy of 1.21 THz and an in-pile Be filter. But despite this improvement, there remained no answer, as far as the transformation in In-Tl was concerned, to the important question of phonon softening as evidenced by either curvature along the T1 branch at low q or the decrease in frequency with decreasing temperature for any measured phonons.

The New Australian Research Reactor

The new Australian research reactor, OPAL (Open Pool Australian Lightwater), was commissioned at Lucas Heights in 2007 and a suite of 14 instruments have now been installed. Two of these, KOALA, a thermal-neutron Laue diffraction instrument, and SIKA, a cold-neutron, triple-axis spectrometer, have the potential to study low-q phonons as a function of temperature. KOALA is similar to the VIVALDI instrument at the Institut Laue-Langevin, Grenoble, where McIntyre et al. had demonstrated the potential of neutron Laue diffraction to study low-q phonons and their temperature dependence [11, 12]. The spectrometer, SIKA [13], was designed and built as the result of an agreement between ANSTO and the National Central University of Taiwan and

Table 1 Details of crystals being discussed in this paper. Alloy composition was estimated from the measured density [8] and hence the expected M_s determined from the phase diagram [5]

Crystal #	Preparation	Flotation density (g cm^{-3})	Composition (at.%Tl)	M_s (K)
1067	Bridgman, 25 mm cylinder	8.489	24.0	250
1072	Bridgman, 10 mm cylinder	8.613	27.0	165
1074	Horiz. Bridgman from seed	8.707	27.5	140
1080	Horiz. Bridgman from seed	8.793	29.0	90

Fig. 3 Thermal diffuse scattering expected around certain Bragg reflections, calculated from the measured phonon dispersion relations [10]

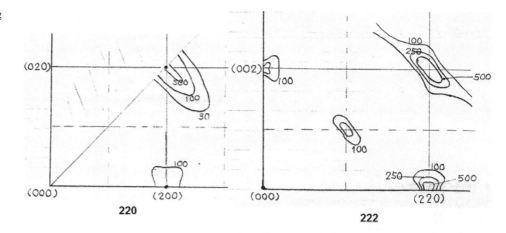

Fig. 4 A neutron Laue diffraction pattern at 260 K for an In-24 at.%Tl crystal, $3 \times 3 \times 2.8$ mm^3, from the KOALA instrument. The indices of the main diffraction spots are given and one example illustrating the temperature dependence of scattering near the $1\bar{1}1$ reflection is shown

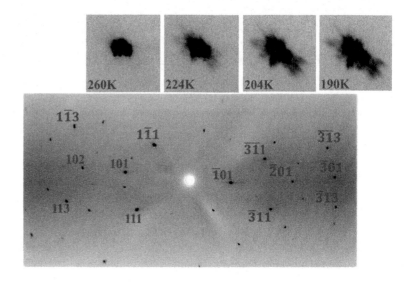

was funded by the National Science Council of Taiwan. The instrument was fully commissioned in 2016.

As a result of these new developments, we were encouraged to re-examine the indium-thallium system in relation to phonon softening. An initial experiment was carried out at KOALA using a small piece of single crystal, $3.0 \times 3.0 \times 2.8$ mm^3, cut from #1067, for which M_s appeared to be around 210 K, on the basis of the first-noticeable changes in the Laue diffraction pattern (Fig. 4). The expected thermal diffuse scattering had been calculated by Wakabayashi (N. Wakabayashi, private communication, 1985), using the measured phonon-dispersion relations [10] as follows:

$$I \propto \sum_{j} \frac{(n + 1/2)}{v_j(\vec{q})} \left(\vec{Q} \cdot \vec{e}_j(\vec{q}) \right)^2$$

where $\vec{Q} = \vec{q} + \vec{G}$, $\vec{G}$ is a reciprocal lattice vector, $v_j(\vec{q})$ is the frequency of the jth mode of eigenvector, $e_j(\vec{q})$, and $n = 1/\left(e^{\frac{hv_j(\vec{q})}{kT}} - 1 \right)$. Typical results showing wing-like contours around Bragg reflections are shown in Fig. 3. However, the high absorption to coherent scattering cross section ratio together with alloy disorder scattering analogous to that for x-rays [14] prevented an observation of the wing-like profiles around certain Bragg reflections, expected for phonon scattering. The diffuse scattering in the vicinity of Bragg

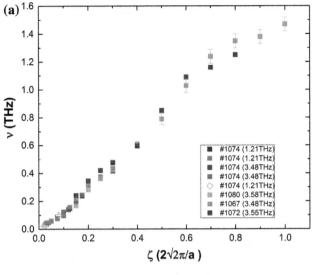

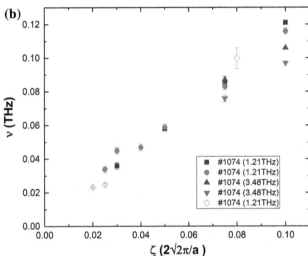

Fig. 5 **a** The complete data set for the T1 phonon branch for all four crystals (#1067, #1072, #1074, and #1080) studied. **b** The low-q region for the T1 branch data. Note the absence of frequencies for certain crystals from **b** as for these crystals phonon groups could not be resolved to as low as $\zeta = 0.10$ rlu, on account of the crystal quality. The recent data from the SIKA spectrometer are shown as open symbols

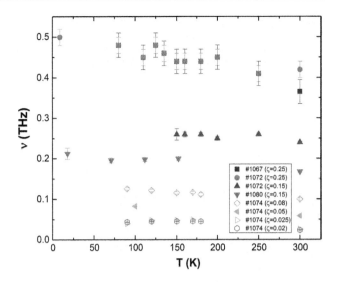

Fig. 6 Temperature dependence of phonon frequencies for a number of ζ values for the different crystals measured. Again the recent data from the SIKA spectrometer are shown as open symbols

complete set of data for the T1 phonon branch are collected in Fig. 5. In producing Fig. 5 in which the T1 data for different crystals having slightly different compositions and therefore slightly different lattice vibrational properties, some normalization procedure should be included to account for these different properties and hence to allow for direct comparison of the T1 frequencies measured. However, this would require the complete lattice dynamics for each of the crystals, which we do not have. Thus, for the purposes of understanding the overall trend along the T1 branch, we have simply plotted the as-measured phonon frequencies. Clearly, as is most evident in Fig. 5b, there is no detectable positive curvature at low q.

The temperature dependences for several T1 phonons are shown in Fig. 6. There is no "softening" found for any T1 phonon frequencies. On the contrary, all frequencies to that for the lowest ζ value of 0.02, increase (i.e., "harden") with decreasing temperature.

An Alternative View for the In-Tl Transformation

Close examination of the dispersion relations for In-Tl (Fig. 1) shows a most unusual feature when compared with the lattice-vibrational behaviour for other fcc crystals. This feature is illustrated in Table 2, where the ratios $v[\zeta\zeta\zeta]T/v[\zeta\zeta\zeta]L$ for the zone-boundary, $[\zeta\zeta\zeta]$ phonons for a number of fcc metals, and alloys have been listed. Clearly, In-Tl with a ratio of 0.226, approximately half the value for all other metals, is unusual.

Geisler [15] proposed a nucleation and growth mechanism along {111} planes of the fcc matrix. Allowing the

reflections observed prior to and through the transformation, (Fig. 4) would appear to arise from disorder scattering.

A second experiment was conducted on the SIKA spectrometer using the same crystal for which the T1 phonons had been resolved to as low as $\zeta = 0.03$ rlu in the earlier experiment at the Brookhaven cold-neutron, triple-axis spectrometer. In this case, the $\zeta = 0.02$ phonon could be resolved and was studied using a fixed final neutron energy of 1.21 THz, as a function of temperature. A clear advantage with the higher thallium concentrations is that the cubic phase can still be detected and the crystal aligned to this cubic phase at temperatures below M_s, as pointed out above in relation to the phase diagram. In this way, the phonons could be measured through the M_s temperature range. The

Table 2 The ratios of the room temperature frequencies for the transverse and longitudinal, zone-boundary, $[\zeta\zeta\zeta]$ phonons, $v(T)/v(L)$, for a number of fcc metal systems. The literature source of the data in each case is given as the reference beside the sample

Metal	$v(T)/v(L)$
Cu [21]	0.458
Ag [22]	0.442
Au [23]	0.396
Ni [24]	0.482
Pd [25]	0.476
Pt [26]	0.495
Pb [27]	0.407
Al [28]	0.429
β_1Cu-Al-Ni [29]	0.493
Ni$_{70}$Pt$_{30}$ [30]	0.373
In-24 at.%Tl	**0.226**

slight change in orientation required for {111} conjugate planes between the fcc and fct structures, as illustrated in Fig. 3 of [15], gives precisely the same x-ray pole-figure results as for the double-shear mechanism [1, 3, 4]. Further, Geisler's mechanism proposed the formation and growth of coherent nuclei along the {111} conjugate planes. When the size of each nucleus reaches the size at which the strain energy becomes intolerable, a twin forms.

Such a mechanism is also consistent with the observation of premartensitic electron diffuse scattering in the form of rods of scattering in $\langle 111 \rangle^*$ directions in reciprocal space. (See Fig. 4 of [16]). The unusual softness to shear for phonons along the $[\zeta\zeta\zeta]$ T branch is also consistent with this mechanism, suggesting that the premartensitic nuclei may be generated by dynamic and static $\langle 111 \rangle \langle 11\bar{2} \rangle$ atomic displacements. According to the theory of Olsen and Cohen [17], the first step in martensite nucleation is the faulting on planes of closest packing, and these faulting displacements are derived from existing defects. When the temperature of the fcc phase is reduced to the transformation temperature, short-range order of $\langle 111 \rangle \langle 11\bar{2} \rangle$ atomic displacements rapidly transforms to long-range order of the martensitic, fct phase.

In addition, these particular atomic displacements have been shown to arise from conduction-electron energy effects associated with strong electron-phonon coupling. Indium-thallium, similar to other dilute indium alloys such as indium-bismuth, indium-cadmium, indium-lead, and indium-tin, exhibits a phase transition which is driven by the electronic contribution to the free energy of the system [18]. This is also evidenced by the rapid variation of M_s with composition and the observed depression of M_s with applied magnetic field [19].

Conclusion

The background literature regarding the transformation mechanism in indium-thallium alloys as involving a double shear in {110} planes driven by phonon softening has been reviewed. More recent measurements directed towards the observation of phonon softening have yielded negative results. The observation of anomalously soft $[\zeta\zeta\zeta]$ T phonons has given rise to an alternative nucleation and growth mechanism involving $\langle 111 \rangle \langle 11\bar{2} \rangle$ defects. Such a mechanism is consistent with electron diffuse-scattering observations and a transition driven by the electronic free energy of the alloy system.

Acknowledgements Access to neutron beamtime at the Australian Centre for Neutron Scattering through proposal nos. P4814 and DB6030 is acknowledged. One of us (TRF) also acknowledges resources provided by The University of Melbourne where he is an Honorary Principal Fellow.

References

1. Bowles JS, Barrett CS, Guttman L (1950) Crystallography of cubic-tetragonal transformation in the indium-thallium system. Trans Metall Soc A.I.M.E 188:1478–1485
2. Chadwick GA (1972) Metallography of phase transformations. Butterworths, London, p 272
3. Burkart MW, Read TA (1953) Diffusionless phase change in the indium-thallium system. Trans Metall Soc A.I.M.E 197:1516–1524
4. Basinski ZS, Christian JW (1954) Experiments on the martensitic transformation in single crystals of indium-thallium alloys. Acta Metall 2:148–166
5. Pollack JTA, King HW (1968) Low temperature martensitic transformations in In-Tl alloys. J Mater Sci 3:372–379
6. Novotny DB, Smith JF (1965) Single crystal elastic constants of f.c.c. Thallium-Indium alloy. Acta Metall 13:881–888
7. Gunton DJ, Saunders GA (1974) The elastic behaviour of In-Tl alloys in the vicinity of the martensitic transformation. Solid State Commun 14:865–868
8. Brassington MF, Saunders GA (1983) Vibrational anharmonicity and the elastic phase transition of Indium-Thallium alloys. Proc R Soc Lond A 387:289–310
9. Gunton DJ, Saunders GA (1973) The soft acoustic phonon mode and its relation to the martensitic transformation in In-Tl alloys. Solid State Commun 12:569–572
10. Finlayson TR et al (1985) Inelastic neutron scattering ftrom a martensitically transforming Indium-Thallium alloy. Solid State Commun 53:461–464
11. McIntyre GJ, Lemée-Cailleau MH, Wilkinson C (2006) High speed neutron laue diffraction comes of age. Phys B 385–386:1055–1058
12. McIntyre GJ, Kohlman H, Willis BTM (2011) Phonons observed by Laue diffraction on a continuous neutron source. Acta Cryst A67:C129–C130
13. Wu CM et al (2016) The multiplexing cold-neutron, triple-axis spectrometer at ANSTO. J. Instrum 11:10009(1)–10009(15)
14. Cowley JA (1950) X-ray measurements of order in single crystals of Cu$_3$Au. J Appl Phys 21:24–30

15. Geisler AH (1953) Crystallography of phase transformations. Acta Metall 1:260–281
16. Finlayson TR et al (1984) An electron diffraction study of a pre-martensitic In-24at.%Tl alloy. Acta Cryst B 40:555–560
17. Olson GB, Cohen M (1976) General mechanism of martensitic nucleation. 3. kinetics of martensite nucleation. Metall Trans A Phys Metall and Mater Sci 7:1915–1923
18. Smith TF (1973) Influence of fermi surface topology on the pressure dependence of T_c for indium and dilute indium alloys. J Low Temp Phys 11:581–601
19. Lashley JC et al (2007) Electronic instabilities in shape-memory alloys: thermodynamic and electronic structure studies of the martensitic transition. Phys Rev B 75:205119-1–205119-6
20. Finlayson TR et al (1988) Studies of the transverse phonon modes in premartensitic indium thallium alloys. Mater Sci Forum 27 (28):107–112
21. Nicklow RM et al (1967) Phonon frequencies in copper at 49 and 298 K. Phys Rev 164:922–928
22. Drexal W, Gläser W, Gompf F (1969) Phonon dispersion in silver. Phys Lett A 28:531–532
23. Lynn JW, Smith HG, Nicklow RM (1973) Lattice dynamics of gold. Phys. Rev. B 8:3493–3499
24. Birgeneau RJ et al (1964) Normal modes of vibration in nickel. Phys Rev A General Phys 136:1359–1365
25. Müller AP, Brockhouse BN (1971) Crystal dynamics and electron specific heats of palladium and copper. Can J Phys 49:704–723
26. Dutton DH, Brockhouse BN, Müller AP (1972) Crystal dynamics of platinum by inelastic neutron scattering. Can J Phys 50:2915–2927
27. Brockhouse BN et al (1962) Crystal dynamics of lead. I. dispersion curves at 100°K. Phys Rev 128:1099–1111
28. Gilat G, Nicklow RM (1966) Normal vibrations in aluminium and derived thermodynamic properties. Phys Rev 143:487–494
29. Hoshino S et al (1975) Phonon Dispersion of the β_1-Phase in Cu-Al-Ni Alloy. Jpn J Appl Phys 14:1233–1234
30. Tsunodo Y et al (1979) Phonon dispersion relations in the disordered $Ni_{1-x}Pt_x$ system. Phys Rev B 19:2876–2885

Author Index

A

Acet, Mehmet, 185
Akama, Daichi, 251
Alves, J.M., 59
Arroyave, Raymundo, 185

B

Babaei, H., 167
Babanli, Mustafa, 115
Bai, Y., 95
Balaev, E.U., 213
Ball, John M., 29
Basak, Anup, 161
Benke, Marton, 73
Blednova, Zh.M., 213
Borza, F., 99
Botelho, R.A., 59
Brandão, L.P.M., 59
Braz Fernandes, F.M., 109
Bujoreanu, L.G., 99

C

Çahır, Asli, 185
Cao, Shanshan, 89, 189, 201, 275
Cardoso, M.C., 59
Carstensen, Carsten, 29
Chen, Hao, 83
Chen, Yiping, 235
Chulist, R., 263
Clemens, Helmut, 247
Collins, P.C., 53
Comăneci, R.I., 99
Czaja, P., 263

D

da Cruz Filho, Bartholomeu Ferreira, 281
Demchenko, Lesya, 115
Dieck, Sebastian, 123
Ding, Hua, 67
Dmitrenko, D.V., 213
Duong, Thien, 185

E

Ebner, Reinhold, 35
Ecke, Martin, 123
Eggbauer (Vieweg), Annika, 35
Entel, Peter, 185
Esfahani, S.E., 53

F

Feng, Biao, 47
Finlayson, T.R., 291
Fitta, M., 263
Freitas, M.C.S., 59
Fujioka, Masaaki, 21, 25, 251
Fujiwara, Kazuki, 143

G

Gao, S., 95
Ge, Y., 257
Ghamarian, I., 53
Gong, Wu, 95, 155
Grigoraş, M., 99
Gruber, Marina, 35
Gruner, Markus E., 185
Gueltig, Marcel, 197

H

Halle, Thorsten, 123
Harjo, Stefanus, 43, 155
Haušild, Petr, 173
Heczko, O., 257
Hemmilä, Mikko, 149
Homma, Ryuichi, 25
Hoshino, Manabu, 21, 25

I

Ishikawa, Kyohei, 25
Ito, Atsushi, 155

Subject Index

Printed by Printforce, the Netherlands